Dept. of Botany
Oregon State Univ.
Corvallis, Ore. 97330

Biology and Control
of Soil-Borne Plant Pathogens

Biology and Control
of Soil-Borne Plant Pathogens

Third International Symposium on
Factors Determining the Behavior of Plant Pathogens in Soil
Held at University of Minnesota, Minneapolis, 5-12 September 1973
in Conjunction with the Second International Congress
of Plant Pathology

Edited by
G. W. BRUEHL

THE AMERICAN PHYTOPATHOLOGICAL SOCIETY
St. Paul, Minnesota: 1975

The American Phytopathological Society
3340 Pilot Knob Road, St. Paul, Minnesota 55121

Copyright © 1975, by
The American Phytopathological Society

ISBN: 0-89054-000-4
Library of Congress Catalog Card Number: 74-79540

Printed in the United States of America

Foreword

R. JAMES COOK—*Agricultural Research Service, U.S. Department of Agriculture, Washington State University, Pullman.*

Ten years have passed since the first international symposium on soil-borne plant pathogens was held on the University of California Campus at Berkeley. I attended that symposium as a Berkeley graduate student, very much a spectator, and was awed by the impressive and unprecedented collection of world experts on soil biology. I can testify unequivocally to the openness, enthusiasm, and scientific precision with which that group went about the task of bringing together information on factors affecting soil-borne plant pathogens.

The book that resulted from that symposium, *Ecology of Soil-Borne Plant Pathogens*, edited by K. F. Baker and W. C. Snyder, is unquestionably among the most useful and comprehensive treatments of an important scientific topic ever assembled. It is impossible, in fact, to fully assess the impact that symposium and resultant book have had on root disease investigations. One probable influence was the inspiration for the First International Congress of Plant Pathology, held 5 years later in London under the organizational leadership of S. D. Garrett. The section of that congress on root diseases and soil-borne plant pathogens was developed as a sequel to the Berkeley symposium, and the resulting book, edited by T. A. Toussoun, R. V. Bega, and P. E. Nelson, now rests properly alongside its successful predecessor on the bookshelves of pathologists the world over.

As the third symposium developed as part of the Second International Congress of Plant Pathology, there was little doubt among the plant pathologists polled that a third volume, companion to the earlier two, must be assembled. We are deeply indebted to The American Phytopathological Society (APS) for agreeing to publish this volume, and to G. W. Bruehl for serving as its editor.

Planning for this volume officially began in August of 1969 in Spokane, Washington, at the 61st annual meeting of APS. C. D. McKeen of Canada and D. W. Burke of the United States served with me as organizers. Working, in part, through the Soil Microbiology Committee of APS, we sought the personal advice and opinions of about 100 pathologists knowledgeable in the ecology of soil-borne plant pathogens and representing 10 countries. There was one mandate in their responses: minimize overlap with or duplication of the preceding two meetings, either of authors or of specific topics, to ensure coverage of new topics by as many different workers as possible. Indeed, identifying 2 years in advance new people with promising research projects and fresh approaches was our biggest

challenge. We were also determined to build on the previous two symposia that emphasized biological control. Moreover, we insisted that the information be sound biology and an accurate representation of what really happens under natural conditions or, better yet, an analysis of a natural or manmade control method already in force. Clearly, the contributors helped us achieve our objectives with their papers on the influence of crop residues and organic wastes, the successful use of host resistance, plant protection by microecology and rhizosphere microorganisms, the existence and importance of soils suppressive to specific pathogens, physical and chemical aspects of the root and soil environment, and the assessment of both aesthetic and economic loss from seed decay, root rots, and wilts.

Two topics in this volume are presented as abstracts grouped under the headings of Soil Fungistasis and Lysis of Fungi and Integration of Pesticide-Induced and Biological Destruction of Soil-Borne Pathogens. These two topics were left intentionally flexible and were handled as colloquia from the start because both defied prediction 2 years in advance and because it was deemed undesirable to restrict their coverage to only four main contributors. With this arrangement, we were able to include the most up-to-date findings on these two rapidly developing and important areas. As a result, at least one recent breakthrough on fungistasis was included: the work of A. M. Smith of Australia, which indicates that ethylene, previously associated only with water logged soils, also occurs in drained agricultural soils at 0.5-1 ppm or more and inhibits fungi at these concentrations. This finding is especially significant in view of still more recent evidence that ethylene is produced by spore-forming bacteria in the anaerobic microsites that form in soils with aerobic microbial consumption of the oxygen. This more recent finding has great relevance to many of the observations and phenomena described herein, but unfortunately was made too late for inclusion in this volume. I am therefore taking the liberty to discuss the findings and their implications.

The existence of anaerobic microsites in soil because of aerobic microbial consumption of oxygen is well established. Apparently, as the volume of these sites expands due to increased aerobic growth, the production of ethylene increases accordingly and arrests growth of the aerobes. Presumably, unless aerobic growth is arrested in time, little or no food would be left in the system for the anaerobes. (Their ubiquitous occurrence in soil at populations of 10^3-10^6/gm is evidence that conditions of both anaerobism and nutrition must be suited for their activity in the tillage layer.) However, as aerobic growth slows due to ethylene, oxygen diffuses back into the anaerobic sites and again limits the anaerobes. The system is thus a balanced one where no one group overruns the other. If it develops, as now seems likely, that ethylene is the cause of widespread soil fungistasis, then pathogens must be recognized as but a small component in the balance of life in soil. However, being particularly sensitive to ethylene, pathogens will have served in history to reveal a phenomenon that might eventually and more properly be termed widespread soil biostasis.

The greater the rate of microbial consumption of oxygen (e.g., that promoted by organic amendments) or the slower the rate of oxygen diffusion (e.g., the limitations imposed by wet or fine-textured soils), the larger is the volume of ethylene-producing anaerobic microsites and the greater is the level of suppression of pathogens. The preference of *Fusarium* for dry acid sandy soils low in organic matter reflects low ethylene tolerance in this genus. In contrast, the Fusarium-suppressive soils are predictably high ethylene producers. Saprophytic actinomycetes are very tolerant of ethylene (Smith, *unpublished*). The early control of potato scab by Millard and Taylor using green plant tissues as amendments and the more recent control of this disease by intensive timely irrigations (Lapwood and Adams, this volume), might be explained in terms of the extraordinary ethylene levels necessary to suppress *Streptomyces scabes*, the former method by increased oxygen consumption and the latter by limitations imposed on oxygen diffusion into the microsites. Pathogens troublesome in wet or waterlogged soils (*Pythium*, *Phytophthora*) are probably also somewhat tolerant of ethylene because of a long evolutionary association with this gas. The Phytophthora-suppressive soils in Queensland, Australia, described in this volume contain 10 ppm ethylene when moist but drained (Smith, *unpublished*), apparently because the growers have incorporated generous amounts of manures and green residues into their orchard soils regularly since the land was first cleared.

The general suppression of pathogens by nonsterile soil, the elevation of this suppression by organic amendments, the lowering of it by tillage and other methods of improved aeration, and the lack of suppression in virgin desert newly reclaimed soils also relates to levels of ethylene (Smith and Cook, *unpublished*). Ethylene apparently is not the cause of specific antagonisms to pathogens such as that associated with take-all decline, but even this must function against a background of ethylene. Nature relies on many systems to maintain biological balance. Nevertheless, because of an apparently general role of ethylene in soil biology, and because pathogens are but a small component in the total balance between aerobes and anaerobes in soil, ethylene management provides a truly optimistic outlook for practical control of soil organisms. In the words of K. F. Baker and W. C. Snyder in the Foreword to *Ecology of Soil-Borne Pathogens*, "It is increasingly clear that mastery of the soil microflora will come only when we understand the obscure complex of nonparasitic organisms, even though the pathogens have seemed the logical point of attack."

Preface

It was a pleasure to assemble the papers of this book. The authors cooperated, and the editorial staff of The American Phytopathological Society did the real work. We are indebted to The American Phytopathological Society for financial sponsorship, and we hope all costs are recovered.

The organization of this volume is the work of the Soil- and Seed-Borne Pathogen Committee of the 2nd International Congress: R. J. Cook, Chairman, D. W. Burke, and C. D. McKeen. They selected the subjects and authors. To them goes the credit for the scope and quality of the contributions. They attempted to minimize duplication of the ideas, subjects, and authors of two previous symposia (Berkeley, California, 1963, and London, England, 1968). R. J. Cook, in his foreword to this volume, states that their biggest challenge was attempting to predict advances 2 years prior to the congress.

This book is in the format of *Root Diseases and Soil-Borne Pathogens*, edited by T. A. Toussoun, R. V. Bega, and P. E. Nelson. We liked their style and felt that its use would facilitate the continuation of a series of such books concluding each congress of plant pathology. In my opinion, assessment of progress or lack of it, made every 5 years, is a worthy objective. It is our hope that support of this volume will justify the next. Such books could be milestones along the path of the study of soil fungi.

Thanks are due Wayne Sinclair for soliciting the papers of F. D. Podger, D. B. Redfern, and E. W. Ross; to Carl J. Eide for indexing; and to Kathleen K. Wolter for copy editing.

G. W. Bruehl, Department of Plant Pathology, Washington State University, Pullman

April 2, 1974

Chairmen of Symposium Sessions

K. F. BAKER, Department of Plant Pathology, University of California, Berkeley

D. W. BURKE, Irrigated Agriculture Research and Extension Center, Prosser, Washington

J. COLHOUN, Department of Cryptogamic Botany, The University, Manchester, England

J. M. KRAFT, Irrigated Agriculture Research and Extension Center, Prosser, Washington

J. L. LOCKWOOD, Department of Botany and Plant Pathology, Michigan State University, East Lansing

C. D. MC KEEN, Canada Agriculture, Central Experimental Farm, Ottawa, Ontario

G. C. PAPAVIZAS, Soilborne Diseases Laboratory, Agricultural Research Service, U.S. Department of Agriculture, Beltsville, Maryland

W. C. SNYDER, Department of Plant Pathology, University of California, Berkeley

G. A. ZENTMYER, Department of Plant Pathology, University of California, Riverside

Contents

PART I

INTRODUCTION

Seed- and Soil-Borne Pathogens: Opening Address

S. D. GARRETT—*Botany School, University of Cambridge, England.*

It is now more than 10 years since an international symposium on "Factors Determining the Behavior of Plant Pathogens in Soil" was held at the University of California at Berkeley. That was the first major international conference devoted wholly to soil-borne plant pathogens; it was organized by a committee headed by W. C. Snyder and K. F. Baker. In 1965, the proceedings of that symposium were published by the University of California Press in a volume edited by Baker and Snyder and entitled *Ecology of Soil-Borne Plant Pathogens* (2). The enthusiasm generated by the Berkeley meeting gave the enterprise a momentum that has been maintained and that shows no sign of flagging. A second meeting was organized within the First International Congress of Plant Pathology, held 5 years later at the Imperial College in London. The proceedings were published in 1970, again by the University of California Press, in a volume entitled *Root Diseases and Soil-Borne Pathogens* (18) and edited by T. A. Toussoun, R. V. Bega, and P. E. Nelson. Now here we are again about to begin the third of these international meetings, a meeting I know will be as successful as its two predecessors. The first symposium volume, published in 1965, was subtitled *Prelude to Biological Control*. This word prelude points up the fact that measures of biological control must be founded on thorough ecological understanding of these pathogens as they live and survive in the soil. Few would deny that the tantalizing prospects of achieving biological control have, in a variety of ways, provided a strong driving force for investigations into microbial ecology. We as plant pathologists know that natural biological control of soil-borne plant pathogens by other soil microorganisms is already giving us substantial help. But we are still trying to learn how to harness these immensely powerful biological forces within the soil so as to secure better control of soil-borne diseases and at the same time to preserve and even to improve the structure and fertility of our soils for future generations.

In this connection, I wish to commend two important books. The first is D. M. Griffin's *Ecology of Soil Fungi* (13), published in 1972. This is a book that should be read by all soil microbiologists, not just by mycologists. The first part is concerned with the relations between species and larger groups of soil fungi and their substrates of plant and animal origin, and with the complex interactions between them. The second part deals with the effects of soil physical factors, an area in which Griffin and his associates have made so many important contributions. The second book is *Biological Control of Plant Pathogens* (1), by K. F. Baker and R. J. Cook, which I have been told will be published by or before mid-1974. I have been privileged to read an advance copy of this book in typescript, and I read it with pleasure and profit. This really is the book we have all been waiting for; its breadth of coverage shows the wide variety of ways in which biological control can be pressed into service and will help to ensure that research workers are aware of all the possibilities for their own particular disease problems.

EXPERIMENTAL MODELS OF SITUATIONS IN THE SOIL.—For the second part of this address, I should like to say something about experimental models of natural situations in the soil. A model is an abstraction from the natural situation, which has thereby been simplified for the purpose of experimentation. An essential preliminary to the creation of a satisfactory model is a thorough study of the natural situation by careful observations of various kinds and by some experimental trials. It is perhaps the most difficult and creative part of any scientific investigation. I propose to discuss only two models. The first example shows the various kinds of difficulty surrounding the choice of a reliable model; the second illustrates admirably the principle of economy in model design.

The first model was designed for a study of factors controlling competitive saprophytic colonization of wheat straw by various species of cereal foot rot fungi. It was developed by F. C. Butler (5, 6) and R. L. Lucas (14) in collaboration with me and has been designated the Cambridge method. The method consists of burying short lengths of autoclaved wheat straw in a progressive series of dilutions of maizemeal-sand inoculum of the fungus to be studied with increasing proportions of a natural soil. By the standard method, the proportion of straws colonized by the inoculant fungus is determined after 1 month of incubation in the inoculum-soil mixtures. The competitive saprophytic ability of the fungus under study can be expressed most concisely by the percentage of natural soil in the inoculum-soil mixture that still permits colonization of 50% of the straws by the inoculant fungus. When describing this method and some of its results in my last book, I expressed some anxieties about the validity of this model (12, p. 111). If the season of cereal harvest has been fine and dry and if a farmer plows in the fresh clean straw soon afterwards, he is creating a situation not unlike that modeled by our Cambridge method. Fortunately for us, Cook and Bruehl (8) and later Cook (7) made a thorough study both of this situation (and of others that depart widely from this model) in the Pacific Northwest region of the United States. When clean bright straw was buried in soils naturally infested by *Fusarium roseum* f. sp. *cerealis* 'Culmorum' and held in laboratory containers at optimum soil temperature and moisture content, more than 90% of the straws were sometimes colonized by this fungus. Under less favorable soil conditions or with a lower soil population of *F. roseum*, correspondingly lower proportions of straws were colonized. But if, under comparable soil conditions, the straws that were buried were dull in color and weathered, i.e., already colonized by air-borne fungi while standing as stubble or lying on the soil for some months after harvest, a much smaller proportion was colonized by· *F. roseum*. In other words, this fungus is most successful as a pioneer colonizer.

An interesting effect of temperature on competitive saprophytic colonization of wheat straw by four species of cereal foot rot fungi was reported by Burgess and Griffin (4): all four fungi colonized a higher percentage of straws at 10° C than at 20° or 30° C. But fresh light has now been thrown on this situation by one of my recent research associates, J. W. Deacon (11), working with both *Ophiobolus graminis* (synonym *Gaeumannomyces graminis*) and *Cercosporella herpotrichoides*. He found that the percentage of straws apparently colonized was much higher if assessed after only 9 days in the inoculum-soil mixtures than if left for the standard period of 28 days. For *O. graminis*, the apparent percentage of straws colonized after 9 days at 21° C was as high as after 28 days at 10° C. From this, Deacon concluded that after 9 days there is a temperature differential in rate of decline of the primary fungal colonizer; it is much more rapid at 21° than at 10° C. So it happens that if percentage of straws colonized by the inoculant fungus is assessed after 28 days of incubation, as by the standard method, the fungus appears to have colonized a higher proportion of straws at 10° than at 21° C. This, however, is not a real effect of temperature on the proportion of straws initially colonized; it is an artifact of the Cambridge method. And so the conclusion of Burgess and Griffin (4) about the effect of temperature on competitive saprophytic colonization by cereal foot rot fungi is now open to serious question. Deacon has further suggested that the effect of nitrogen (5) in increasing the percentage of straws apparently colonized by *O. graminis* and, conversely, in decreasing the percentage apparently colonized by *Cochliobolus sativus* may in reality be an effect not on colonization but on subsequent survival of the respective fungi in the straws during the latter part of the 28-day incubation period. It has repeatedly been found that excess nitrogen prolongs saprophytic survival in straw of *O. graminis*, but shortens that of *C. sativus* (12, p. 152). We can here recall the finding by Papavizas and Davey (15) that high percentages of segments cut from mature dry stems of buckwheat were competitively colonized by *Rhizoctonia solani* after burial in various naturally infested soils for 4 days. When the segments were left for longer than 4 days, recovery of *R. solani* on antibiotic-supplemented water agar declined sharply.

In the light of these comments, do I now think the Cambridge method has sufficient validity as a model for situations in natural soil? I will evade this question by asking an easier one: has the model given us information about natural situations that we could not have obtained without it or without a model of any kind? I think it has, because it has not only answered our questions but has prompted other questions that have been answered by Deacon (11). His work and the work of Papavizas and Davey (15) have emphasized that the microbial equilibrium within a substrate is not static but changes more rapidly than we had realized. Their work has shown that if a primary fungal colonizer has secured only a partial occupation of a virgin substrate in the course of saprophytic competition, its tenancy of that substrate will be insecure and possibly brief. This contrasts with the situation in which a fungus achieves complete occupation of a substrate as a pioneer, either ahead of saprophytic competitors by parasitic invasion of living tissue or by being inoculated onto an autoclaved substrate. This prior colonization, as Bruehl and Lai (3) have so aptly called it, gives the fungus a secure hold on its substrate, a hold it may still retain after months of burial in the soil.

The second model is that employed by W. C. Snyder and associates to elucidate the behavior of chlamydospores of *Fusarium solani* f. sp. *phaseoli* when brought into contact with germinating seeds, roots, and hypocotyls of bean seedlings. The papers describing this work are by Toussoun and Snyder (19), Schroth and Snyder (16), Schroth et al. (17), Cook and Schroth (9), and Cook and Snyder (10). They found that germinating bean seeds were rarely infected by f. *phaseoli*, whereas seedling hypocotyls were regularly infected in the same soils. Exudates from germinating seeds provided a complete nutrient medium, with sugars balanced by amino acids. This promoted a high percentage germination of chlamydospores, but the germlings soon lysed, probably due to development of a large population of zymogenous bacteria. Infection of germinating seeds thus failed to occur. Hypocotyl exudates contained sugars but only a trace of amino acids. Such exudates stimulated a lower percentage of chlamydospores to germinate, but there was no lysis of germlings, and so hypocotyl infection followed in due course.

This successful investigation merits close study within the context of my particular theme. Snyder and his associates began by making a close study of the natural situation, by employing microscopic observation of the events leading up to infection, and by analyzing the sugar and amino acid composition of exudates from germinating seeds, root lengths of various ages, and hypocotyls. They followed this with a model study of the effects of various synthesized exudates upon the behavior of chlamydospores in natural soil, with respect both to chlamydospore germination and to the fate of the germlings. Although at one stage they made use of porous-porcelain seed models impregnated with a synthetic exudate, the visual resemblance of this particular model to a bean seed really added little to its value, except perhaps to give some additional aesthetic satisfaction. And so my final words are these. There is no need for a model to look like the real thing; it only has to function like it.

LITERATURE CITED

1. BAKER, K. F., and R. J. COOK. 1974. Biological control of

plant pathogens. W. H. Freeman, San Francisco. 433 p.

2. BAKER, K. F., and W. C. SNYDER. 1965. Ecology of soil-borne plant pathogens. Univ. Calif. Press, Berkeley and Los Angeles. 571 p.

3. BRUEHL, G. W., and P. LAI. 1966. Prior colonization as a factor in the saprophytic survival of several fungi in wheat straw. Phytopathology 56:766-768.

4. BURGESS, L. W., and D. M. GRIFFIN. 1967. Competitive saprophytic colonization of wheat straw. Ann. Appl. Biol. 60:137-142.

5. BUTLER, F. C. 1953. Saprophytic behaviour of some cereal root-rot fungi. I. Saprophytic colonization of wheat straw. Ann. Appl. Biol. 40:284-297.

6. BUTLER, F.C. 1953. Saprophytic behaviour of some cereal root-rot fungi. II. Factors influencing saprophytic colonization of wheat straw. Ann. Appl. Biol. 40:298-304.

7. COOK, R. J. 1970. Factors affecting saprophytic colonization of wheat straw by Fusarium roseum f. sp. cerealis 'Culmorum'. Phytopathology 60:1672-1676.

8. COOK, R. J., and G. W. BRUEHL. 1968. Relative significance of parasitism versus saprophytism in colonization of wheat straw by Fusarium roseum 'Culmorum' in the field. Phytopathology 58:306-308.

9. COOK, R. J., and M. N. SCHROTH. 1965. Carbon and nitrogen compounds and germination of chlamydospores of Fusarium solani f. phaseoli. Phytopathology 55:254-256.

10. COOK, R. J., and W. C. SNYDER. 1965. Influence of host exudates on growth and survival of germlings of Fusarium solani f. phaseoli in soil. Phytopathology 55:1021-1025.

11. DEACON, J. W. 1973. Behaviour of Cercosporella herpotrichoides and Ophiobolus graminis on buried wheat plant tissues. Soil Biol. Biochem. 5:339-353.

12. GARRETT, S. D. 1970. Pathogenic root-infecting fungi. Cambridge Univ. Press, London. 294 p.

13. GRIFFIN, D. M. 1972. Ecology of soil fungi. Chapman and Hall, London. 193 p.

14. LUCAS, R. L. 1955. A comparative study of Ophiobolus graminis and Fusarium culmorum in saprophytic colonization of wheat straw. Ann. Appl. Biol. 43:134-143.

15. PAPAVIZAS, G. C., and C. B. DAVEY. 1961. Saprophytic behaviour of Rhizoctonia in soil. Phytopathology 51:693-699.

16. SCHROTH, M. N., and W. C. SNYDER. 1961. Effect of host exudates on chlamydospore germination of the bean root rot fungus, Fusarium solani f. phaseoli. Phytopathology 51:389-393.

17. SCHROTH, M. N., T. A. TOUSSOUN, and W. C. SNYDER. 1963. Effect of certain constituents of bean exudate on germination of chlamydospores of Fusarium solani f. phaseoli. Phytopathology 53:809-812.

18. TOUSSOUN, T. A., R. V. BEGA, and P. E. NELSON. 1970. Root diseases and soil-borne pathogens. Univ. Calif. Press, Berkeley and Los Angeles. 252 p.

19. TOUSSOUN, T. A., and W. C. SNYDER. 1961. Germination of chlamydospores of Fusarium solani f. phaseoli in unsterilized soils. Phytopathology 51:620-623.

PART II

DAMAGE CAUSED BY SEED- AND SOIL-BORNE ROOT-INFECTING FUNGI

Incidence of Root Rots—Their Prediction and Relation to Yield Losses: A Glasshouse Study

MARION H. EBBEN and F. T. LAST—*Glasshouse Crops Research Institute, Littlehampton, Sussex, and Institute of Tree Biology, Edinburgh, United Kingdom.*

In recent years, great strides have been made in studying the epidemiology of foliar pathogens. Because of their nature, the macroscopic symptoms of damage caused by these pathogens can usually be recorded without destructive sampling procedures. Consequently, it has often been possible to use successive observations of the same set of host plants to construct progress curves with a greatly decreased error component, changes of slope being related to differing environmental conditions. Root diseases, however, are less amenable, and as yet pathologists have not solved the problem of obtaining satisfactory estimates of root disease without destroying entire root systems.

Our understanding of the epidemiology of root pathogens is rudimentary compared with that of foliar pathogens, particularly when we consider the important group of soil fungi that debilitate, rather than kill, their hosts.

To some extent, our lack of understanding of roots-in-disease reflects the dearth of knowledge concerning roots-in-health. Most treatises concerned with quantitative aspects of spore dispersal and subsequent deposition on "susceptible" sites are concerned with air-borne propagules; fewer are concerned with those depending on splash dispersal, and at the bottom of the list are those concerned with an edaphic milieu. Recently, Baker et al. (2) described their approach to (a) motile and nonmotile propagules of soil-borne or soil-living pathogens and (b) motile and nonmotile infection sites on roots. This attractive essay on root pathology could perhaps be made more meaningful if we were able to predict the searching ability of root systems. Is it possible to construct the spread of a root system from relatively few parameters as is being done for the canopies of trees? From the models produced, could the probability of susceptible roots contacting viable propagules of pathogens, variously distributed within soil, be predicted? Are all sections of a root system equally important in terms of nutrient absorption, as anchors, as sources and sinks of growth regulators? As information is assembled on such factors, the significance of debilitating pathogens might be put into perspective.

In Western Europe, glasshouse tomatoes are affected by a range of soil-borne pathogens; the damage done by species of *Verticillium* and *Fusarium* was recognized many years ago, whereas that done by *Pyrenochaeta lycopersici* (8), formerly considered to be a grey sterile fungus (GSF), was only recently identified (10). *Pyrenochaeta lycopersici* affects the development of tomato roots by causing the development of (a) brown lesions on small roots and (b) brown furrowed bark, "corkiness", on large roots, these symptoms being associated with variable amounts of root pruning (Fig. 1). These developments debilitate the host, causing yield decreases if the incidence of symptoms, collectively known as brown root rot (BRR), exceeds 20% within 12 weeks of planting (6).

Before the recent search for disease resistance in wild species and in the absence of an acceptable rotation system, soil-borne pathogens of glasshouse crops were commonly controlled by partial sterilants of soil. How frequently should partial sterilants be applied to effectively minimize yield losses caused by *P. lycopersici*? Such a decision depends upon the nature of the sterilant used and on the characteristics of the edaphic environment, which, together with the inherent properties of the pathogen, combine to influence the subsequent population dynamics of the surviving propagules of *P. lycopersici*.

Having arranged experiments to study the epidemiology of BRR, we gradually realized that glasshouses are the commercial equivalent of controlled environment chambers. In them, it is possible to control soil moisture and regulate the supply of nutrients. With this greater control, it is not surprising to find that the incidence of *P. lycopersici* is less erratic than that of *Gaeumannomyces graminis* (syn. *Ophiobolus graminis*) attacking cereals and causing take-all. Perhaps open-ended polyethylene "glasshouses" with trickle irrigation systems might prove useful in field experiments with root diseases of agricultural crops; they could enable at least soil moisture to be largely controlled.

This paper describes an extension of earlier work done on unheated glasshouse tomato crops grown at Luddington Experimental Horticulture Station, Warwicks., from where the increasing incidence of BRR on the roots of five successive tomato crops grown in untreated soil was reported by Last et al. (7). The severity of symptoms increased within each crop and in successive crops, the pattern defined by regression analyses accounting for 72.6% of the variation with a coefficient of variation of 24.3%.

CULTIVATION DETAILS.—Late-planted crops of Moneymaker were grown for 5-6 months in unheated glasshouse dates varied from 12 May to 12 June. The crop 3 months. From 1963 to 1971, sowing dates varied between 21 March and 26 April, and transplanting to the glasshouse dates varied from 12 May 59 12 June. The crop was harvested between 14 July and 1 November. Within six houses, 48 plots (each 4.9 m² and with 24 plants) were separated by concrete slabs buried to a depth of 0.45 m. The results detailed in this paper were obtained from plants grown in the four replicate plots that remained untreated by any method of soil sterilization between 1963 and 1971. Other treatments in the experiment involved the use of partial soil sterilants, including steam and metham sodium (1963-1965) and methyl bromide (1966-1971).

Base dressings were applied to all plots before planting; rates per m² varied slightly from year to year but were ca. 70 g hoof and horn, 70 g superphosphate, 70-140 g sulfate of potash, and 130-200 g magnesium sulfate. Soluble nutrients were applied when watering through a ground level trickle irrigation system with one nozzle per plant. Amounts were based on recommendations in *Liquid*

Fig. 1. Moderately affected roots of glasshouse tomatoes grown for approximately 5.5 months in soil infested with *Pyrenochaeta lycopersici*, the cause of brown root rot. Inset shows details of corkiness on most roots and brown lesions on others.

Feeding of Tomatoes, Ministry of Agriculture, Fisheries and Food, AL 520.

OBSERVATIONS.—Two roots per plot were sampled for disease assessment at approximately monthly intervals after planting between 1963-1967 and at two monthly intervals in 1968-1971. They were lifted in blocks of soil 400 cm³, when root systems were freed of excess soil before being washed. Disease incidence was assessed as % BRR, or the percentage of a surviving root system with brown lesions (Table 1).

TABLE 1. Mean incidence of percentage brown root rot (*Pyrenochaeta lycopersici*) on nine annual tomato crops grown successively in the same untreated glasshouse soil

Cropping season	Weeks after planting								
	5	8	9	13	16	17	21	23	24
1963	3		4	7		11	15		
1964	7		11	14		20	18		
1965	7		9	14		25	41		
1966	9		15	22		32	49		
1967	8		20	34		46	60		
1968		17			45				58
1969		16			37			58	
1970			18		40			54	
1971		18			45				58

Multiple regression equations were calculated using data for the 9-year period 1963-1971, the equations describing changes in observed amounts, "y", of BRR at increasing intervals from planting within seasons, "x_1", and in different seasons, "x_2". The fitted equations were of the form:

$$y^1 = b_o + b_1 x_1 + b_2 x_2 + b_{12} x_1 x_2 + b_{11} x_1^2 + b_{22} x_2^2$$

where y = proportion of root diseased, expressed as a fraction sensu van der Plank (11).

x_1 = weeks after planting; coded 0 for week 13, − or + 0.125 per week earlier or later,

x_2 = year of experiment; coded 0 for 1966, − or + 0.4 per year earlier or later.

Two transforms of y were sometimes tested (a) $y^1 = 100$ y, converting proportions to percentages, and (b) $y^1 = $ arc sin$\sqrt{y}$, the angular transform. With the data for the period 1963-1971, analyses using angular transforms and percentages of y accounted for 75.3% and 72.7% of the variation and gave coefficients of variation 23.6 and 36.0, respectively. A similar result was obtained earlier using data for 1963-1967. From the 1963-1971 data, the following equation was derived:

$$y^1 = b_o + b_1 x_1 + b_2 x_2 + b_{12} x_1 x_2 + b_{22} x_2^2$$

$$y = 29.275 + 10.476 x_1 + 8.037 x_2 + 2.181 x_1 x_2 - 3.354 x_2^2.$$

Although the incidence of % BRR increased progressively within all nine seasons, comparisons between seasons suggest that maximal levels of infection were reached in 1969/70; i.e., after 7 years of continuous cropping (Fig. 2). Thus a mean maximum of ca. 63%

BRR was reached in the 8th year and was detected 24 weeks after planting. For early-season infections recorded 8 weeks after planting, a peak of ca. 19% was recorded two or three seasons earlier in 1968/69; the midseason peak occurred in 1969. Notwithstanding the indication that disease incidence was no longer significantly increasing above the amounts recorded for the seventh and eighth seasons, there is a reasonable relation ($P = 0.001$) between the midseason incidence of BRR in one year (x) with that in the next (y):

$$y = 14.44 + 0.63x.$$

This equation indicates that 10% BRR recorded 16 weeks after planting in one year will increase to ca. 21% in the succeeding year, but 60% would instead decrease to 52% (Fig. 3), with 40% being the break-even point. Interest in the buildup of BRR is associated with effects on yield. To calculate % yield decreases (Fig. 4), the four untreated plots were paired with four treated plots, three of which had been partially sterilized annually with metham sodium or methyl bromide; the fourth plot was not treated in 1966, but was treated annually before and after. Yield losses estimated when picking ceased at the end of the season were significantly related to amounts of BRR

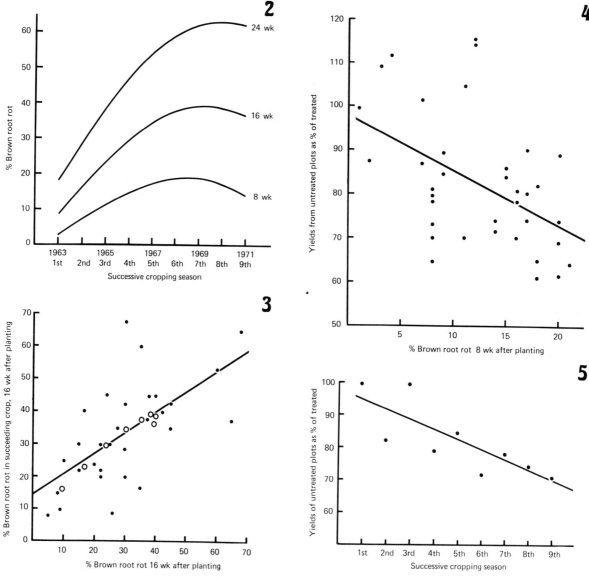

Fig. 2-5. Fig. 2. Fitted curves showing the effects of repeated cropping on the incidence of tomato brown root rot (BRR) 8, 16, and 24 weeks after planting. Fig. 3. Relation between the incidence of tomato BRR 16 weeks after planting with that at a similar stage in the development of the immediately previous crop. o = data from fitted curves. Fig. 4. Relation between tomato yields and the incidence of BRR recorded 8 weeks after planting. Yields from the experimental untreated plots are expressed as percentages of those from tomatoes growing in "sterilized" soil. Fig. 5. Effects of continuous cropping on tomato yields.

recorded 8 and 16 weeks after planting, but not with those observed after 24 weeks. Increasing the amount of BRR 8 weeks after planting from 10 to 15% decreased yields by 7%.

Because amounts of BRR recorded 8 weeks after planting seemed to have reached a peak of ca. 20% in the sixth and seventh crops, it might be expected that yield losses would not continue to increase in severity. In practice, however, crops grown in the eighth season yielded less than those in the seventh, and those in the ninth yielded less than those in the eighth (Fig. 5), suggesting that repeated cropping was influencing the host's response to infection.

DISCUSSION.—The increase of tomato brown rot in a glasshouse was not subject to the irregularities commonly found in recording the incidence of debilitating root diseases of field crops, e.g., take-all of cereals. This difference is probably attributable to the effects of contrasting soil conditions, e.g., soil moisture, which remain relatively unvarying during glasshouse cropping but are subject to the vagaries of weather in the field.

One of the aims of epidemiology is to acquire the knowledge that will enable the incidence of disease to be predicted. In this instance, attempts were made after 5 and 7 years of continuous cropping, but, because the rate of increase lessened after six seasons, cropping

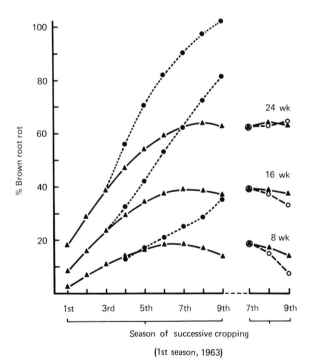

Fig. 6. Within and between season changes in the incidence of brown root rot (BRR) caused by *Pyrenochaeta lycopersici* when tomato crops were cultivated in 9 successive years. Curves fitted using 1963-1971 data (▲—▲); the predictions of BRR incidence from its occurrence during 1963-1967 (●.....●) and 1963-1969 (○–○). Disease incidence curves are given for 8, 16, and 24 weeks after planting.

predictions made using data from five seasons were totally inappropriate. Predictions made after 7 years tended to exaggerate the magnitude of the subsequent decline. The theoretical curve calculated from data for the first 5 years suggested that there would be 35% BRR after 8 weeks in the 9th year; the curve using data for years 1-7 predicted 7.5%, whereas the fitted curve for years 1-9 gave 14% (Fig. 6).

The increase of disease through the first 7 years of cropping followed the pattern postulated for increasing concentrations of nonmotile inocula of root-infecting pathogens (1). The subsequent decreases in the 8th and 9th years were small compared with the declines commonly recorded for take-all of cereals. Although soil-living actinomycetes antagonistic to *P. terrestris*, the pink rot pathogen of onions, have been isolated (4), it is likely that the curtailed increase of *P. lycopersici* should be sought in terms similar to those explaining the population dynamics of the potato cyst nematode, *Heterodera rostochiensis* (9). A stage is reached when the host's root system is damaged so severely and at such an early stage that it never attains a size to enable populations of *P. lycopersici* to be self-perpetuating, always supposing that (a) its infective propagules are primarily formed as a result of its association with living susceptible hosts and (b) saprophytic phases are of little consequence.

Although the evidence from crops grown at Luddington does not suggest that repeated cropping increases the significance of biological control agents, they could be responsible, in part, for site differences. *Pyrenochaeta lycopersici* increased more rapidly at Luddington than at the Experimental Horticulture Station at Hoddesdon, Herts.

Because *P. lycopersici* causes root pruning and because estimates of BRR are understandably restricted to the roots that survive, it would be desirable to take into account the effects of *P. lycopersici* on root system size. This should be done bearing in mind that many plants can tolerate appreciable root loss without their profitable yields being adversely affected (5). In the series of crops described in this paper, seasonal tomato yields continued to decrease even when the incidence of BRR lessened after the seventh season. Such a result may reflect the inadequacy of our system for recording the effect of root disease on root size. On the other hand, it could suggest that continuous cropping may, by other means, influence the ability of plants to tolerate attack. But basically the trouble lies in our inadequate background knowledge of the nature and function of the roots of healthy plants. Roots are in dynamic equilibrium with other components of growth; to be able to infer damaging relationships, the nature of this relation needs to be better understood. Would % BRR recorded 8 weeks after planting be of equal significance to crops harvested for periods of 3 and 5 months? Similarly, would the length of cropping greatly affect the significance of % BRR recorded 16 weeks after planting as an indicator of the amounts of inocula jeopardizing succeeding crops?

Interest in the mathematically defined progress of BRR on this site must lie in the probability that the same pattern is repeated on other soils. Factors that could be expected to affect the rate of increase on different sites would be (a) initial inoculum levels, (b) virulence of the pathogen, e.g., the proportion of highly virulent strains may be increased when a susceptible crop has been grown previously, and (c) rate of host growth. Soil temperature, plant nutrition, and shoot growth could all influence these factors.

On this site, disease levels significantly damaging to crop yields occurred during the log phase of disease increase when BRR 16 weeks after planting was ca. 30%. Data from fumigation treatments using methyl bromide at different concentrations on soils with different initial inoculum levels show that high dosages or annual fumigation at lower rates can decrease BRR incidence to levels below those damaging to yields (3). However, disease incidence was always higher than the very low levels found in the first 2 years of cropping on the virgin site, and the subsequent increase of disease on fumigated soils has been observed to be more rapid than that on untreated plots of the original site.

As with other debilitating diseases, total eradication of the pathogen appears to be neither necessary nor in practice possible, and an increasing knowledge of epidemiology should enable control measures to be more logically determined.

ACKNOWLEDGEMENTS.—We thank E. A. Turner and the director and staff of Luddington E.H.S. for their assistance and cooperation, and J. H. Gisborne, G.C.R.I., for mathematical assistance.

LITERATURE CITED

1. BAKER, R. 1968. Mechanisms of biological control of soil-borne pathogens. Annu. Rev. Phytopathol. 6:263-294.

2. BAKER, R., C. L. MAURER, and RUTH A. MAURER. 1967. Ecology of plant pathogens in soil. VII. Mathematical models and inoculum density. Phytopathology 57:662-666.

3. EBBEN, MARION H. 1971. Tomato brown root rot: the build-up of soil inoculum and its control by fumigation, p. 243-250. Proc. 6th Br. Insecticide Fungicide Conf. (1971).

4. FREEMAN, T. E., and E. C. TIMS. 1955. Antibiosis in relation to pink root of shallots. Phytopathology 45:440-442.

5. LAST, F. T. 1971. The role of the host in the epidemiology of some nonfoliar pathogens. Annu. Rev. Phytopathol. 9:341-362.

6. LAST, F. T., and M. H. EBBEN. 1966. The epidemiology of tomato brown root rot. Ann. Appl. Biol. 57:95-112.

7. LAST, F. T., M. H. EBBEN, R. C. HOARE, E. A. TURNER, and A. R. CARTER. 1969. Build-up of tomato brown root rot caused by Pyrenochaeta lycopersici Schneider and Gerlach. Ann. Appl. Biol. 64:449-459.

8. SCHNEIDER, R., and W. GERLACH. 1966. Pyrenochaeta lycopersici nov. spec., der Erreger der Korkwurzelkrankhreit der Tomate. Phytopathol. Z. 56:117-122.

9. SEINHORST, J. W. 1970. Dynamics of populations of plant parasitic nematodes. Annu. Rev. Phytopathol. 8:131-156.

10. TERMOHLEN, G. P. 1962. Onderzoekingen over kurkwortel van tomaat en over de kurkwortelschimmel. Tijdschr. Plantenziekten 68:295-367.

11. VAN DER PLANK, J. E. 1963. Plant diseases: epidemics and control. Academic Press, New York and London. 349 p.

Gnotobiotic Assessment of Plant Health

WILLIAM A. KREUTZER AND RALPH BAKER—Department of Botany and Plant Pathology, Colorado State University, Fort Collins.

It has long been the desire of plant pathologists to understand the interacting systems of higher plants and their soil-borne pathogens. The principal drawback has been a lack of effective methodology. Since the days of de Bary and Kühn, the procedure used in proving pathogenicity of a soil-borne organism has been infestation of steamed rooting medium with the suspected pathogen before planting. Modifications of this method are used to study the interrelations of microorganisms as well as their effects on higher plants. Although much information has been obtained with this technique, future research in higher plant-soil microbial interactions will require more precise tools. Biological control of the environment must be added to physical control. One can remove the biological components of the environment, then put them back, one by one. In considering this problem, Braun (12) wrote, "Environment if broken into its component parts is no longer environment." In principle we agree; but, unfortunately, this leads to an experimental cul-de-sac. We must violate this truism. We must invoke the science of gnotobiology.

DEFINITIONS.—Reyniers et al. coined the term gnotobiotic to describe an environment consisting of only known organisms (23). The term is derived from two Greek words: γνωτος (gnotos), meaning known or understood, and βιος (bios), meaning life; hence, "known life." The term axenic is from the Greek ξενος (xenos), meaning foreign or strange; hence, "without foreign (life)," and is used to describe an environment in which only a single species is present. This terminology was extended to include monoxenic (one known organism), dixenic (two organisms), trixenic (three organisms), etc. (23). Thus far, the idea has not caught on.

In this paper we use four terms: sterile (no organisms present in the environment), axenic (only one organism present), gnotobiotic (more than one organism present, all known), and agnotobiotic (more than one organism present, one or more unknown).

HISTORY.—In the first recorded axenic experiment, Boussingault (9) grew legumes. Pure culture techniques were later adopted. Until the turn of the century, gnotobiotic research on higher organisms was focused on the role of microorganisms in the nitrogen nutrition of higher plants (41). Because axenic methods involving higher plants were cumbersome and relatively expensive, their use declined over the next 30 years. Axenic culture became the foundation for microbiological investigations, however, and was used in tissue culture (116).

Increased funding for medical research revived interest in gnotobiotic techniques for work on higher animals (64). With the advantage of improved and less expensive methods developed in these animal investigations, research on higher plants was renewed (59, 61, 66, 113).

METHODOLOGY.—For over 100 years, investigators have constructed axenic-gnotobiotic chambers (isolators) for plants. Isolators were of two types (Fig. 1, 2): those for the maintenance of axenic root systems only (8, 11, 49, 96, 98, 119), the apparatus of current choice (Fig. 1E) being that designed by Stotzky et al. (96); and those designed for the axenic maintenance of entire plants (1, 6, 15, 19, 28-30, 34, 41, 46, 48, 49, 57, 59, 73, 99, 109).

Modern techniques in axenic research (37, 59, 60, 66, 113) employ plastic isolators (Fig. 3) adapted from those used in animal studies (64). Some unfavorable effects were presumably caused by vapors from the polyvinyl sheeting used in isolator manufacture. Our observations confirm those of Fujiwara et al. (31), who noted abnormal growth in some plants and attributed it to diisobutyl phthalate vapors from polyvinyl sheeting. Some isolator plastics are autoclavable (57) to avoid the use of chemical sterilants. Isolators and introduced equipment are usually sterilized with peracetic acid, decomposing to acetic acid and water (5).

Autoclaved nutrients and water are introduced into chambers through sterilized locks. For prolonged plant culture, it is desirable to have reservoirs of sterile nutrients external to the plant isolators (19, 28, 58, 99, 113). We supplied continuous sterile nutrients from an external reservoir. Success depends on housing the reservoir and its outlet tubing within larger sterile isolators.

As a rooting medium, most investigators prefer synthetic products like vermiculite, perlite, or BR-8 blocks (59, 107). Soil is not suitable because heat treatments alter its structure, increase its soluble constituents, change its organic composition, and lower its water-holding capacity (108).

Germ-free propagules may be obtained either by surface sterilization or by removal of sterile tissues from embryos, buds, or plant organs. Seed sterilization usually is accomplished with liquid or volatile "germicides."

These include mercuric chloride, Formalin, hydrogen peroxide (109), calcium or sodium hypochlorite (96, 118), propylene oxide (104), peracetic acid (59), and sodium mercuricthiosalicylate (36). Because of its superior germicidal properties, ease of application, and relative nonphytotoxicity, sodium hypochlorite is preferable. Even the most effective germicide, however, is not capable of eliminating microorganisms when contamination is within plant tissues (60, 76, 104, 113,

117). We found it difficult to remove resident bacteria from propagative material (60, 113). Tyner (104) had some success by fumigating moistened seed with propylene oxide or heating to 50°-60° C. However, such treatments usually injure embryonic tissues. In some cases, seed may be freed of internal contamination if produced in an environment devoid of free water during flowering and fruiting, e.g., in a greenhouse.

To detect external contamination, we flooded all 12-

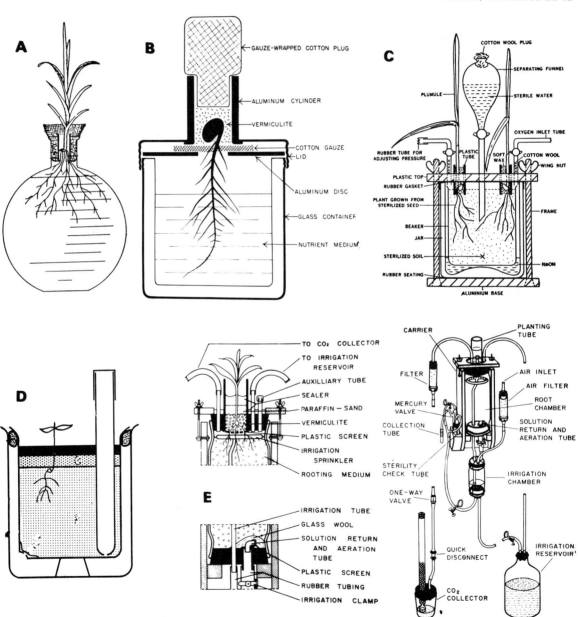

Fig. 1. Isolating systems and accessory units designed to maintain root systems axenically. *A)* Early apparatus used to grow "sterile" root systems with cheesecloth and cotton barriers in neck of flask (119). *B)* Jar unit containing surface-sterilized Lima beans; roots are forced down into sterile medium, while top is pushed into air (8). *C)* Apparatus used to compare oxygen consumption of sterile roots with roots containing rhizosphere flora (98). *D)* Growth chamber consisting of a 400-ml beaker within a 600-ml beaker, with cotton seal between beaker lips. Paraffin-coated sand topped with mercury resin (smaller beaker) provides contamination barrier. Sterile nutrient is added through filler tube (11). *E)* Planting tube detail (left) and plant culture unit (right) of apparatus designed to grow axenic root systems (96). (Copies of diagrams by permission.)

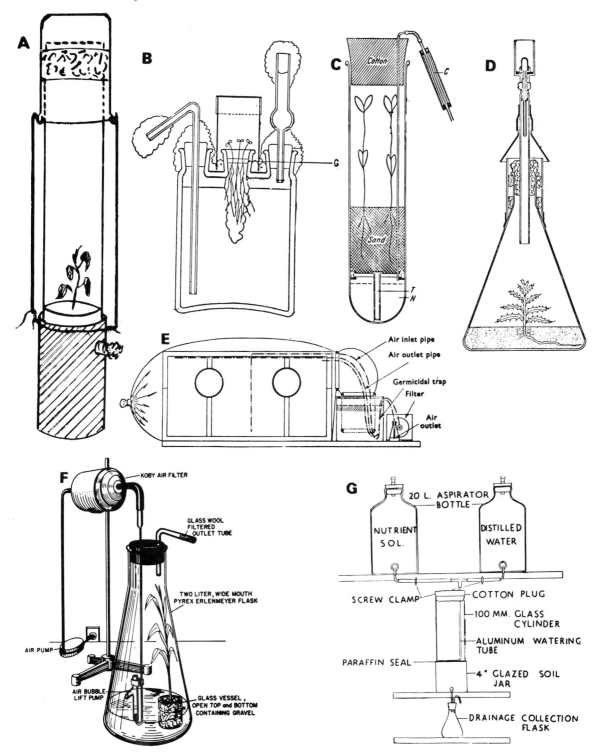

Fig. 2. Isolating systems and accessory units designed to maintain plants axenically. *A)* Early apparatus consisting of loose interconnecting glass cylinder barriers (29). *B)* Modified three-necked Wolff bottle for growing germ-free plants; nutrient agar layer (G) monitors for presence or absence of microorganisms (6). *C)* Small tube unit, in which nutrient solution (N) at bottom can be forced into sand through tube (T) by air pressure (73). *D)* Berkefeld filter for gas exchange fitted over an Erlenmeyer flask (1). *E)* Autoclavable modern plastic unit designed originally for large animal research (57). *F)* Simple apparatus for introduction of fresh air into a germ-free environment (34). *G)* Culture apparatus designed to permit introduction of nutrient solution to axenic plants (46). (Copies of diagrams by permission.)

day-old seedlings with nutrient broth and incubated submerged plants for 96 hours before rinsing with sterile water (113). After axenic seedlings were established in pots, the contents were monitored on a weekly basis by incubating debris and plant tissues in fluid thioglycollate, peptone broth, Sabouraud-dextrose broth, and Bacto AC liquid medium at 22° and 35° C for 35 days.

Additional details of axenic-gnotobiotic methods for plants are available in the literature (59-61, 66, 67, 113).

HIGHER PLANT AXENIC SYSTEMS.—*Resident bacterial forms in plant tissues.*—Reports of bacteria or bacteria-like forms within tissues of healthy plants are numerous (20, 42, 75, 86-88, 100, 101). Perotti (75) and Tonsig and Bracci-Orsenigo (101) noted bacteria within stems, leaves, tubers, and roots of healthy herbaceous plants. Hennig and Villforth (42) discovered spore-forming bacteria within tissues of 28 different plants. Sanford (87) reported bacteria in apparently healthy potato and green bean stems, in alfalfa and sweet clover taproots, and in potato tubers. Unidentified bacterial forms have been found within seeds and plants of certified bean varieties (88) and within healthy carnation seeds, stems, and leaves (20). *Bacillus subtilis* has been found in healthy soybean root, stem, and leaf tissues (26), peanut cotyledons (76), and peanut tissues (113). *Bacillus megaterium* has been found in healthy potato tuber tissues (43) and in *Trifolium* taproots (77).

Although Colorado workers (60, 113) grew presumed axenic peanuts from seedlings to mature plants in isolators, most plants between 50 and 90 days of age had endogenous bacteria-like microflora of two general kinds: pleomorphic forms and *B. subtilis* types. Pleomorphic forms were isolated from mature stem sections in Bacto AC liquid medium or fluid thioglycollate broth, manifested as tenuous, amorphous, and cloudy growths. At first, growths were plant tissue-dependent and anaerobic, forming no surface pellicles. Microorganisms resembled spheroidal L-forms, or appeared as coccoid to rod-like forms and tenuous filamentous forms. These could be established only in liquid broth containing peanut tissues. Later, independent colonies were grown on solid media. Isolates were coccoid and rod-like; representative forms were identified as *B. subtilis*. *Bacillus subtilis* produces potent antibiotics. Pettit et al. (76) isolated a strain of *B. subtilis* from interiors of peanut cotyledons antagonistic to species of *Macrophomina, Choanephora, Chaetomium,* and *Penicillium*.

By aseptically removing cotyledons from surface-sterilized peanut embryos, Hale (37) succeeded in eventually producing axenic peanut plants. Using similar techniques, we also grew axenic plants; however, they were always stunted and abnormal.

Practical aspects.—Colorado's carnation industry is based on axenic methodology (3, 78).

With the exception of Mangin's early techniques (69), use of pathogen-free propagative material (chrysanthemums) by culture indexing was probably first suggested by Dimock (21). This system was adopted by the carnation industry; however, presence of virus diseases and infections from *Fusarium roseum* f. *cerealis*

disseminated on plant surfaces forced a reassessment of disease control methods. An approach for disease control devised by Baker and Phillips (3) was based upon increasing knowledge of tissue culture (116, 117) and the lead supplied by Quak (80), who first showed that axenic carnation plants could be obtained by culturing shoot tips. Heat-treated plants grown axenically in test tubes of nutrient media by the newly devised procedures were pathogen free (3, 78). Here, tissue culture techniques have been applied to benefit agriculture.

The next improvement for obtaining pathogen-free propagative units appears to be axenic culture of plantlets in quantity derived from cell cultures (7).

Basic aspects.—Nitrogen assimilation.—Axenic methods have clarified the mode of nitrogen absorption and assimilation. Prior to 1825, it was believed that plants obtained their nitrogen directly from the atmosphere (56). By 1855, Boussingault (9), using modified axenic methods, demonstrated that this was not true. Since that time, axenic techniques have been used to determine the types of nitrogenous compounds utilized by plants. In 1911, Hutchinson and Miller (45) demonstrated that a wide range of nitrogenous compounds could be absorbed by axenic root systems. Later investigators (64) showed that nitrate and ammonium salts absorbed through roots are principal sources of nitrogen. Although amino acids are absorbed by axenic root systems (e.g., tobacco), they apparently are not utilized to any extent as nitrogen sources (35). Indeed, amino acids actually can be phytotoxic (94).

The so-called carnivorous plants are unique because of their alleged need for raw protein. Bladderwort (*Utricularia* spp.), sundew (*Drosera* spp.), and Venus flytrap (*Dionaea* spp.) entrap and ingest small insects. It

Fig. 3. Individual polyvinyl plant isolator in current use in the Gnotobiotic Laboratory, Colorado State University. Technician shown in process of monitoring for surface contaminations.

follows that axenic growth techniques should be ideal for nutrient studies. Pringsheim and Pringsheim (79) grew the bladderwort axenically on inorganic nutrients, but it did not flower unless beef extract was added. Although the butterwort (*Pinguicula lusitanica*) grew axenically and flowered in a mineral medium, leaf and flower development were enhanced following "feeding" with *Drosophila* and egg yolk (39).

Carbon assimilation.—Contrary to Duclaux's initial contention (25) that plants could not absorb carbohydrates through their roots, Loo (63) found that sucrose absorbed through the roots of axenically grown *Baeria* accelerated initiation and production of flowers. Langridge (55) grew the crucifer, *Arabidopsis thaliana*, in axenic culture and found it could absorb and utilize different sugars. However, in axenic studies, Knudson and Smith (53) confirmed Duclaux's finding that corn and pea roots would not absorb starch.

Phosphorus and trace elements.—Schulow (91) grew maize and pea root systems in "sterilen Kulturen" and studied the absorption and assimilation of inorganic and organic phosphorus compounds. Steinberg (93) used axenically grown duckweed (*Lemna minor*) to determine deficiencies of Fe, Mn, and Ca, and recommended the technique employed to study trace element nutrition in crop plants.

Root exudations.—Within recent years, comprehensive reviews of root exudations and their influence on soil-borne plant disease have been published (83, 90). In 1920, Knudson first demonstrated that axenic Canada field pea and maize roots formed reducing sugars from sucrose, indicating that roots secrete invertase. Organic nitrogen and organophosphate substances also were found in root excretions of axenically grown seedlings and mature plants. Later, amino acids, enzymes, carbohydrates, vitamins, and nucleotides were also detected. Roots from axenically grown field peas secreted 22 different amino-type compounds; oats secreted 14 similar substances. Components of exudates may vary qualitatively and quantitatively with environment and/or plant type, age, or health. Most recently, low levels of calcium (not deleterious to growth) have been found to increase peanut cell membrane permeability, resulting in increased exudation of sugars (92).

Clearly, axenic roots excrete many diverse biochemical substances.

AXENIC-GNOTOBIOTIC SYSTEMS.—*Higher plant-unspecialized microbiological interactions.*—Before the French Academy in 1885, Duclaux (25) considered the possibility of an obligate relationship between soil microorganisms and higher plants: "La destruction, par les microbes, de la matière organique du sol, et la production sur ce sol d'une vegetation nouvelle, sont deux phénomènes qui s'accompagnent constamment. Ont-ils une relation necessaire l'un avec l'autre?"

Duclaux reported that axenically grown peas and beans could not utilize casein, starch, or sugars, as enzymes essential for the degradation of these substances apparently were not secreted by plant roots. These roots behaved like those grown by Boussingault in distilled water 30 years earlier (9).

In assessing the importance of Duclaux's presentation, Pasteur (74) correctly concluded that microbial degradation of Duclaux's amendments was essential for their ulitization by higher plants, but incorrectly concluded that plants could not live in the absence of microorganisms: "... avec la pensée preconcue que la vie, dans ces conditions, deviendrait impossible...."

The effects of saprophytic soil microorganisms on the growth and development of higher plants have been the subject of speculation. For example, Kreutzer (54) classified all soil organisms as either plant sustainers or plant inhibitors. Most information is on the effects of saprophytes on plant pathogens. Although many microbial antagonists of plant pathogens are known, their activities and importance in plant disease control have been questioned (2).

The direct effects of saprophytic microorganisms on higher plant health are even less understood. The complexity of this problem was emphasized by Domsch (22), who determined influences of 25 fungal soil saprophytes on representative crop plants agnotobiotically. All stages from marked stimulation to marked inhibition were seen in test plants. Clearly, investigations of this type are best conducted in gnotobiotic systems.

Results obtained from gnotobiotic studies using individual or mixed soil saprophytes with higher plants indicated that axenic root systems were generally superior to their infested counterparts. This was especially true where roots were grown in a liquid medium, and if organic material was lacking or limited in quantity.

Maize plants grew better axenically in the absence of organic amendments than with *B. subtilis*; however, where organic supplements were added, plants grew better in its presence (13). Duckweed growth in an inorganic sterile medium was superior to that in a gnotobiotic medium (16). In contrast, growth and reproduction rates of *Lemna* were better in a gnotobiotic system containing *Escherichia coli* or *B. subtilis* than in an axenic system with organic materials. Red clover and sunflower grown in axenic nutrient solutions had a "stärkere wurzelbildung" than infested control plants (114). Apparently, such depression in gnotobiotic plant growth is caused by root damage (4, 10, 38, 81). Root damage and reduced plant size resulted when gnotobiotic squash root systems were infested with a mixture of soil microorganisms (4). In similar studies, Bowen and Rovira (10) observed reduced root lengths and reduction in numbers and lengths of root hairs in tomato, pine, and subterranean clover seedlings. They concluded that "a predominance of microorganisms causing root and root-hair suppression in the field could materially influence crop yield."

A likely explanation for detrimental effects on root growth is that sloughed root detritus and accompanying root exudations (90) supply a favorable substrate for bacteria, resulting in oxygen depletion and an increase of phytotoxic bacterial metabolites. Other unfavorable results were reported by Reuszer (81), who found that sunflower roots in gnotobiotic liquid media containing mixtures of soil microorganisms became brown and flaccid. Even more root damage was observed by Hameed and Couch (38) when washed conidia of the saprophyte *Penicillium simplicissimum* was added to axenic cultures of marigold. Here is an instance of a nonpathogenic organism becoming a plant pathogen under gnotobiotic conditions.

One indirect effect was reported. Using radiotracer-labeled nutrients, Subba-Rao et al. (97) found that axenic tomato roots infested with *Trichoderma viride* or *Fusarium* sp. suppressed the uptake of phosphate and sulfate ions.

A few reports indicate that plants exposed to microorganisms are superior to axenic control plants. With selected organic materials in the rooting medium, maize grew better gnotobiotically (*B. subtilis* infestation) than axenically (13). In the presence of organic material, duckweed also grew better when microorganisms were added (16).

There have been a few claims (without reference to organic material in the rooting system) that gnotobiotic plants were superior to complementary axenic controls. However, in such instances, root systems usually were grown in a solid artificial medium (sand, gravel, perlite-sand, etc.). Gawel (33) increased top and root weights of a fescue grass by adding an *Arthrobacter* sp. to the axenic rooting medium. Lindsey (59) grew dwarf tomatoes axenically and agnotobiotically by adding 0.05 g soil to each pot. Agnotobiotic plants were larger, taller, and had greater dry weights than axenic controls. Lindsey and Baker (61) found that dwarf tomatoes grown in a *T. viride*-infested environment were significantly taller than those grown in a *Chaetomium* or axenic environment; however, there was no significant difference between plant heights in the *Chaetomium* and germ-free environments. In addition, dwarf tomato plants infested with *Rhizopus nigricans* or *Fusarium roseum* f. sp. *cerealis* had significantly greater dry weights than axenic controls. Lindsey and Baker concluded that fungi growing in the rhizosphere may have growth-regulative effects on tomatoes.

Fungi may be the most beneficial microorganisms in a gnotobiotic system. Two exceptions are indicated. In addition to Gawel's claim (33) of plant stimulation by *Arthrobacter*, Welte and Trolldenier (115) reported that mixed soil bacteria, as well as pure cultures of *Bacillus mycoides*, *Mycobacterium phlei*, and *Pseudomonas fluorescens*, increased the dry weight of gnotobiotically grown red clover over that of axenic controls.

Higher plant-specialized microbiological interactions.—Rhizospheres.—Literature on the rhizosphere is abundant (83, 90); therefore, we shall consider only that resulting from axenic-gnotobiotic research. What knowledge we have about the underlying causes of rhizosphere formation has arisen primarily from axenic experiments. From such research, Schroth and Hildebrand (90) concluded that root exudation is the principal reason for rhizosphere formation.

Gnotobiotic research has indicated that rhizosphere microorganisms may have a direct beneficial effect on higher plants. Cullimore and Woodbine (17) noted that an alga was stimulated by proximity to axenic pea rootlets. Subba-Rao et al. (97) observed that although *T. viride* and *Fusarium* sp. increased glucose assimilation in the tomato rhizoplane, these fungi suppressed the uptake of phosphate and sulfate ions.

Literature on the postulated role of microorganismal antagonists within rhizospheres in the control of soil-borne plant diseases is voluminous (2). There are a few examples in which gnotobiotic methods have been utilized. Tomato rhizosphere-inhabiting strains of *Arthrobacter* sp. lysed hyphae of the pathogens *Pythium debaryanum* and *Fusarium oxysporum* f. *lycopersici* (71). *Bacillus subtilis* var. *niger* from sugar beet rhizospheres prevented growth of the seed-borne pathogen *Phoma betae* in a modified gnotobiotic system (89).

The belief that most microorganismal antagonists exert their effects through the production of antibiotics has been questioned, and there are few examples of significant antibiotic effects in natural higher plant-plant pathogen systems (2). Axenic-gnotobiotic methods could be employed to clarify the role of antagonists.

Pathogens.—Gnotobiotics have been used to establish etiological relationships in higher plant-plant pathogen systems. Tseh-an et al. (102) studied the parasitism of *Pratylenchus penetrans* on axenically grown white clover seedlings. DuCharme and Hanks (24) produced axenic citrus seedlings with the intent of inoculating their roots with germ-free burrowing nematodes (*Radopholus similis*). Steinberg (95) demonstrated the etiological nature of the tobacco disease called "frenching." Using gnotobiotic techniques, he found that the hormonal symptoms could be caused by amino acid diffusates from *B. cereus*, *B. pumilus*, *Corynebacterium simplex*, and, occasionally, *Erwinia carotovora* (94, 95).

In gnotobiotic studies, Lindsey (60) noted that although the toxin-forming *Aspergillus flavus* could colonize peanut pericarps and testae, the fungus was nonpathogenic to peanut tissues. In subsequent studies, Wells et al. (113) and Wells and Kreutzer (112) concluded that under their defined gnotobiotic conditions the fungus was capable of establishing itself first in senescent petals and later in developing gynophores. This food-base principle is well known, having been observed in plant tissue invasion by species of *Botrytis*, *Cladosporium*, *Alternaria*, *Curvularia*, and *Helminthosporium* (32). A similar effect may be seen in seed laboratories where there is frequent massive growth of saprophytic fungi within germination blotters. Because of high inoculum densities (32), these fungi frequently invade and destroy otherwise resistant seedlings. In large measure, what is and what is not a pathogen depends on the nature of the biological environment.

Chain-reaction effects also may be identified by gnotobiotics. Wells et al. (113) found that colonization of immature and mature peanut pericarps by *A. flavus* was reduced in the presence of *T. viride*. However, when *Penicillium funiculosum* was added to the system, it not only nullified this antagonistic effect, but it appeared to stimulate pericarp tissue colonization by *A. flavus*. Although *T. viride* invaded mature and immature pericarps, the fungus was seldom established in peanut testae or seeds.

The effect of soil saprophytes on fungal root pathogens of great economic importance is an interesting area of gnotobiotic investigation. Eaton and Rigler (27) inoculated both axenically grown and agnoto-biotically grown maize with the cotton root rot fungus *Phymatotrichum omnivorum*, and summarized their findings thus: "An important interaction was demonstrated between root-surface saprophytes and the parasitic activity of *Phymatotrichum omnivorum* .... After inoculation with *P. omnivorum*, maize plants growing on sterile sand-bentonite substrates were rapidly attacked and killed. The roots of maize plants on otherwise similar but nonsterile substrates remained healthy."

Normal resistance of certain lines of cotton to *P. omnivorum* also was considered to be the result of fluorescent pseudomonads present on root surfaces.

Gnotobiotic methods have been used to investigate the influence of the host's nutritional state on higher plant-plant pathogen interactions. Alfalfa grown under axenic conditions showed no adverse effects from severe clipping; however, 25 to 57% plant mortality occurred in severely clipped alfalfa inoculated with the suspected root rot pathogen *Fusarium tricinctum* in a gnotobiotic system (67, 68).

Disease resistance and susceptibility are final facets of higher plant-plant pathogen interaction investigated with gnotobiotic methods. Gnotobiotically grown alfalfa resistant to the wilt pathogen *Corynebacterium insidiosum* was obtained by treating with cells of an avirulent strain of the same microorganism (14). Only the test organism, not other microorganisms naturally present in root tissues, was found to be responsible for this induced resistance.

Mycorrhizas.—Much remains to be learned in the field of mycorrhizas. Besides the need for studying effects on the host plant, Harley (40) stresses the need for knowledge on (a) the mechanism of exchange of materials between host and fungus, (b) the problems involved in the establishment of the association, and (c) the host's effects upon the activities of the fungus. Only a beginning has been made in each of these research areas.

Probably the most critical work on mycorrhizal associations was that done over 40 years ago by Knudson (51, 52) who, in elegant but simple axenic experiments, clarified two classical mycorrhizal relationships. In the first study (51), he grew hybrid orchids in glass isolators "free of contamination" for 4 years. Adding glucose aseptically induced flowering without the natural fungal symbiont. Later (52), he demonstrated that heather seedlings without mycorrhizas grew well in the presence of sugar supplements.

Mosse (72) grew clover seedlings axenically in a nitrogen-deficient medium and found that *Endogone* sp. established itself in clover tissues only if a *Pseudomonas* sp. were added to the system. By use of supplementary gnotobiotic techniques to produce inoculum, Ross and Harper (82) were able to precisely evaluate the stimulatory effect of *Endogone* on field-grown soybeans. In investigating another aspect of the problem, Turner (103) found that exudates from some mycorrhizal fungi stimulated elongation of excised *Pinus sylvestris* axenic root radicles.

Marx (70) has reviewed the role of mycorrhizas as biological deterrents of root disease. In gnotobiotic studies, ectomycorrhizas of *Pinus* acted as biological barriers in preventing root infections by *Phytophthora cinnamomi*.

Nodule formers.—Over 80 years ago, Lawes and Gilbert (56) summed up the significance of pioneer axenic plant research on the nitrogen nutrition of higher plants thus: "... it is just about half a century since Boussingault commenced experiments to determine whether plants assimilate free nitrogen. From his results he concluded that they did not; and those obtained at Rothamsted about thirty years ago confirmed the conclusions of Boussingault."

Lawes and Gilbert emphasized the then recent researches of Hellriegel and Wilfarth (41) who, using gnotobiotic techniques, demonstrated that certain soil bacteria were the cause of nodules on leguminous plant roots, bringing about the assimilation of atmospheric nitrogen.

Even though current knowledge of nitrogen fixation by legumes is based on the early use of gnotobiotic methods, little emphasis has been placed on the use of these techniques within recent years. Nevertheless, over the past 40 years a few investigators have employed axenic-gnotobiotic methods in studies of "Knöllchenbakterien" and the remarkable effects of their parasitism. Gnotobiotic techniques were used by Virtanen et al. (105), who found that amino acids were excreted by legume root nodules 20 years before this same phenomenon was discovered in nonparasitized root systems. Later, Weir (111) found that free carbohydrates stimulate nodulation of gnotobiotically grown red clover and alfalfa to the limit set by existing soil nitrogen. Dart and Mercer (18) employed similar methods to study the invasion of barrel medic root hairs and rootlets by *R. meliloti*. More recently, axenic techniques were used by Ljunggren (62) in discovering that exudates essential for root invasion by nodule-forming Rhizobia stimulated the excretion of polygalacturonase from legume roots.

Axenic-gnotobiotic methodology should be valuable in future research even in this well-trodden area.

SYNTHESIS.—*The axenic-gnotobiotic concept: strengths and weaknesses.*—The state of being axenic (germ-free) is a negative concept. In the words of Luckey (64), "The microbe is absent until proven present." Herein lies the weakness of this approach. The assumed axenic condition, which must precede any gnotobiotic condition, is based upon our inability to demonstrate the presence of a foreign organism.

Foreign organisms in a presumed axenic-gnotobiotic system can be found in two major ways, by physical detection and by physiological detection. A third way, chemical detection, although not currently used, is theoretically possible. Physical detection is based either on macrodetection or microdetection. In macrodetection, a microorganism makes its presence known by its external signs (e.g., colonies and sporulating masses) and, if pathogenic, by the symptoms induced in its host. In microdetection, the presence of the microorganism may be determined by signs within plant tissues (e.g., spores, hyphae, and cellular inclusion bodies). In physiological detection, classical culture methodology and optimum incubation conditions are employed.

The general method now used to monitor foreign microorganisms is cultural (physiological) detection. Both strengths and weaknesses are inherent in this method. Assuming the use of reasonably wide ranges of incubation periods and media, cultural methods will detect either obligate saprobes or facultative parasites. Inasmuch as most potential contaminants are saprobic microorganisms readily grown in culture media, it is possible to monitor a broad organismal spectrum. However, by definition, such physiological methods will not reveal the presence of obligate parasities. Nor, with the common media employed, will they detect mycoplasma-like forms. Finally, no true virus can be detected by such methods.

In a given system, being axenic is a concept rigidly restricted by the parameters of the methodology employed. Thus, it cannot be proven that any environment is (a) truly sterile, (b) truly axenic, or (c) truly gnotobiotic. This raises the question of whether "germ-freeness" is the normal state in plant tissues. Is the higher plant a monobiotic system? Some of the evidence reviewed in this paper indicates this may not be so. Although such a concept is counter to all hallowed botanical principles, higher plants could actually be polybiotic systems.

Future uses.—In this paper, we stress the need for methods to clarify higher plant-soil-borne pathogen interactions. More facts relating to the higher plant and its subterranean environment are needed. The entire maze of plant roots, plant pathogens, and symbionts, with its associated specialized and unspecialized saprophytes, can be charted using axenic-gnotobiotic techniques. This research will be expensive and time-consuming. But it must be done.

Looking further into the future, axenic-gnotobiotic methodology will be needed to back up our space probes. No technique previously developed for other purposes has found more unexpected but appropriate applications than gnotobiotics in problems associated with lunar and proposed planetary quarantines. Risks associated with back contamination from extraterrestrial samples are unforeseeable, and extrapolations involving each unexplored extraterrestrial body present a new equation for which there is no model.

Whereas most speculations indicate that extraterrestrial samples would not be harmful to Earth's biosphere, any nation committed to spatial exploration must accept the social responsibility for seeing that no injurious agents escape as a result of the project. For this reason, the United States set up the Lunar Receiving Laboratory for the quarantine and testing of both returned astronauts and the samples collected (47).

Use of axenic-gnotobiotic techniques in such a quarantine is opportune. For example, if symptoms appear on terrestrial organisms challenged with extraterrestrial samples, it is imperative that possible causal agents from Earth be absent. It follows that any disease manifestation must be caused by an alien pathogen. Gnotobiotic techniques insure just this sort of environment and possess the added advantage of confining possible extraterrestrial forms within appropriate biobarriers. There is reason to be more fearful of a *Pseudofusarium martianus* from the Syrtis Major than an intelligent bifurcated tetrapod from the Mare Erythraeum.

Lunar samples returned to Earth by our astronauts after the Apollo missions contained no organisms or substances harmful to terrestrial life. However, according to several reports, increases in size and pigmentation occurred when various species of plants were exposed to lunar material (106, 107, 110). This phenomenon is currently being investigated through a National Aeronautic and Space Administration grant at Colorado State University.

Axenic-gnotobiotic techniques have also been applied to avoid possible planetary contamination (44). To forestall confusing signals from life detection devices (50) or to avoid planetary contamination (65), it is necessary to apply stringent measures for sterilizing spacecraft. The discipline reaches staggering dimensions in projecting the concept of gnotobiotic planets (65) in which a programmed ecology is designed to change the character of a planet's surface or its atmosphere. Such a scheme has been advanced by Sagan (85) for Venus.

ACKNOWLEDGEMENT.—The authors are grateful to Eleanor J. Baker for her assistance in the translation of pertinent literature. They also wish to thank Nina Smith for editorial assistance.

LITERATURE CITED

1. ÅSLANDER, A. 1930. A method for growing plants under sterile conditions. Svensk. Bot. Tidskr. 24:111-112.
2. BAKER, R. 1968. Mechanisms of biological control of soil-borne pathogens. Annu. Rev. Phytopathol. 6:263-294.
3. BAKER, R., and D. J. PHILLIPS. 1962. Obtaining pathogen-free stock by shoot tip culture. Phytopathology 52:1242-1244.
4. BARKER, H. A., and T. C. BROYER. 1942. Notes on the influence of microorganisms on growth of squash plants in water culture with particular reference to manganese nutrition. Soil Sci. 53:467-477.
5. BARRETT, J. P. 1958. Sterilizing agents for Lobund flexible film apparatus. Proc. Animal Care Panel 9:127-133.
6. BEAUMONT, A. B., and G. J. LARSINOS. 1928. Method for growing small-seeded plants under sterile conditions. Science 67:350-351.
7. BEN-JAACOV, J., and R. W. LANGHANS. 1972. Rapid multiplication of chrysanthemum plants by stem-tip proliferation. Hortscience 7:289-290.
8. BLANCHARD, F. A., and VIOLET M. DILLER. 1950. Technique for growing plants with roots in a sterile medium. Plant Physiol. 25:767-769.
9. BOUSSINGAULT, J. B. 1855. Recherches sur la végétation, enterprises dans le but d'examiner si les plantes fixent dans leur organisme l'azote qui est à l'état gazeux dans l'atmosphère. Ann. Chim. Phys. 43:149-228.
10. BOWEN, G. D., and A. D. ROVIRA. 1961. The effects of microorganisms on plant growth. I. Development of roots and root hairs in sand and agar. Plant Soil 15:166-188.
11. BRADFUTE, O. E., R. A. LUSE, L. BRAAL, and A. D. MC LAREN. 1962. Growth of sterile plant roots in sand or soil in an inexpensive growth chamber. Proc. Soil Sci. Soc. Amer. 26:406-408.

12. BRAUN, E. L. 1958. The development of association and climax concepts. Their use in interpretation of the deciduous forest, p. 329-339. *In* W. C. Steere (ed.), Fifty years of botany. Golden jubilee volume Bot. Soc. Amer. McGraw-Hill Co., New York.

13. BRIGHAM, R. O. 1917. Assimilation of organic nitrogen by Zea mays and the influence of Bacillus subtilis on such assimilation. Soil Sci. 3:155-196.

14. CARROLL, R. B., and F. L. LUKEZIC. 1972. Induced resistance in alfalfa to Corynebacterium insidiosum by prior treatment with avirulent cells. Phytopathology 62:555-564.

15. CLARK, N. A. 1932. Technique for the growth of Lemna under sterile conditions with controlled temperature and light. Iowa State J. Sci. 7:13-16.

16. CLARK, N. A., and E. M. ROLLER. 1931. The stimulation of Lemna major by organic matter under sterile and non-sterile conditions. Soil Sci. 31:299-308.

17. CULLIMORE, D. R., and M. WOODBINE. 1963. A rhizosphere effect of the pea root on soil algae. Nature (London) 198:304-305.

18. DART, P. J., and F. V. MERCER. 1964. The legume rhizosphere. Arch. Mikrobiol. 47:344-378.

19. DEROPP, R. S. 1946. Apparatus for the prolonged sterile culture in vitro of whole plants or excised plant tissues. Science 104:371-373.

20. DICKEY, R. S., and P. E. NELSON. 1970. Pseudomonas caryophylli in carnation. IV. Unidentified bacteria isolated from carnation. Phytopathology 60:647-653.

21. DIMOCK, A. W. 1943. A method of establishing Verticillium-free clones of perennial plants. Phytopathology 33:3 (Abstr.).

22. DOMSCH, K. H. 1963. Der Einfluss saprophytischer Bodenpilze auf die Jugendentwicklung höherer Pflanzen. Z. Pflanzenkrankh. 70:470-475.

23. DOUGHERTY, E. C. 1953. Problems of nomenclature for the growth of organisms of one species with and without associated organisms of other species. Parasitology 42:259-261.

24. DUCHARME, E. P., and R. W. HANKS. 1961. Gnotobiotic techniques and the study of Radopholus similis on citrus. Plant Dis. Reptr. 45:742-744.

25. DUCLAUX, E. 1885. Physiologie végétale. Sur la germination dans un sol riche en matières organiques, mais exempt de microbes. Compt. Rend. Acad. Sci. 100:66-68.

26. DUNLEAVY, J., J. F. KUNKEL, and J. J. HANWAY. 1966. High populations of Bacillus subtilis associated with phosphorus toxicity in soybeans. Phytopathology 56:83-87.

27. EATON, F. M., and N. E. RIGLER. 1946. Influence of carbohydrate levels and root-surface microfloras on Phymatotrichum root rot in cotton and maize plants. J. Agr. Res. 72:137-161.

28. ESTEY, R. H., and T. H. SMITH. 1962. Note on the construction and operation of equipment for providing a sterile environment for the growth of plants from seed to maturity. Can. J. Plant Sci. 42:386-389.

29. FRED, E. B. 1919. The growth of higher plants in soils free of microorganisms. J. Gen. Physiol. 1:623-629.

30. FUJIWARA, A., K. OHIRA, K. CHIBA, and I. KONNO. 1968. Facilities for sterile culture of higher plants, p. 387-391. *In* M. Miyakawa and T. D. Luckey (eds.), Advances in germfree research and gnotobiology. The Chemical Rubber Co., Cleveland, Ohio.

31. FUJIWARA, A., K. OHIRA, K. CHIBA, and I. KONNO. 1969. Harmful effects on plant growth of the vinyl sheet utilized in the isolator for germfree higher plant culture, p. 75-82. *In* M. Miyakawa and B. S. Wostmann (eds.), Technology in germfree and gnotobiotic life research.

Academic Press of Japan, Tokyo.

32. GARRETT, S. D. 1960. Inoculum potential, p. 23-56. *In* J. G. Horsfall and A. E. Dimond (eds.), Plant pathology—an advanced treatise. Vol. III. Academic Press, New York and London.

33. GAWEL, L. J. 1961. The role of rhizosphere bacteria in the growth of tall fescue. M.S. Thesis, Oreg. State Univ., Corvallis. 67 p.

34. GERMAN, F., and V. T. BOWEN. 1951. A technique for growing plants under sterile conditions. Plant Physiol. 26:840-842.

35. GHOSH, B. P., and R. H. BURRIS. 1950. Utilization of nitrogenous compounds by plants. Soil Sci. 70:187-203.

36. GILMORE, A. E. 1950. A technique for embryo culture of peaches. Hilgardia 20:147-169.

37. HALE, M. G. 1969. Loss of organic compounds from roots. I. Cultural conditions for axenic growth of the peanut, Arachis hypogaea L. Plant Soil 31:463-472.

38. HAMEED, K. M., and H. B. COUCH. 1972. Effects of Penicillium simplicissimum on growth, chemical composition, and root exudation of axenically grown marigolds. Phytopathology 62:669 (Abstr.).

39. HARDER, R., and I. ZEMLIN. 1967. Förderung der Entwicklung und des Blühens von Pinguicula lusitanica durch Fütterung in axenischer Kultur. Planta 73:181-193.

40. HARLEY, J. L. 1965. Mycorrhiza, p. 218-230. *In* K. F. Baker and W. C. Snyder (eds.), Ecology of soil-borne plant pathogens. Univ. Calif. Press, Berkeley and Los Angeles.

41. HELLRIEGEL, H., and H. WILFARTH. 1888. Untersuchungen über die Stickstoffnahrung der Gramineen und Leguminosen. Beilageheft. Z. ver. Rübenzucker-Ind. deut. Reich (Beren). 234 p.

42. HENNIG, K., and F. VILLFORTH. 1940. Experimentelle Untersuchungen zur Frage der Bacteriensymbiose in höheren Pflanzen und ihrer Beeinflussung durch "Leitelemente." Biochem. Zeitschrift 305:299-309.

43. HOLLIS, J. P. 1951. Bacteria in healthy potato tissue. Phytopathology 41:350-366.

44. HOROWITZ, N. H., R. P. SHARP, and R. W. DAVIES. 1967. Planetary contamination. I: The problem and the agreements. Science 155:1501-1505.

45. HUTCHINSON, H. B., and N. H. J. MILLER. 1911. The direct assimilation of inorganic and organic forms of nitrogen by higher plants. Centrabl. Bakt., II. Abt., Bd. 30:513-547.

46. KATHREIN, H. R. 1951. A technique for the cultivation of higher plants under sterile conditions. Plant Physiol. 26:843-847.

47. KEMMERER, W. W., JR., J. A. MASON, and B. C. WOOLEY. 1969. Physical, chemical and biological activities at the Lunar Receiving Laboratory. Bioscience 19:712-715.

48. KHAN, I. U. 1948. A technique for growing citrus seedlings under aseptic conditions of culture. Phytopathology 38:756-757.

49. KLEIN, G., and J. KISSER. 1922. Die sterile Kultur der höheren Pflanzen. Botanische Abhandlungen 2:1-64. Verlag von Gustav Fisher, Jena.

50. KLEIN, H. P., J. LEDERBERG, and A. RICH. 1972. Biological experiments: the Viking Mars lander. Icarus 16:139-146.

51. KNUDSON, L. 1930. Flower production by orchid grown non-symbiotically. Bot. Gaz. 89:192-199.

52. KNUDSON, L. 1933. Non-symbiotic development of seedlings of Calluna vulgaris. New Phytologist 32:115-127.

53. KNUDSON, L., and R. S. SMITH. 1919. Secretion of amylase by plant roots. Bot. Gaz. 68:460-466.

54. KREUTZER, W. A. 1960. Soil treatment, p. 431-476. *In* J.

G. Horsfall and A. E. Dimond (eds.), Plant pathology–an advanced treatise. Vol. III. Academic Press, New York and London.

55. LANGRIDGE, J. 1957. The aseptic culture of Arabidopsis thaliana (L.) Heynh. Austral. J. Biol. Sci. 10:243-252.

56. LAWES, J. B., and J. H. GILBERT. 1891. The sources of the nitrogen of our leguminous crops. Ser. 3, J. Roy. Agr. Soc. Engl. 2:657-702.

57. LEV, M. 1962. An autoclavable plastic unit for rearing animals under germ free conditions. J. Appl. Bacteriol. 25:30-34.

58. LEV, M. 1964. A device for the external supply of sterile water and a simple air sterilizing filter for germ free units. J. Appl. Bacteriol. 27:41-44.

59. LINDSEY, D. L. 1967. Growth of beans, tomatoes, and corn under gnotobiotic conditions. Phytopathology 57:960-964.

60. LINDSEY, D. L. 1970. Effect of Aspergillus flavus on peanuts grown under gnotobiotic conditions. Phytopathology 60:208-211.

61. LINDSEY, D. L., and R. BAKER. 1967. Effect of certain fungi on dwarf tomatoes grown under gnotobiotic conditions. Phytopathology 57:1262-1263.

62. LJUNGGREN, H. 1969. Mechanism and pattern of Rhizobium invasion into leguminous root hairs. Physiol. Plant. 5:1-82.

63. LOO, S. W. 1946. Preliminary experiment on the cultivation of Baeria chrysostoma under sterile conditions. Amer. J. Bot. 33:382-389.

64. LUCKEY, T. D. 1963. Germfree life and gnotobiology. Academic Press, New York and London. 512 p.

65. LUCKEY, T. D. 1968. Gnotobiology and aerospace systems, p. 317-353. In M. Miyakawa and T. D. Luckey (eds.), Advances in germfree research and gnotobiology. The Chemical Rubber Co., Cleveland, Ohio.

66. LUKEZIC, F. L., J. R. BLOOM, and R. B. CARROLL. 1969. Development of Fusarium root rot and its effect on root and crown carbohydrate levels of clipped alfalfa plants grown in a gnotobiotic environment. Phytopathology 59:1575-1579.

67. LUKEZIC, F. L., J. R. BLOOM, and ROSLYN G. LEVINE. 1969. Influence of top removal on the carbohydrate levels of alfalfa crowns and roots grown in a gnotobiotic environment. Can. J. Plant Sci. 49:189-195.

68. LUKEZIC, F. L., ROSLYN G. LEVINE, and R. B. CARROLL. 1968. The influence of Fusarium root rot and harvesting on the carbohydrate levels of alfalfa roots grown in a gnotobiotic environment. Phytopathology 58:401 (Abstr.).

69. MANGIN, L. 1899. Sur une maladie nouvelle des oeillets. Compt. Rend. Acad. Sci. (Paris) 129:731-734.

70. MARX, D. H. 1972. Ectomycorrhizae as biological deterrents to pathogenic root infections. Annu. Rev. Phytopathol. 10:429-454.

71. MITCHELL, R., and ESTHER HURWITZ. 1965. Suppression of Pythium debaryanum by lytic rhizosphere bacteria. Phytopathology 55:156-158.

72. MOSSE, BARBARA. 1962. The establishment of vesicular-arbuscular mycorrhiza under aseptic conditions. J. Gen. Microbiol. 27:509-520.

73. NILSSON, P. E. 1957. Aseptic cultivation of higher plants. Arch. Mikrobiol. 26:285-301.

74. PASTEUR, L. 1885. Observations relatives à la Note précédente de M. Duclaux. Compt. Rend. Acad. Sci. (Paris) 100:68.

75. PEROTTI, R. 1935. Attraverso i meandri della biosfera agronomica. Ann. Soc. Agr. (Bologna) 63:1-34.

76. PETTIT, R. E., RUTH A. TABER, and B. G. FOSTER. 1968. Occurrence of Bacillus subtilis in peanut kernels. Phytopathology 58:254-255.

77. PHILIPSON, M. N., and J. D. BLAIR. 1957. Bacteria in clover root tissue. Can. J. Microbiol. 3:125-129.

78. PHILLIPS, D. J. 1968. Carnation shoot tip culture. Colo. State Univ. Exp. Stn. Tech. Bull. 102. 22 p.

79. PRINGSHEIM, E. G., and OLGA PRINGSHEIM. 1962. Axenic culture of Utricularia. Amer. J. Bot. 49:898-901.

80. QUAK, F. 1957. Meristeemcultur, gecombineerd met warmtebehandeling voor het vekrijgen van virusvrije anjerplanten. Tijdschr. Plantenziekten 63:13-14.

81. REUSZER, H. W. 1962. Axenic techniques in the determination of root functions and interrelationship of microorganisms and plant roots. Soil Sci. 93:56-61.

82. ROSS, J. P., and J. A. HARPER. 1970. Effect of Endogone mycorrhiza on soybean yields. Phytopathology 60:1522-1556.

83. ROVIRA, A. D. 1965. Plant root exudates and their influence upon soil microorganisms, p. 170-186. In K. F. Baker and W. C. Snyder (eds.), Ecology of soil-borne plant pathogens. Univ. Calif. Press, Berkeley and Los Angeles.

84. ROVIRA, A. D. 1969. Plant root exudates. Bot. Rev. 35:35-57.

85. SAGAN, C. 1961. The planet Venus. Science 133:849-858.

86. SAMISH, Z., and R. ETINGER-TULCZYNSKA. 1963. Distribution of bacteria within the tissue of healthy tomatoes. Appl. Microbiol. 11:7-10.

87. SANFORD, G. B. 1948. The occurrence of bacteria in normal potato plants and legumes. Sci. Agr. 28:21-25.

88. SCHNATHORST, W. C. 1954. Bacteria and fungi in seeds and plants of certified bean varieties. Phytopathology 44:588-592.

89. SCHÖNBECK, F., and W. A. KREUTZER. 1971. Nullification of antagonism of Phoma betae by Bacillus subtilis var. niger in soil and in a simulated rhizosphere. Phytopathology 61:1447-1450.

90. SCHROTH, M. N., and D. C. HILDEBRAND. 1964. Influence of plant exudates on root-infecting fungi. Annu. Rev. Phytopathol. 2:101-132.

91. SCHULOW, I. 1913. Versuche mit sterilen Kulturen höherer Pflanzen. Ber. Deut. Bot. Ges. 31:97-121.

92. SHAY, F. J., and M. G. HALE. 1973. Effect of low levels of calcium on exudation of sugars and sugar derivatives from intact peanut roots under axenic conditions. Plant Physiol. 51:1061-1063.

93. STEINBERG, R. A. 1941. Use of Lemna for nutrition studies on green plants. J. Agr. Res. 62:423-430.

94. STEINBERG, R. A. 1947. Growth responses to organic compounds by tobacco seedlings in aseptic culture. J. Agr. Res. 75:81-92.

95. STEINBERG, R. A. 1947. Growth responses of tobacco seedlings in aseptic culture to diffusates of some common soil bacteria. J. Agr. Res. 75:199-206.

96. STOTZKY, G., W. CULBRETH, and L. B. MISH. 1962. Apparatus for growing plants with aseptic roots for collection of root exudates and CO_2. Plant Physiol. 37:332-341.

97. SUBBA-RAO, N. S., R. G. S. BIDWELL, and D. L. BAILEY. 1961. The effect of rhizoplane fungi on the uptake and metabolism of nutrients by tomato plants. Can. J. Bot. 39:1759-1764.

98. SWABY, R. J., and BEVERLEY I. PASSEY. 1953. A simple macrorespirometer for studies in soil microbiology. Austral. J. Agr. Res. 4:334-339.

99. SZEMBER, A. 1959. Providing aseptically cultivated plants with water through bacteria-tight glass filters. Plant Soil 11:392-394.

100. TERVET, I. W., and J. P. HOLLIS. 1948. Bacteria in storage organs of healthy plants. Phytopathology 38:960-967.

101. TONSIG, S., and L. BRACCI-ORSENIGO. 1936. Sulla

prenza di batteri nei vari organi delle piante superiori. Nuovo Giorn. Botan. Ital. (Nuova Ser.) 42:1-8.

102. TSEH-AN, C., R. A. KILPATRICK, and A. E. RICH. 1961. Sterile culture techniques as tools in plant nematology research. Phytopathology 51:799-800.

103. TURNER, P. D. 1962. Morphological influence of exudates of mycorrhizal and non-mycorrhizal fungi on excised root cultures of Pinus sylvestris L. Nature (London) 194:551-552.

104. TYNER, L. E. 1958. The effect of water on the partial sterilization of barley seed by propylene oxide and by heat. Phytopathology 48:177-178.

105. VIRTANEN, A. I., T. LAINE, and S. V. HANSEN. 1936. Excretion of amino acids from the root nodules of leguminous plants. Nature (London) 137:277.

106. WALKINSHAW, C. H., and P. H. JOHNSON. 1971. Analysis of vegetable seedlings grown in contact with Apollo 14 lunar surface fines. Hortscience 6:532-535.

107. WALKINSHAW, C. H., H. C. SWEET, S. VENKETESWAREN, and W. H. HORNE. 1971. Results of Apollo 11 and 12 quarantine studies on plants. Bioscience 20:1297-1302.

108. WARCUP, J. H. 1957. Chemical and biological aspects of soil sterilization. Soils Fert. 20:1-5.

109. WARIS, H. 1958. Simple devices for aseptic culture of seed plants. Physiol. Plant. 11:627-630.

110. WEETE, J. D., and C. H. WALKINSHAW. 1972. Apollo 12 lunar material: effects on plant pigments. Can. J. Bot. 50:101-104.

111. WEIR, JEAN B. 1958. Investigations on the influence of certain carbohydrates on the nodulation of Medicago sativa L. Phyton 10:153-163.

112. WELLS, T. R., and W. A. KREUTZER. 1972. Aerial invasion of peanut flower tissues by Aspergillus flavus under gnotobiotic conditions. Phytopathology 62:797 (Abstr.).

113. WELLS, T. R., W. A. KREUTZER, and D. L. LINDSEY. 1972. Colonization of gnotobiotically grown peanuts by Aspergillus flavus and selected interacting fungi. Phytopathology 62:1238-1242.

114. WELTE, E., and G. TROLLDENIER. 1962. Der Einfluss der Bodenmikroorganismen auf Trockensubstanzbildung und Aschegehalt in Nährlösung wachsender Pflanzen. Arch. Mikrobiol. 43:138-147.

115. WELTE, E., and G. TROLLDENIER. 1963. Einfluss von Mikrofloren verschiedener Böden und von Bakterienreinkulturen auf das Pflanzenwachstum. Arch. Mikrobiol. 47:42-56.

116. WHITE, P. R. 1936. Plant tissue cultures. Bot. Rev. 2:419-437.

117. WHITE, P. R. 1943. Germ-free plants and plant parts as material for physiological and pathological studies, p. 188-204. In J. A. Reyniers (ed.), Micrurgical and germ-free methods. Charles C. Thomas, Springfield, Ill.

118. WILSON, J. K. 1915. Calcium hypochlorite as a seed sterilizer. Amer. J. Bot. 2:420-427.

119. WILSON, J. K. 1920. Device for growing large plants in sterile media. Phytopathology 10:425-429.

Appraisal of Loss from Common Root Rot in Wheat

R. D. TINLINE, R. J. LEDINGHAM, and B. J. SALLANS[1]—*Plant Pathologists and Retired Plant Pathologist, Research Station, Research Branch, Agriculture Canada, Saskatoon, Saskatchewan.*

The need for loss assessment is widely recognized. It has been stressed by various authors, including Chester (2), Chiarappa et al. (3), Grainger (10), Le Clerg (17), and Ordish and Dufour (22), and requires no reiteration here. The purpose of this paper is to present some specific work on loss assessment; we have chosen to discuss loss in spring wheat caused by common root rot as it occurs in western Canada. Initially, the term common root rot was used for a complex of diseases in cereals characterized by necrosis of basal stems, crown, subcrown internodes, and roots (1, 27). In this broad context, it includes among others crown rot of wheat caused by *Fusarium graminearum* such as occurs in Australia (24) and Fusarium foot and root rot caused primarily by *Fusarium culmorum* as described by Cook (6) in the U.S. Pacific Northwest, and which we suspect sometimes occurs in western Canada. As Cook (6) pointed out, symptoms of the diseases differ markedly from those caused by *Cochliobolus sativus*, the dominant incitant of common root rot in the Canadian west. The usual symptoms there are brown lesions on lower leaf sheaths, crowns, and subcrown internodes; root lesions often occur, but not extensively. Infected plants usually develop to maturity, and, although stunting occurs, visible signs of disease frequently are not manifest on aboveground parts. In isolations, *C. sativus* generally is obtained from 80% or more, and fusaria, including *F. avenaceum*, *F. acuminatum*, and *F. culmorum*, alone or together with *C. sativus*, is generally obtained from a relatively low but variable percentage of lesioned tissues.

The interrelationship of the causal organisms appears to be complex. In greenhouse tests, the subcrown internodes of wheat plants previously infected with *C. sativus* were successfully invaded by *F. culmorum* or *F. acuminatum*; *C. sativus*, however, was seldom a successful challenger to the fusaria when they prepossessed the tissue (Tinline, *unpublished data*). If a similar situation occurs in the field, these findings may help explain the *C. sativus-Fusarium* spp. succession in plant bases observed by Hoes (13) and Oswald (23), and the failure of Statler and Darlington (33) to recover *C. sativus* from plants inoculated with both organisms.

Disease assessment.—Disease assessment constitutes an essential step in loss appraisal work. Chester (2) and Large (15) have stressed the importance of quantifying disease, and Moore (21) has cited disease assessment as one of three aspects in the measurement of disease losses. In western Canada, common root rot is endemic and is present annually in almost all wheat fields. Frequently it first appears a few weeks after plants emerge. Apparently, most infections originate from propagules of the pathogens in the soil; Chinn et al. (4), using a flotation method, found an average of 118 conidia of *C. sativus* per g of soil in 100 cultivated wheat fields. The incidence of disease increases during the season, and Verma (35) found the increase was that of a "simple interest disease," as described by van der Plank (34). Severity of the disease on plants also increases with time. Generally, diseased plants are scattered throughout the host population. A comparison of common root rot-diseased plants and healthy ones was made by Simmonds et al. (32) during two seasons. They observed that one of the early symptoms was the lesioning of the subcrown internode. Lesions of the external crown near the soil line also may represent primary infections. Their occurrence was correlated with disease intensities based on subcrown internode lesions (26). We found direct relationships between intensity of subcrown internode lesioning and reductions in fresh weight and grain yield, but not between intensity of external crown lesions and these attributes (*unpublished data*). Since the same plants were used in this dual classification, it seems clear that the subcrown internode lesions provided the better index of disease. Further, the results corroborated earlier work (31), and most disease assessments have been based on lesions of the subcrown internode.

In root rot studies, categories such as clean, slight, moderate, and severe are frequently employed to characterize severity of disease on plants or plant parts. Although the terms are somewhat descriptive, they are imprecise for comparative analysis, and they do not reflect the incidence of disease. When the categories are assigned numerical values, however, a disease rating, here referred to as "disease intensity" may be calculated using McKinney's "infection index" or a modification of it that integrates both the incidence and severity of disease. For example, disease intensity (%) =

$$\frac{\Sigma \text{ (Category value} \times \text{no. plants in category)}}{\text{Total no. plants} \times \text{maximum category value}} \times 100.$$

An obvious difficulty with arbitrary disease categories is the lack of uniformity with which observers assign plants to them. In recent years in western Canada, an

---

[1]Deceased, 16 February 1974.

attempt to promote greater uniformity has involved the use of photographs depicting the disease categories and the classification of plants on the basis of the area of subcrown internode lesioned. Thus, clean plants have no lesions, slight have 1-25%, moderate have 25-50%, and severe have over 50% of the area lesioned (18). The categories are illustrated in Figure 1.

An expeditious method of comparing disease in large populations of plants also has been used in some loss studies. It is simply a two-class (healthy and diseased) system; the percentage incidence of moderately and severely diseased plants together constitute the diseased category. Sallans (*unpublished data*) obtained a good linear relationship between intensity based on four disease categories and the two-class disease incidence; the coefficient of correlation in each of 4 years was 0.98. Ledingham et al. (19) obtained a similar high correlation.

Disease has been assessed on the basis of subcrown internode lesions several times during plant development (35). However, an optimum time for it has not been established, and most loss studies have been conducted with a single date of sampling, about growth stage 11.3, when plants are nearly mature.

Disease intensity and losses.—Comparison of yields of individual healthy and diseased plants from the same population has been the method used most frequently to assess loss. In such studies, plants were sorted into disease classes and reductions in yield obtained. Almost invariably, reduction per plant was higher in the moderate than slight disease category and highest in the severe disease category, demonstrating a relationship between disease intensity and plant yield. Within categories, however, the magnitude of reduction varied considerably; e.g., whereas average reductions of 6.0, 12.5, and 28.2% were obtained from many survey fields by Ledingham et al. (19), 26, 38, and 59% were recorded from one location by Verma (35) in the slight, moderate, and severe disease categories, respectively. Many factors undoubtedly influence the degree of yield reduction;

consequently, derived quantitative values should be used with discretion.

Using the individual plant method, Machacek (20) calculated loss in survey fields as a percentage of the actual to the potential yield of samples, where the potential yield was derived by multiplying the average yield of healthy plants by the total density. His samples were taken from 60 wheat fields in the Province of Manitoba annually for 3 years. He estimated mean provincial losses of 8.0, 16.4, and 12.1% for the years 1939, 1940, and 1941, respectively, and an average loss of 12.1%.

Ledingham et al. (19) conducted an extensive survey of losses for the 3-year period 1969-1971 in the Prairie Provinces of Canada. The area devoted there to spring wheat production in the 3 years ranged from 4.9 million to 9.9 million ha. In all, 509 fields were sampled. As in Machacek's study, losses in individual fields were derived, and these were averaged by crop reporting districts. The percentage loss was applied to actual production in each district to obtain the potential production. Potential production for the provinces and the entire region was determined. Provincial and total losses then were obtained from the differences between potential and actual production. The overall loss estimate was 5.7%.

During this same 3-year period, Verma (35) determined reduction in dry weight at intervals throughout each season and loss in yield at the final date of sampling in Manitou wheat at one location in Saskatchewan. Various treatments such as fertilizers were applied to some crop areas to ascertain their effect on disease. In most treatments, reductions in dry weight assessed as percentage of potential were larger at each subsequent sampling date. In 1970, however, in untreated areas in both fallow and stubble crop, the largest decreases were recorded prior to plant maturity, suggesting that environmental factors may differentially affect the dynamic processes of plant growth and disease development. He found that losses in yield of grain were very similar in magnitude to the reductions in dry weight recorded at the final sampling. Yield loss estimates were

Fig. 1. Common root rot severity classes based on extent of lesioning in subcrown internodes (left to right): clean, slight, moderate, and severe.

30, 38, and 22% for the 3 years, respectively. Losses of comparable magnitude (about 29%) previously had been reported by Sallans and Ledingham (28) for a small zone in the Province of Saskatchewan in which severe disease had occurred in patches. The authors estimated loss from the yields of samples from diseased and adjacent healthy-appearing patches in fields.

Both Ledingham et al. (19) and Verma (35) found that reduction in grain yield from disease was due primarily to a reduction in the number of heads (tillers) per plant, and, to a lesser degree, to the number of kernels per head and the weight of the kernels. Severely diseased and healthy plants did not differ appreciably in percentage of crude fiber, phosphate, and total calories per g (35). Diseased plants (35) and grain (19) contained slightly more protein than did healthy ones.

We have used the extent of lesioning on subcrown internodes in the field extensively to classify cultivars and lines of wheat for resistance. Ranking of cultivars between locations seems quite uniform (30). We found that yield losses in moderately resistant cultivars were less than in moderately susceptible ones (*unpublished data*). For example, in replicated tests in 1971 and 1972, losses calculated for the moderately resistant cultivar Neepawa were 10 and 3%, respectively, and for the moderately susceptible cultivar Chinook were 17 and 5%, respectively. The method of ascertaining loss was similar to that used by Verma (35) and Ledingham et al. (19).

Correlations have been found between disease intensities and yields by various investigators (12, 25, 29). In most of these studies, regression methods were used to estimate loss. Greaney and Machacek (12) induced different disease intensities by inoculating seed with *C. sativus* and *F. culmorum* and showed significant negative correlations between disease intensities and yields of plots. Sallans (25) demonstrated a relationship between wheat yields and disease intensity over a 10-year period. Intensities for each of nine crop-reporting districts in the Province of Saskatchewan were obtained from annual surveys of fields within the districts; yields and data on four other variables were from provincial annual reports. Next to June-July rainfall, common root rot had the most significant effect of the variables on yield. The calculated average loss from disease was about 30% (5.14 ± 1.79 bu/acre). Together the five variables accounted for 78% of the variance in yield; consequently, the lower limit of loss, approximately 10%, appears to be pertinent.

Disease incidence-yield relationships were found in the majority of field tests at various locations in which a number of cultivars and lines (16 the 1st year, 6 the 2nd year) differing in resistance were grown (29 and *unpublished data*). The two-class system was used to rate plants for disease. Plot yields were obtained in the tests. Where complicating factors like immaturity, stem or leaf rust, and stand were apparent, partial regression methods were used. The regression coefficients that were significant by analyses of variance were used in multiple regression equations from which losses in the cultivars and lines from common root rot were computed. Although the test materials may have had different yielding abilities because some of the included lines had been selected solely for their resistance to disease, losses appeared to be least in the more resistant lines and highest in the most susceptible ones. Interestingly, using this method, the losses in cultivar Manitou in 10 tests over 2 years in the Province of Saskatchewan ranged from 2.4 to 9.7% and averaged 5.7%. The average loss estimated there by Ledingham et al. (19) over 3 years, during which Manitou occupied 60% of the area devoted to wheat, was 6.6%. The two methods of appraising loss gave fairly similar results.

We examined 1969 and 1970 survey data obtained for the Province of Saskatchewan for a relationship between crop yield and disease intensity. We did not find a simple one; it likely was obscured by the differences in field productivity. A highly significant negative correlation occurred, however, between yield expressed as a percentage of the potential yield per unit-area and the disease intensity. From linear regressions that best fit the data, a disease increase of 10% was associated with a yield decrease of 2.6% in 1969 and about 4% in 1970. The average disease intensities for these years, 21.0 and 21.3%, respectively, represented losses of 5.5 and 8.5%. Ledingham et al. (19) reported losses of 5.7 and 10.2%. Apparently, disease intensities by themselves do not indicate the magnitude of loss. As Ledingham et al. stated, "The disease rating must be multiplied by a conversion factor to give an approximation of loss." They indicated that a conversion factor is determined by simply dividing the calculated loss from a group of fields by their average disease intensity. Once a conversion factor has been determined, it may be used, "within limits," for other fields for which disease intensities are known.

The question of competition effects.—Loss in yield derived from the relationship of diseased to healthy plants within the same population may be an inflated rather than a realistic value if healthy plants compensate in part for the loss incurred by infected ones. Although Chester (2) was of the opinion that where diseased plants normally are in competition with healthy ones loss determined by the individual plant method would closely reflect the loss actually experienced in practice, various investigators of losses from common root rot have expressed concern about competition effects (19, 20, 29, 35). Probably due to the lack of adequate controls, the question of competition remains unresolved. In greenhouse tests in which we grew four plants per container and inoculated all, two, or none of the plants per unit at one time with *C. sativus*, competition effects were evident between the inoculated and uninoculated plants in the same container. In the field, however, incidence of disease increases during the season (35). Consequently, it may be assumed that yield compensation is most likely to occur late in the season, perhaps at a time when the healthy plants are least able to benefit from reduced competition.

We tried several approaches in the field to ascertain whether compensation is an important variable in loss appraisal. One involved a calculation of loss in varied densities of one wheat cultivar. Although Greaney (11) had reported previously that increased rates of seeding increased incidence and severity of common root rot, we did not obtain similar results. Another approach involved a comparison of the yield per plant in moderately resistant and moderately susceptible lines that were grown singly and intermixed in varying proportions, and a third involved determining percentage loss in spaced plants. Results were inconclusive; we found no evidence of a compensatory yield in healthy plants. Conversely, an indication that compensation may be important appeared in some of our survey data for the Province of Saskatchewan. The percentage reduction in yield of diseased plants (calculated on a per plant basis) was 29 and 22% greater in 1969 and 1970, respectively, for fields with a high rather than a low average disease intensity.

Grainger (10) considered yield compensation a factor in loss assessment; plant density is a variable in its expression. He indicated that compensation may be negligible with some diseases like root diseases, where the relationship is linear between percentage loss of roots and percentage loss of yield. Additional investigation seems requisite, however, to ascertain the significance of compensation in loss appraisals of common root rot.

DISCUSSION.—Obviously, all the loss figures reported in the investigations reviewed here are estimates and as such their usefulness has limitations. The individual plant and regression methods employed in their derivation have attendant difficulties; the former may exaggerate loss because of competition effects, the latter is imperfect in that many other variables besides disease are involved in yield variation. Although relationships were found between disease intensity or incidence and yield, disease was assessed primarily on the subcrown internode, and this must be considered less reliable than measurement of disease on all affected parts. Despite probable large errors of estimates, a comparison of some figures seems pertinent. A presumptive decline in magnitude of loss in potential production is apparent over time. Whereas Ledingham et al. (19) recently found mean losses in the Provinces of Saskatchewan and Manitoba of 6.6 and 2.7%, respectively, Sallans (25) and Machacek (20) previously had estimated losses of at least 10 and 12%, respectively, for the same regions.

Although it is tempting to assign this tentative reduction in loss to the use of more resistant cultivars, a myriad of other changes also has occurred in such factors as cultural practices (weed control, fertilizer usage, tillage, etc.) and new crops which precludes any unequivocal assignment. In view of the prevalence and annual recurrence of the disease, even a low percentage loss indicates a sizable total loss in unrealized production. In some other wheat-growing areas, the disease may be more damaging. Fenster et al. (7) reported that date of seeding greatly influenced root and crown rot intensity of winter wheat in Nebraska. Planting dates to minimize disease resulted in increased yields estimated at 336 kg/ha (5 bu/acre). Root rot caused chiefly by *C. sativus* has been reported by Goloshchapov and Lopatin (8), Gorodilova (9), and Korshunova (14) to cause losses of considerable magnitude in the Soviet Union.

One of the major difficulties in assessing disease or loss in root rots such as common root rot arises from the multiplicity of seed- and soil-borne pathogens that may be involved. Although isolations may indicate that one fungus among several is dominant, we really do not know what proportion of the total disease or loss relates to it. This imprecision, recognized by Colhoun (5) and others, has obvious significance in the search for improved methods of control. In our opinion, it emphasizes the importance of following the course of disease in detail, as advocated by Large (15) and Last (16), and including sites and identification of pathogens throughout plant development before we can hope to apportion loss to the individual components of a complex.

LITERATURE CITED

1. BUTLER, F. C. 1961. Root and foot rot diseases of wheat. N.S.W. Dept. Agr. Sci. Bull. 77. 98 p.
2. CHESTER, K. STARR. 1950. Plant disease losses: their appraisal and interpretation. Plant Dis. Reptr. Suppl. 193:190-362.
3. CHIARAPPA, L., H. C. CHIANG, and R. F. SMITH. 1972. Plant pests and diseases: assessment of crop losses. Science 176:769-773.
4. CHINN, S. H. F., R. J. LEDINGHAM, and B. J. SALLANS. 1960. Population and viability studies of Helminthosporium sativum in field soils. Can. J. Bot. 38:533-539.
5. COLHOUN, J. 1969. Seed and soil borne pathogens of cereals. Proc. 5th Br. Insect. Fung. Conf., 1969. Vol. 3:620-625.
6. COOK, R. J. 1968. Fusarium root and foot rot of cereals in the Pacific Northwest. Phytopathology 58:127-131.
7. FENSTER, C. R., M. G. BOOSALIS, and J. L. WEIHING. 1972. Date of planting studies of winter wheat and winter barley in relation to root and crown rot, grain yields and quality. Nebr. Agr. Exp. Stn. Res. Bull. 250. 32 p.
8. GOLOSHCHAPOV, A. P., and M. I. LOPATIN. 1969. The spread and diagnosis of and losses from Cochliobolus sativus on spring wheat. Sb. Nauchn. Rab. kurgan. s.-kh. Inst. 16:105-112. (Rev. Appl. Mycol. 50:1177. 1971.)
9. GORODILOVA, L. M. 1967. Root rot of wheat in N. Kazakhstan. Vestn. Sel'skokhoz. Nauki, Alma Ata 10:45-50. (Rev. Appl. Mycol. 47:120. 1968.)
10. GRAINGER, J. 1967. Economic aspects of crop losses caused by disease, p. 55-98. FAO Symp. on Crop Losses, FAO, UN, Rome.
11. GREANEY, F. J. 1946. Influence of time, rate and depth of seeding on the incidence of root rot in wheat. Phytopathology 36:252-263.
12. GREANEY, F. J., and J. E. MACHACEK. 1934. Studies of the control of root-rot diseases of cereals caused by Fusarium culmorum (W. G. Sm.) Sacc. and Helminthosporium sativum P., K., & B. I. Field methods with root-rot diseases. Sci. Agr. 15:228-240.
13. HOES, J. A. 1966. A possible Bipolaris sorokiniana-Fusarium nivale sequence in roots and crowns of winter wheat in Washington. Plant Dis. Reptr. 50:154-157.
14. KORSHUNOVA, A. F. 1968. Root rots of wheat. Zashch. Rast., Mosk. 13:12-14. (Rev. Appl. Mycol. 48:120. 1969.)
15. LARGE, E. C. 1966. Measuring plant disease. Annu. Rev. Phytopathol. 4:9-28.
16. LAST, F. T. 1971. The role of the host in the epidemiology of some nonfoliar pathogens. Annu. Rev. Phytopathol. 9:341-362.
17. LE CLERG, E. L. 1964. Crop losses due to plant diseases in the United States. Phytopathology 54:1309-1313.
18. LEDINGHAM, R. J. 1970. Special method No. 29. Crop loss assessment methods. FAO Manual on the evaluation and prevention of losses by pests, diseases and weeds. FAO, UN, Rome.
19. LEDINGHAM, R. J., T. G. ATKINSON, J. S. HORRICKS, J. T. MILLS, L. J. PIENING, and R. D. TINLINE. 1973. Wheat losses due to common root rot in the Prairie Provinces of Canada, 1969-71. Can. Plant Dis. Surv. 53:113-122.
20. MACHACEK, J. E. 1943. An estimate of loss in Manitoba

from common root rot in wheat. Sci. Agr. 24:70-77.

21. MOORE, J. 1967. A review of current methods applicable to measuring crop losses caused by plant diseases, p. 275-287. FAO Symp. on Crop Losses, FAO, UN, Rome.

22. ORDISH, G., and D. DUFOUR. 1969. Economic bases for protection against plant diseases. Annu. Rev. Phytopathol. 7:31-50.

23. OSWALD, J. W. 1950. Etiology of cereal root rots in California. Hilgardia 19:447-462.

24. PURSS, G. S. 1969. The relationship between strains of Fusarium graminearum Schwabe causing crown rot of various gramineous hosts and stalk rot of maize in Queensland. Austral. J. Agr. Res. 20:257-264.

25. SALLANS, B. J. 1948. Interrelations of common root rot and other factors with wheat yields in Saskatchewan. Sci. Agr. 28:6-20.

26. SALLANS, B. J. 1960. Common root rot (Helminthosporium sorokinianum and Fusarium spp.), p. 10-12. 38th Annu. Rept. Can. Plant Dis. Surv., 1958.

27. SALLANS, B. J. 1965. Root rots of cereals. III. Bot. Rev. 31:505-536.

28. SALLANS, B. J., and R. J. LEDINGHAM. 1943. An outbreak of common root rot in southwestern Saskatchewan in 1942. Sci. Agr. 23:589-597.

29. SALLANS, B. J., and R. D. TINLINE. 1968. The experimental approach in assessing disease losses in cereals: root diseases. Can. Plant Dis. Surv. 48:68-70.

30. SALLANS, B. J., and R. D. TINLINE. 1969. Consistency of reaction in wheat lines to common root rot. Can. J. Plant Sci. 49:197-201.

31. SIMMONDS, P. M. 1935. Study of the root-rot problem of wheat and barley caused by Helminthosporium sativum in Saskatchewan, p. 24-25. Prog. Rept. Dom. Botanist, 1931-1934.

32. SIMMONDS, P. M., R. C. RUSSELL, and B. J. SALLANS. 1935. A comparison of different types of root rot of wheat by means of root excavation studies. Sci. Agr. 15:680-700.

33. STATLER, G. D., and L. C. DARLINGTON. 1972. Resistance of hard red spring wheat and durum wheat to seedling blight and crown rot. Plant Dis. Reptr. 56:788-791.

34. VAN DER PLANK, J. E. 1963. Plant diseases: epidemics and control. Academic Press, New York and London. 349 p.

35. VERMA, P. R. 1973. Studies of common root rot (Cochliobolus sativus) in Manitou wheat: epidemiology, loss assessment, inoculum density, and effects of phosphate. Ph.D. thesis, Univ. Saskat., Saskatoon. 386 p.

The Role of Phytophthora Cinnamomi in Dieback Diseases of Australian Eucalypt Forests

F. D. PODGER—*Forest Research Institute, Forestry and Timber Bureau, Canberra, Australia.*

Australia's estate of intensively managed forest is dominated by the indigenous genus *Eucalyptus*, upon which the country's growing forest industry is heavily dependent. During this century, particularly in the past two decades, large areas of eucalypt forests have been severely damaged by one or another of a group of diseases known as eucalypt diebacks. In extreme cases, almost complete destruction of all strata in floristically complex forests has resulted. In others, damage has been restricted to the eucalypt overstory. These diebacks cause serious loss of growth to some of Australia's potentially most productive forests. No national inventory of the total area affected is yet available, but it may already exceed 150,000 ha, and the area affected is increasing.

The best known and most extensive of the diebacks is jarrah dieback. It destroys endemic jarrah (*Eucalyptus marginata*) forests (Fig. 1, 2) in southwestern Australia, and has been shown to be due to root rot caused by *Phytophthora cinnamomi* (39). This fungus is distributed widely in eastern Australia (45, 46). Although it has been associated with several diebacks there, including some almost identical to jarrah dieback, it is also found beneath extensive tracts of eucalypt forest that are apparently quite unaffected (45). There are several other diebacks in eastern Australia that are almost certainly not due to *P. cinnamomi*.

Before discussing these diseases, I will outline briefly the broad floristic and environmental patterns in Australian forests. The relative susceptibility of the various floristic elements to *P. cinnamomi* and the influence of varied environments upon their vulnerability determine the characteristics of the diebacks.

THE FOREST COMMUNITIES AND THEIR ENVIRONMENTS.—The approximately 38 million ha of potentially productive forest land in Australia are restricted almost entirely to a narrow fringe, seldom more than 100 miles wide, around the eastern and southeastern seaboards. This zone, which begins in the Queensland tropics near 11°S latitude, reaches its southern limit near 43°S latitude in the island state of Tasmania.

Each of the three principal elements of the Australian flora, the Indo-Malayan, the Antarctic, and the Australian, is represented in the forest flora. The Indo-Malayan element is predominant in tropical and subtropical rainforests of the north, with a few species extending well to the south, usually in sheltered coastal gullies on well-drained, fertile soils. The Antarctic element reaches its maximum development in the highlands of Tasmania, where *Nothofagus* dominates extensive areas of cool temperate rainforest. Disjunct relics of the Antarctic flora extend at increasingly higher elevation into subtropical northern New South Wales. These elements are of minor significance for the forest industry compared to the Australian element. Most forested landscapes in Australia are dominated by sclerophyllous evergreen forests of *Eucalyptus*. In the far southwest of the continent the approximately 2 million ha of commercial forests are entirely dominated by this genus. These forests are separated from the east coast forests by almost 3,000 km of semiarid steppe and plains and have been isolated since at least the Pleistocene. Although the two regions share no native tree species, their eucalypt forests are floristically and structurally similar.

Eucalyptus has adapted to habits of great diversity ranging from subalpine to semidesert. It forms tall closed forests where rainfall is high in subtropical regions of summer rainfall and in temperate regions of uniform or winter rainfall. About 200 of the approximately 550 taxa in the genus are found in the forested zone. Some 80 are utilized for various purposes by the forest industry.

Of the eight subgenera in *Eucalyptus* (51), the two largest, *Symphyomyrtus* (about 347 taxa) and *Monocalyptus* (about 106 taxa), are of particular importance both commercially and for this review. *Symphyomyrtus* (S) occurs throughout the forest zone, reaching its maximum importance for forestry in the subtropics. It has been outstandingly successful in exotic plantings abroad, but takes second place to *Monocalyptus* (M) in Australia's domestic forest industry. *Monocalyptus* has reached its maximum development in the temperate winter rainfall zone, where it supplies almost the entire intake of short-fibered pulpwood for paper manufacture. *Monocalyptus* rapidly declines in importance northward through the subtropics and is represented by only a few species in the tropics. Generally speaking, it has not been a successful exotic abroad and in Australia is much more subject to dieback diseases than is *Symphyomyrtus*. Species of the various subgenera often co-occur, but *Symphyomyrtus* occupies water gaining sites more frequently than does *Monocalyptus*.

THE EUCALYPT DIEBACKS. — *Terminology.* — Australian usage of the term dieback is loose and not restricted to conventional definition ["Progressive dying back from the tips of twigs, branches or tops" (56)].

Some degree of dieback in this conventional sense is characteristic of many native stands considered to be in generally good health. The primary leaf bud systems in *Eucalyptus* are naked and allow rapid expansion of foliage whenever conditions are favorable for growth. They are also vulnerable to defoliation, particularly by insects, but may rapidly recover as epicormic shoots develop from secondary bud systems ordinarily suppressed by inhibitors produced in the primary bud systems. These mechanisms are useful adaptations to environments in which defoliation by drought, fire, and insect plagues is recurrent. Thus, dieback in the conventional sense is a feature of many eucalypt forests that are able to maintain their stand structure and broad species composition in dynamic balance with the environment.

The diseases which Australian foresters refer to as diebacks and with which this review is concerned result in spectacular damage and often widespread mortality. The symptoms in individual trees are not always the conventional symptoms of dieback. They may vary from sudden and lethal wilt of apparently healthy trees to chronic dieback over many years prior to death. Root rot is a component of the syndrome in all these diseases. In most cases, *P. cinnamomi* has been implicated at one stage or another during the investigations of their cause.

Jarrah dieback.—In southwestern Australia (Fig. 3), extensive areas in the dry sclerophyll forests of *Eucalyptus marginata* (M) are dead and dying (3, 39). The disease occurs as a mosaic of patches that have gradually increased in size and number (see Fig. 2, p. 974, Phytopathology 62). The problem was first recorded in 1921, when only a few small patches were known. It now affects an area estimated at up to 100,000 ha. Interpretation on time sequences of aerial photographs has indicated that the area affected has increased by 4% per annum on several large sample areas (3).

The occurrence of the disease in relation to environmental factors, its symptoms, and its effects upon plant communities have been described elsewhere (38, 39, 43). Although it is best known for its effects in the jarrah forest, it also severely damages dry sclerophyll woodland and heath communities (39). In the overstory of eucalypt forests, it severely affects only *E. marginata* (M). Other overstory species are either mildly affected, e.g., *E. calophylla* (subgenus *Corymbia*), or are field resistant, e.g., *E. wandoo* (S), *E. gomphocephala* (S), *E. diversicolor* (S), *E. rudis* (S), and *E. patens* (M). In dry sclerophyll communities, many species of trees and shrubs in the understory and shrub layers are killed, particularly those in the families Proteaceae, Epacridaceae, Xanthorrhoeaceae, Papilionaceae, Dilleniaceae, and Myrtaceae. The disease does not cause damage to understory species of the wet sclerophyll Karri (*E. diversicolor*) forests where the families Rhamnaceae, Rutaceae, and Mimosaceae are prominent (38).

Because the incidence of jarrah dieback was seen to be closely associated with timber cutting and roadmaking, the disease was at first thought to be due to deterioration of the forest environment brought about by disturbance

of sites already marginal for the susceptible species (18, 37). But by 1963 there was strong circumstantial evidence that indicated jarrah dieback was due to a soil-borne pathogen. In 1964, *P. cinnamomi* was isolated from diseased forest (43) and later was shown to be constantly associated with the disease (39). It has never been isolated from unaffected forest despite repeated attempts (3, 39). By 1971, field inoculation trials made in 1966 produced proof that the introduction of *P. cinnamomi* to previously unaffected forest was sufficient to initiate the entire sequence of disease symptoms (39). This result, considered along with (a) the spreading mosaic pattern of occurrence, (b) the association of the disease with logging and roading, (c) the apparent absence of the fungus in unaffected forest, and (d) the extreme susceptibility of so much of the flora, has lead to the conclusion that *P. cinnamomi* is an introduced fungus in Western Australia (39).

For both economic and practical reasons, it has been impossible to implement an eradication program or to attempt to contain the epidemic within the boundaries of existing infestation (3). However, a marked reduction in the dispersal of infested soil has been effected by reducing traffic between diseased and healthy forest, sanitizing vehicles where such traffic is unavoidable, and banning

Fig. 1. A healthy stand of second growth jarrah (*Eucalyptus marginata*) approximately 60 years old and approximately 80 ft (25 m) mean dominant height with an understory of *Banksia grandis* and *Persoonia* spp. No isolates of *Phytophthora cinnamomi* were detected in tests on a total of 3,025 soil samples taken beneath this stand on 45 sampling occasions between 10 May 1965 and 1 May 1967. (Photo: F. White)

the use of gravels from diseased areas for roadbuilding in unaffected areas. These measures are expected to sharply reduce the rate at which new centers of infestation are established (3). They are practicable only because in most instances the boundaries of infestation can be readily recognized. On most terrain in the jarrah forest, linear rates of autonomous spread by *P. cinnamomi* are so slow that the zone of infestation closely corresponds to the presence of severe symptoms in the understory and shrub layers. Attempts to reforest devastated areas with jarrah have been unsuccessful, but *E. saligna* (S), *E. globulus* (S), and *E. microcorys* (S), all from eastern Australia, have shown early promise (3).

The severity of the disease is attributed to the coincidence of the high susceptibility of the flora and its extreme vulnerability (35, 39). This vulnerability is due to a combination of circumstances, including (a) a large food base for inoculum buildup provided by the high proportion of susceptible species in the flora, (b) low fertility of the soils and their low levels of microbial activity, which result in a low level of antibiosis, (c) the gravelly and sandy textures of soils whose physical structure allows extensive rewetting of large volumes of soil by relatively light rainfall, so favoring zoospore formation and dispersal, and (d) the severity of the summer drought, which places severe stress upon plants whose root systems are damaged during periods of spring rain (35, 39, 43).

The great susceptibility of the jarrah forest, its vulnerability, and the seemingly inexorable spread of *P. cinnamomi* suggest that, given long continuation of present conditions, this fungus could destroy jarrah as a forest community and deprive Western Australia's wood-using industry of its prime source of raw material (3, 39, 54).

Dieback on the Brisbane and Otway Ranges, Victoria.—Perhaps the most dramatic disease development in Australian forests has been the explosive epidemic of *P. cinnamomi* infection in the Brisbane Ranges 60 km west of Melbourne, Victoria. In 1969 the

Fig. 2. Recent dieback in a stand of jarrah of mixed age and mean dominant height of approximately 75 ft (24 m). All age classes have been affected. Note the dead understory plants, which include *Banksia grandis*, *Persoonia elliptica*, and *Xanthorrhoea preissii*. *Phytophthora cinnamomi* has been constantly associated with this disease. (Photo: D. Darling)

fungus was isolated from four small patches of dying ground vegetation under low sclerophyll forest (41). Although there were no signs of ill-health in the eucalypt overstory, the symptoms and the range of genera affected in the ground flora were strikingly similar to those of the initial stages of jarrah dieback. There was such remarkable similarity of soils and climate and of composition and structure of the communities with those of the jarrah forest that these forests seemed likely to be vulnerable to *P. cinnamomi* (41). Since then, the following has been reported: (a) rapid extension of the disease (up to 175 m per annum on flat terrain and 400 m per annum down drainage lines) (61, 63), (b) close association of the disease with *P. cinnamomi* (63), (c) severe dieback and mortality in *E. obliqua* (M), *E. baxteri* (M), and *E. macrorhyncha* (M), (d) field tolerance in *E. sideroxylon* (S), *E. viminalis* (S), and *E. aromaphloia* (S), and apparent field immunity in *E. ovata* (S) and *E. goniocalyx* (S) (61), (e) no success in isolating *P. cinnamomi* from unaffected stands in the same district (61), (f) proof of the pathogenicity of pure cultures of local isolates within 12 months of inoculation into previously unaffected forest (60), and (g) strong circumstantial evidence of the introduction of the fungus on earthmoving equipment to both the Brisbane Ranges (62) and Wilson's Promontory National Park, approximately 180 km southeast of Melbourne (11, 59, 62).

The effect of the disease upon species composition is striking. Two species of the Papilionaceae, *Dillwynia sericea* and *Hovea heterophylla*, were eliminated on the infested areas studied, and 21 other species of woody shrubs in the families Brunoniaceae, Dilleniaceae, Epacridaceae, Myrtaceae, Proteaceae, Papilionaceae, and Xanthorrhoeaceae seem likely to be eliminated (61). By contrast, the numbers of most monocotyledons, particularly grasses and sedges, sharply increased, as did some native herbs and exotic weeds (61). It is expected that within a few years the disease will have converted a two-layered, closed forest of dry sclerophyll shrubs and trees dominated by *Monocalyptus* to a floristically impoverished savannah of sedges, grasses, and herbs carrying scattered specimens of *Symphyomyrtus*.

In the Otway Ranges, about 200 km southwest of Melbourne, G. C. Marks (*personal communication*) has associated *P. cinnamomi* with dieback virtually identical to that in the Brisbane Ranges.

Although control measures similar to those employed with jarrah dieback would seem to be suitable, the practical problems of defining the limits of infestation are greatly exacerbated by rapid spread of the fungus, which may occur ahead of symptoms for variable distances up to 30 m (61, 63).

Recapitulation.—The two diebacks treated above are the only eucalypt diebacks for which there is unequivocal evidence that *P. cinnamomi* is the primary cause; in both cases, Koch's postulates have been satisfied under field conditions. Opinion on the role of *P. cinnamomi* in other diebacks, to which we shall shortly turn, is largely a matter of inference. This inference is based upon more or less satisfactory information concerning the distribution of the fungus in relation to the diseases and upon the responses of the various components of the communities. The classic pattern of *P. cinnamomi* diebacks (i.e., susceptibility of *Monocalyptus* and tolerance in *Symphyomyrtus* accompanied by susceptibility in the Proteaceae, Epacridaceae, and Xanthorrhoeaceae and resistance of grasses and sedges) has been observed only in the presence of *P. cinnamomi*.

It might seem reasonable, then, to suppose that *P. cinnamomi* is the cause of any dieback exhibiting this particular array of effects. Further, where these criteria might be met but are not, it would seem reasonable to assume that *P. cinnamomi* is either absent or that its activity is suppressed. However, the situation in the diebacks of Tasmania, New South Wales, and eastern Victoria is rather more complex and puzzling than the above model suggests.

EUCALYPT DIEBACKS IN TASMANIA.—Four diebacks are currently recognized in Tasmania (58): east coast dieback, gully dieback, regrowth dieback, and high altitude dieback. All are imperfectly known, but are now subject to detailed investigation. *Phytophthora cinnamomi*, which is reported to be widespread in Tasmania (45), plays a role in three of these diebacks.

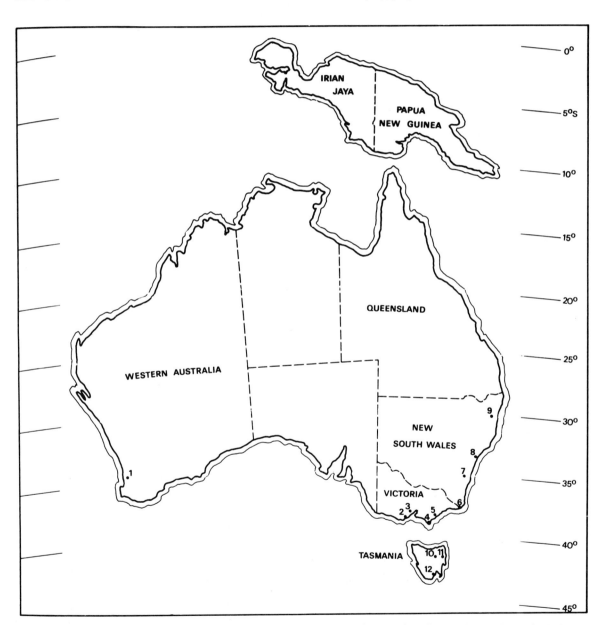

Fig. 3. Map of Australia showing locations referred to in the text. *1)* Jarrah forest region of south Western Australia, *2)* Otway Ranges, *3)* Brisbane Ranges, *4)* Wilson's Promontory, *5)* Rosedale, *6)* Genoa-Mallacoota, *7)* Berimbamala Creek, *8)* Ourimbah State Forest, Gosford-Wyong district, *9)* Nymboida-Dorrigo district, *10)* main occurrence of high altitude dieback, *11)* main occurrence of gully dieback, and *12)* main occurrence of southern regrowth dieback.

Recent work (23, 36), however, indicates that it is probably the prime cause of only one.

East Coast Dieback.—In the lowland and foothill forests of Tasmania's east coast, a dieback that meets the classic criteria for diagnosing *P. cinnamomi* infection occurs. *Eucalyptus sieberi* (M), *E. obliqua* (M), *E. amygdalina* (M), *Banksia marginata*, *Epacris impressa*, *Xanthorrhoea australis*, and *Hibbertia* sp. are killed, but *E. ovata* (S), *E. globulus* (S), and the sedges are field tolerant (36). *Phytophthora cinnamomi* has been isolated from the diseased areas, but has not been detected in unaffected forest nearby (36). Inoculation of pure culture of *P. cinnamomi* onto feeder roots of trees growing in the forest has indicated susceptibility to root rot in *E. obliqua* (M), *E. sieberi* (M), *E. amygdalina* (M), *E. viminalis* (S), and *B. marginata*. The tests do not constitute proof of pathogenicity under field conditions, however, since the excavated root systems, although attached to the parent trees, were cultured and inoculated in containers with moist vermiculite (36). Field inoculation trials were established more recently (58).

No estimates of the area affected, rates of spread, or history of the problem have been reported.

Gully Dieback.—This dieback, which affects dry sclerophyll forest on the eastern slopes of Tasmania's northeastern highlands, was first described in 1972 (15). Comparison of aerial photographs taken in 1950 and 1969 showed that both the acreage affected and the intensity of the disease increased markedly during the period. In an area of 32,000 ha sampled in 1969, at least 2,400 ha were affected (16). The disease occurs principally on lower slopes and in gullies and occasionally on ridges. In the first areas examined, there was strong evidence that *P. cinnamomi* caused this disease (15); *E. obliqua* (M), *E. amygdalina* (M), *E. delegatensis* (M), and *E. viminalis* (S) were killed, but *E. ovata* (S) was unaffected. There was also mortality among species of the dry sclerophyll ground cover, but not among the understory of wet sclerophyll gully species that are mainly from the families Compositae and Mimosaceae. *Phytophthora cinnamomi* was isolated from three areas. Since then, 21 diseased gullies have been tested, but *P. cinnamomi* has been found in only two, both close to roads (36). The negative samples were baited, rebaited, sown with susceptible plants that grew well, and were then rebaited. It was concluded (36) that gully dieback is not due to *P. cinnamomi*. However, in some places, the effects of this fungus are superimposed upon another disease whose cause is unknown. The problem may involve a complex of pathogens, including *Pythium*, *Phytophthora*, and *Armillaria*, as well as insect defoliators and drought (36, 58).

Regrowth Dieback.—This disease was first reported in 1968 and has since been described in more detail (5, 15, 16). It affects regrowth eucalypts in tall wet sclerophyll forests ranging in age from 30 to 170 years. The tall understory of species in the families Rhamnaceae, Pittosporaceae, Compositae, and Mimosaceae is unaffected. *Eucalyptus obliqua* (M), *E. regnans* (M), and *E. simmondsii* (M) are affected, but *E. globulus* (S) is not.

The extent of the problem is difficult to assess because dead and dying trees are diffusely distributed among apparently healthy trees and there are no marked effects on the understory. The disease is common on some 16,000 ha in southern Tasmania, and there are smaller occurrences in northern parts of the island (16).

Evidence from growth plots established long before the disease was recognized suggests that this dieback is a very recent phenomenon that has not occurred previously in the life of the existing stands (5). Once decline has begun, it seems to be irreversible, even though it may take up to 10 years for a tree to die.

Phytophthora cinnamomi has been isolated from beneath regrowth dieback areas, but constant association has not been demonstrated. It is unlikely that it plays a major role, as soil temperatures appear to limit fungus activity for the greater part of the year (23). *Armillaria mellea* is intimately associated with the problem, but insect defoliation (22) and drought may be involved as well.

High Altitude Dieback.—This dieback kills *E. delegatensis* (M) over large tracts of country above 700 m in northeastern Tasmania, but does not affect *E. dalrympleana* (S) or the understory flora, which includes *Persoonia* (Proteaceae) and *Pultenaea* (Papilionaceae) as well as wet sclerophyll and rainforest elements. In 1964, it was estimated (12) that approximately 1,800 of the 5,000 ha sampled on the Mt. Maurice Plateau were severely affected. The cause of the disease is unknown, and efforts to isolate *P. cinnamomi* have produced negative results (58). Ellis (12) believes this dieback is a natural part of the successional sequence toward climax rainforest of *Nothofagus*. It develops in fire-sere eucalypt communities when fire is excluded for long periods. Ellis has speculated on several possible causal mechanisms (12, 13). The disease is unknown in stands of *E. delegatensis* on mainland Australia, even where understories of rainforest species occur.

EUCALYPT DIEBACKS IN NEW SOUTH WALES.—Despite the widespread occurrence of *P. cinnamomi* in native forests in New South Wales (45, 46), there are no extensive areas of dieback that can be attributed solely to this fungus. Only one dieback has damaged significant areas of eucalypt forest, and its effects differ markedly from those of other eucalypt diebacks. The disease results in defoliation and death of *E. saligna* (S), *E. propinqua* (S), *E. paniculata* (S), *E. deanei* (S), *E. microcorys* (S), *E. acmenioides* (M), *E. umbra* (M), *E. pilularis* (M), and *Angophora intermedia* (30, 31). There are no detailed accounts of the effects of the disease upon the understory although a few understory plants were affected at Ourimbah in 1972 (17). In the two areas of severely damaged forest near Nymboida on the Dorrigo Plateau and at Ourimbah in 1966, no marked effects were evident in the understories, even though they contained *Persoonia* sp. and *Orites excelsa* (both Proteaceae) and *Trichocarpa laurina* (Epacridaceae).

The disease occurs in scattered patches throughout coastal New South Wales, but is best known on Ourimbah State Forest in the Gosford-Wyong district.

Moore (31) reported that in 1959 there were 150 separate occurrences varying in size up to 1,500 ha and that the disease had increased from minor importance in 1949 to local economic significance in 1958.

The cause of the dieback is unknown, but it is almost certainly complex. Most trees had been continually defoliated for some years by leaf-sap sucking psyllids and had been attacked by borers, particularly *Xyleborus truncatus* (30), with which an *Ambrosiella* sp. is associated (Stahl, *personal communication*). Stahl found that in pure culture this fungus produced a water soluble toxin that rapidly wilted cuttings of soft-leaved plants. Secondary borers of at least five other genera may also contribute to the decline (30). The disease occurs mainly on shallow soils over clay subsoils, and its major development followed the unusually wet years between 1949 and 1956 (30). It has been suggested that these factors debilitated the trees, making them more attractive to psyllids and more susceptible to *Xyleborus. Armillaria mellea* is associated with the disease in some places. More recently, isolations of several pathogens in the Pythiaceae have been reported; these included *P. cinnamomi* and *P. citricola* (46). However, there are no published accounts of detailed surveys to indicate whether these fungi are constantly associated with the disease. Some evidence suggests they may not be. In 1966, I isolated *P. cinnamomi* from beneath dying *E. saligna* on farmland near Ourimbah, but I was unable to isolate it from beneath dying forest both at Ourimbah and at Nymboida, where *Pythium* spp. were abundant.

The role of *P. cinnamomi* in this disease is far from clear. Perhaps it has an occasional distribution overlaying a disease due entirely to other causes. Or perhaps the lack of damage to the understory and the susceptibility of both *Symphyomyrtus* and *Monocalyptus* may be an expression of an unusual array of pathogenic variants in local populations of *P. cinnamomi*. For this, however, we have no direct evidence.

EUCALYPT DIEBACKS IN EASTERN VICTORIA.—The first account of rapid decline and death of groups of eucalypts in eastern Victoria was given in 1959 (24). According to Marks et al. (28), Lee's unpublished records showed a total of 20 ha affected in patches not exceeding 2 ha. On the basis of air-photo records and local lore, it has been suggested that the disease originated in the late 1930's (28). It extended rapidly during the unusually wet seasons in the late 1950's and is now estimated to affect about 15,000 ha, of which 5,000 ha are severely damaged (28).

The species now affected include *E. obliqua* (M), *E. sieberi* (M), *E. muellerana* (M), *E. globoidea* (syn. *E. scabra*) (M), and *E. radiata* (M), but *E. bosistoana* (S) and *E. sideroxylon* (S) are field resistant (24, 28). The understory species *B. marginata*, *X. australis*, and *Daviesia latifolia* are also severely affected (28).

Armillaria mellea occurs in some patches, but it is not constantly associated with the dieback (24). *Phytophthora cinnamomi* has been isolated from each of the areas sampled, and the pattern of disease occurrence and spread reportedly is consistent with the behavior of an introduced pathogen (28).

Although in many places the disease bears a striking resemblance to jarrah dieback, recent observation indicates that the disease situation and the role of *P. cinnamomi* in this region are somewhat more complex than had previously been suggested (35, p. 303). There are classic symptoms near Rosedale, where *E. consideniana* (M), *B. marginata*, *Epacris impressa*, and *X. australis* are killed en masse and *E. bridgesiana* (S) is healthy. Recently, however, I have seen other areas where the symptoms are not consistent with the classic response to *P. cinnamomi* infection. Immediately west of the Mallacoota airfield, there is an extensive area of advanced dieback in *E. sieberi* (M), *E. muellerana* (M), *E. globoidea* (M), and *E. gummifera* (*Corymbia*). Beneath this stand, a wide range of understory species ordinarily considered to be susceptible to *P. cinnamomi* is apparently unaffected. The range includes *B. marginata*, *B. spinulosa*, *Persoonia* spp., *Hibbertia* sp., *Xanthorrhoea* sp., and *Callistemon citrinus*. Yet nearby, on the Betka track, D. Cameron (*personal communication*) has observed typical early symptoms of *P. cinnamomi* infection where *B. spinulosa* and *Hakea ulicina* had died en masse beneath apparently healthy *E. sieberi*. Further, dieback of susceptible species does not always occur in the presence of *P. cinnamomi*, as the fungus occurs beneath extensive tracts of apparently healthy forest (28). Although it has been suggested that *P. cinnamomi* probably causes loss of growth in such places (26), it is not an invariable consequence (21).

The dieback situation in eastern Victoria is further complicated by the occurrence of Gosford-Wyong type dieback, notably around Genoa, where psyllids of the genera *Glycaspis* and *Eucalyptolyma* (Yen, *personal communication*) occur in plague proportions on eucalypts over an apparently healthy understory.

SYNOPSIS.—The distribution of *P. cinnamomi* in coastal Australia is widespread, but discontinuous and extending. The fungus is intimately associated with several new and serious diseases on extensive areas in southern Australia. In east coast forests north of latitude 37°S, it has caused remarkably little damage. These differences in response among forests containing inherently susceptible species and in which the fungus is widespread have attracted considerable attention, giving rise to a number of questions concerning the variability of the fungus and the influence of environment upon the activity of *P. cinnamomi* and upon the vulnerability of its hosts, and to the question of the introduced or indigenous origin of the Australian populations of the fungus. These questions were last reviewed in 1971 (35), but they remain essentially unresolved. However, enough new information has since appeared to make a reappraisal of the evidence worthwhile.

THE PATHOGEN.—Until recently, Australian populations of *P. cinnamomi* were thought to be remarkably uniform (9, 10, 43). Only the A_2 mating strain was known, and the morphology and pathogenicity of isolates from various parts of Australia and New Zealand seemed to be very similar. However, there is now considerable evidence of variation (8, 27, 50, 55), especially in growth response to temperature (28, 55).

There are marked regional differences in the relative proportions of these variants, but no published accounts show how this nutrient dependent variability (55) influences the competitive ability, the pathogenicity, or the survival capacity of local populations. Although variations in pathogenicity to lupines have been demonstrated among some Victorian isolates, the differences could not be related to pathogenicity toward *Eucalyptus* (28).

Recent work has indicated several possible mechanisms for maintaining variation in Australian populations of *P. cinnamomi*. Both the sexually compatible A_1 and A_2 types have been isolated. The A_2 type is widespread, and the A_1 has now been detected at five locations in Western Australia, New South Wales, and Queensland from among 372 isolates examined (48). Cultural studies on single zoospore isolates have provided evidence of heterokaryosis, and hyphal anastomosis has been observed in vitro (55). Other possible mechanisms for the formation and survival of heterokaryons include the formation of oospores between hyphae of sexually incompatible isolates that have been exposed to volatiles produced by *Trichoderma viride* (6). Similar response has been reported for Australian isolates exposed to volatiles produced by *T. koningii* (49). Several species of *Phytophthora* that are known pathogens of *Eucalyptus* and that have been isolated from forest soils in Australia (38, 46) "mate" with *P. cinnamomi* to form oospores in vitro. It is possible that they may hybridize in nature.

Although it is now apparent that Australian populations of *P. cinnamomi* are more variable than hitherto appreciated, the present state of knowledge falls far short of that required (14) to understand how the populations of this pathogen react with their hosts in the great variety of environments in which they co-occur.

THE IMPORTANCE OF ENVIRONMENT.—The powerful effects that environmental factors exert upon disease development and expression are now widely recognized (27, 28, 32-35, 39, 45, 54). The variety of factors involved and the complexity of their interactions are becoming even more apparent. It has been demonstrated that Australian soils vary markedly in the extent to which they exert fungistatic effects upon *P. cinnamomi* (7). The relative susceptibilities to infection of *E. marginata* and *E. calophylla* have been shown to be influenced by their distinctive rhizosphere floras (25), and, in vitro, some basidiomycetes that may form mycorrhizae on eucalypts exhibit antagonism to *P. cinnamomi* (44). Soil nutrient status influences the host-pathogen interaction (19); calcium content of lupine roots affects the rate of invasion by *P. cinnamomi* (4); and boron deficiency can increase susceptibility of lupine roots to infection (Jehne and Snowdon, *unpublished*). The significance of unusually wet periods, poorly drained soils, and soil temperature in influencing epidemic development in eucalypt forests has been confirmed by recent observations (23, 28, 54, 63). The complex relationships between soil flooding and *P. cinnamomi* infection have been partly unraveled. It has been shown, for example, that changes in the balances of metabolites that occur in root exudate when plants are flooded (29) may markedly influence the chemotaxis of zoospores (1).

These effects may in part account for the apparent correlation between tolerance to waterlogging in some plants and their field tolerance of *P. cinnamomi* (42).

Examples of the importance of forest structure and species composition upon preinfection buildup of inoculum and of climate and stand structure upon postinfection stress on the host have been given by Newhook and his colleagues (32-35, 39).

Despite these advances, there have been no reports of in vivo tests of any of the several hypotheses that have been advanced to account for the damage caused by *P. cinnamomi* in some forests and its failure to cause serious disease in others. So multitudinous are the components of environment that may influence control of *P. cinnamomi* and its effects that the emergence of an integrated theory of environmental control in the near future seems unlikely.

PHYTOPHTHORA CINNAMOMI INTRODUCED OR INDIGENOUS.—Although the relevance of this matter for forest practice has been questioned (2), most workers now agree that it has an important bearing upon the choice of methods that might be adopted to control the disease (28, 35, 40, 47).

That *P. cinnamomi* must have been introduced into Western Australia some time after the first European settlement in 1827 is widely accepted (35, 47, 54). However, two quite distinct explanations of the origins of the eastern Australian populations are current. One group of investigators believes that the fungus was introduced (27, 35, 63) and that the low incidence of disease in northern forests and at higher altitudes in the south is due to environmental controls. Another group has suggested that *P. cinnamomi* is a native fungus that is in balance with its hosts and their environments. Where the activities of European man have disturbed the environmental balance so that natural restraints on the activity of *P. cinnamomi* are removed, disease may develop (45). This debate has been discussed in detail elsewhere (35). New arguments have now been advanced in favor of the hypothesis of indigenous origin modified to accommodate earlier counterarguments. According to these arguments (45, 47): (a) *P. cinnamomi* entered Australia from the north during the Tertiary, along with elements of the Indo-Malayan flora, (b) it occupied moist situations in the landscape where reciprocal adaptation for avirulence in the fungus and tolerance in the flora occurred, (c) it failed to colonize drier sites on which inherently susceptible elements of the Australian flora have survived, and (d) European man has inadvertently transferred the fungus to new localities and so disturbed natural balances that new sets of environmental conditions in both new and old habitats favor the pathogen more frequently than before. The following observations have been cited in support of the hypothesis: (a) the variability of the fungus, particularly the co-occurrence of the two mating strains, (b) the apparent adaptation of the flora on those sites most favorable to the fungus, and (c) the isolation of *P. cinnamomi* from a remote and undisturbed area [sic] in New South Wales.

These arguments, however, do not exclude the alternative hypothesis.

First, it is possible that both mating strains and all their variants were introduced. In the 100 years or more between the first European settlement and the effective implementation of quarantine to prevent the movement of soil into Australia, a vigorous import trade in balled planting stock flourished. Many of the whole plants traded are known to be hosts of *P. cinnamomi*. The proponents of the hypothesis of indigenous origin of *P. cinnamomi* evidently accept the possibility of multiple origin, as they record variation in the Western Australian populations of the fungus (46, 48) and accept that they were probably introduced (45). There seems to be no reason why there should not have been multiple introduction on the longer settled and more heavily populated east coast.

Secondly, the supposed entry of *P. cinnamomi* to eastern Australia along with the Indo-Malayan flora is open to question. Theories of the origin of *P. cinnamomi* must take account of the biogeographical history of the area (47). The proponents of the indigenous hypothesis accept the theory of immigrant origin of the Indo-Malayan element (52), but an alternative theory of local origin of these floristic elements has been suggested (20). Nevertheless, there is considerable evidence that past migrations of animals and plants took place across a former land-bridge between New Guinea and Australia. It might seem reasonable to suppose that, irrespective of the origins of the Indo-Malayan elements, *P. cinnamomi* entered through that portal. But this is not consistent with the apparent absence of the fungus from the lowlands of New Guinea (57). Despite wide search, only the A_1 isolate has been recovered and then only in the highlands, where the circumstances of its occurrence and its effects (53) are not inconsistent with recent introduction into New Guinea.

Thirdly, the evidence of the tolerance to *P. cinnamomi* of Australian native plants is not entirely consistent with the proposal that on moist sites in eastern Australia there has been reciprocal adaptation of fungus and flora. Many plants that are tolerant of moist soils are also tolerant of *P. cinnamomi*, but this is probably not causally related. If it were, particular explanations would be required for the great susceptibility to *P. cinnamomi* of a number of species that are well adapted and sometimes restricted to sites that are seasonally waterlogged. Examples include the narrow-leaved peppermint group in *Eucalyptus*, *Aotus ericoides*, *Leucopogon australis*, *B. marginata*, and *Sprengelia incarnata*.

Finally, the area described as remote and undisturbed by logging (47) was exposed to the chance of infestation by *P. cinnamomi* for many years prior to isolation of the fungus. The Bimberamala catchment is rugged and little disturbed, but it forms part of the Budawang escarpment for the settled plateau of the Southern Tablelands district. Local settlers report that cattle have been transferred back and forth between their farms and the Bimberamala Valley for more than 80 years. The western slopes were logged for timber with heavy equipment more than 15 years ago, and stumps and skid trails have been found within 50 m of the watershed above the Bimberamala catchment.

The question probably cannot be answered unequivocally. Meanwhile, decisions on control procedures are required. In these circumstances, the most suitable control strategy is to take precautions against the alternative that is potentially most serious; i.e., that *P. cinnamomi* is an introduced pathogen.

Managers of threatened wildlands are adapting their practices so that further spread of the fungus is reduced to a minimum. They look to research to devise means by which the fungus can be controlled on lands it will infest. This will require a greater understanding than we now possess concerning the fungus and its interaction with its hosts and their environments. Until this is achieved, it seems unlikely that the role of *P. cinnamomi* in Australian forests will be reduced to one of minor significance.

LITERATURE CITED

1. ALLEN, R. N., and F. J. NEWHOOK. 1973. Chemotaxis of zoospores of Phytophthora cinnamomi to ethanol using capillaries of soil pore dimensions. Trans. Br. Mycol. Soc. 61:287-302.
2. ANONYMOUS. 1972. Eucalypt dieback threat to Australian forests. Appita 25:251-253.
3. BATINI, F. E., and E. R. HOPKINS. 1972. Phytophthora cinnamomi Rands—a root pathogen of the jarrah forest. Austral. For. 36:57-68.
4. BELLANY, G., W. A. HEATHER, and B. H. PRATT. 1971. The effect of calcium chloride treatment in reducing the rate of spread of Phytophthora cinnamomi within roots of Lupinus angustifolius. Paper to Austral. Plant Path. Conf., Hobart, Tasmania.
5. BIRD, T. 1973. Regrowth dieback. Submission to Eucalypt Crown Dieback Seminar, Lakes Entrance, Victoria.
6. BRASIER, C. M. 1971. Induction of sexual reproduction in single A_2 isolates of Phytophthora species by Trichoderma viride. Nature New Biol. 231:283.
7. BROADBENT, P., K. F. BAKER, and Y. WATERWORTH. 1971. Bacteria and actinomycetes antagonistic to fungal root pathogens in Australian forests. Austral. J. Biol. Sci. 24:925-944.
8. CHANG, S. T., C. J. SHEPHERD., and B. H. PRATT. 1973. Sexual behaviour in the genus Phytophthora—peculiarities and problems. Abstr., 45th ANZAAS Cong., Perth, Western Australia, Sec. 12. p. 20-21.
9. CHEE, K. H., and F. J. NEWHOOK. 1965. Variability in Phytophthora cinnamomi Rands. N.Z. J. Agr. Res. 8:95-103.
10. CHEE, K. H., and F. J. NEWHOOK. 1965. Variability in sexual reproduction of Phytophthora cinnamomi Rands. N.Z. J. Agr. Res. 8:947-950.
11. EDWARDS, G. P. 1973. The cinnamom fungus in national Parks, Submission to Eucalypt Crown Dieback Seminar, Lakes Entrance, Victoria.
12. ELLIS, R. C. 1964. Dieback of alpine ash in northeastern Tasmania. Austral. For. 28:75-90.
13. ELLIS, R. C. 1971. Dieback of alpine ash as related to changes in soil temperature. Austral. For. 35:152-163.
14. ERWIN, D. C., G. A. ZENTMYER, J. GALINDO, and J. S. NIEDERHAUSER. 1963. Variation in the genus Phytophthora. Annu. Rev. Phytopathol. 1:375-396.
15. FELTON, K. C. 1972. Eucalypt diebacks in Tasmania. Appita 26:207-208.
16. FELTON, K. C., and T. BIRD. 1972. Economic effects of eucalypt diebacks in Tasmania. Paper to 44th ANZAAS Cong., Sydney, Australia. 7 p.
17. GERRETTSON-CORNELL, L. 1973. A preliminary study on the morphology of Phytophthora cinnamomi Rands

from Ourimbah State Forest. Informatore Botanico Italiano 5:78-80.

18. HAMILTON, C. D. 1951. The dying of jarrah (Eucalyptus marginata) in Western Australian forests. Progress on work done to 1948. Unpub. rept., For. and Timber Bur., Canberra.

19. HEATHER, W. A., G. BELLANY, and B. H. PRATT. 1971. The effect of mineral nutrition of soil on the infection of roots of Lupinus angustifolius by Phytophthora cinnamomi. Paper to Austral. Plant Pathol. Conf., Hobart, Tasmania.

20. HERBERT, D. A. 1966. Ecological segregation and Australian phytogeographic elements. Proc. Roy. Soc. Qld. 78:101-111.

21. INCOLL, W. D., and P. C. FAGG. 1973. The effect of Phytophthora cinnamomi on the basal area increment of natural stands of Eucalyptus sieberi. Submission to Eucalypt Crown Dieback Seminar, Lakes Entrance, Victoria.

22. KILE, G. A. 1973. Insect defoliation in the eucalypt regrowth forests of southern Tasmania. Austral. For. Res. 6:9-18.

23. KILE, G. A. 1973. Regrowth dieback—possible contributory factors. Submission to Eucalypt Crown Dieback Seminar, Lakes Entrance, Victoria.

24. LEE, H. M. 1962. Death of Eucalyptus spp. in East Gippsland. For. Comm. Vic. Tech. Paper 8:14-18.

25. MALAJCZUK, N., A. J. MC COMB, and C. A. PARKER. 1973. The possible role of rhizosphere microorganisms in conferring resistance to infection by Phytophthora cinnamomi. Abstr., 45th ANZAAS Cong., Perth, Western Australia, Sec. 12. p. 72-73.

26. MARKS, G. C., and D. T. HARTIGAN. 1972. The problems of learning to live with Phytophthora cinnamomi in infected forest soils. Paper to 44th ANZAAS Cong., Sydney, Australia.

27. MARKS, G. C., F. Y. KASSABY, and P. C. FAGG. 1972. Die-back tolerance in eucalypt species in relation to fertilization and soil populations of Phytophthora cinnamomi. Austral. J. Bot. 21:53-65.

28. MARKS, G. C., F. Y. KASSABY, and S. T. REYNOLDS. 1971. Die-back in the mixed hardwood forests of eastern Victoria: a preliminary report. Austral. J. Bot. 20:141-154.

29. MC MANMON, M., and R. M. M. CRAWFORD. 1971. A metabolic theory of flooding tolerance: the significance of enzyme distribution and behaviour. New Phytologist 70:299-306.

30. MOORE, K. M. 1959. Observations on some Australian forest insects. 4. Xyleborus truncatus Erichson 1842 (Coleoptera: Scolytidae) associated with dying Eucalyptus saligna Smith (Sydney Blue Gum). Proc. Linn. Soc. N.S.W. 84:186-193.

31. MOORE, K. M. 1962. Entomological research on the cause of mortalities of Eucalyptus saligna Smith (Sydney Blue Gum). N.S.W. For. Comm. Res. Notes No. 11. 8 p.

32. NEWHOOK, F. J. 1960. Climate and soil type in relation to Phytophthora attack on pine trees. Proc. N.Z. Ecol. Soc. 7:14-15.

33. NEWHOOK, F. J. 1970. Phytophthora cinnamomi in New Zealand, p. 173-176. In T. A. Toussoun et al. (ed.), Root diseases and soil-borne pathogens. Univ. Calif. Press, Berkeley and Los Angeles.

34. NEWHOOK, F. J. 1973. The importance of environment in the reaction of Australasian forests to Phytophthora cinnamomi. Abstract No. 0984 in Abstracts of Papers, 2nd Int. Cong. Plant Pathol., Minneapolis, Minn.

35. NEWHOOK, F. J., and F. D. PODGER. 1972. The role of Phytophthora cinnamomi in Australian and New Zealand forests. Annu. Rev. Phytopathol. 10:299-326.

36. PALZER, C. R. 1973. The relationship of Phytophthora cinnamomi to gully dieback in the Fingal district. Unpub. rept., For. Comm., Tasmania.

37. PODGER, F. D. 1959. A review of the disorder jarrah dieback. Unpub. rept., For. and Timber Bur., Canberra.

38. PODGER, F. D. 1968. Aetiology of jarrah dieback. M.S. For. Thesis, Univ. Melbourne. 292 p.

39. PODGER, F. D. 1972. Phytophthora cinnamomi, a cause of lethal disease in indigenous plant communities in Western Australia. Phytopathology 62:972-981.

40. PODGER, F. D. 1972. Phytophthora cinnamomi—an invader from abroad or a native gone wrong? Paper to 44th ANZAAS Cong., Sydney, Australia.

41. PODGER, F. D., and D. H. ASHTON. 1970. Phytophthora cinnamomi in dying vegetation on the Brisbane Ranges, Victoria. Austral. For. Res. 4:33-36.

42. PODGER, F. D., and F. E. BATINI. 1971. Susceptibility to Phytophthora cinnamomi root rot of thirty-six species of Eucalyptus. Austral. For. Res. 5:9-20.

43. PODGER, F. D., R. F. DOEPEL, and G. A. ZENTMYER. 1965. Association of Phytophthora cinnamomi with a disease of Eucalyptus marginata forest in Western Australia. Plant Dis. Reptr. 49:943-947.

44. PRATT, B. H. 1971. Isolation of basidiomycetes from Australian eucalypt forest and assessment of their antagonism to Phytophthora cinnamomi. Trans. Br. Mycol. Soc. 56:243-250.

45. PRATT, B. H., and W. A. HEATHER. 1972. The origin and distribution of Phytophthora cinnamomi Rands in Australian native plant communities and the significance of its association with particular plant species. Austral. J. Biol. Sci. 26:559-573.

46. PRATT, B. H., and W. A. HEATHER. 1972. Recovery of potentially pathogenic Phytophtora and Pythium spp. from native vegetation in Australia. Austral. J. Biol. Sci. 26:575-582.

47. PRATT, B. H., W. A. HEATHER, and C. J. SHEPHERD. 1972. Recovery of Phytophthora cinnamomi from native vegetation in a remote area of New South Wales. Trans. Br. Mycol. Soc. 60:197-204.

48. PRATT, B. H., W. A. HEATHER, and C. J. SHEPHERD. 1972. Transcontinental occurrence of A_1 and A_2 strains of Phytophthora cinnamomi in Australia. Austral. J. Biol. Sci. 25:1099-1100.

49. PRATT, B. H., J. H. SEDGELY, W. A. HEATHER, and C. J. SHEPHERD. 1971. Oospore production in Phytophthora cinnamomi in the presence of Trichoderma koningii. Austral. J. Biol. Sci. 25:861-863.

50. PRATT, B. H., and C. J. SHEPHERD. 1972. Variation in Australian isolates of Phytophthora cinnamomi Rands. Austral. Plant. Pathol. Soc. Newsletter 1(4):25-26.

51. PRYOR, L. D., and L. A. S. JOHNSON. 1971. A classification of the Eucalypts. Austral. Nat. Univ. Press, Canberra. 102 p.

52. ROBBINS, R. 1971. On the biogeography of New Guinea. Austral. External Territories Bull. 11:31-37.

53. SHAW, D. E., E. G. CARTLEDGE, and D. J. STAMPS. 1972. First records of Phytophthora cinnamomi in Papua New Guinea. Papua N. G. Agr. J. 23:46-48.

54. SHEA, S. R., and E. R. HOPKINS. 1973. Environmental factors in relation to distribution, intensity and control of jarrah dieback. Paper to 45th ANZAAS Cong., Perth, Australia.

55. SHEPHERD, C. J. 1972. The ecology of Phytophthora cinnamomi. Paper to 44th ANZAAS Cong., Sydney, Australia.

56. SOCIETY OF AMERICAN FORESTERS. 1950. Forestry Terminology. 2nd ed. Munns (ed.). Soc. Amer. For., Washington, D.C. 93 p.

57. STAMPS, D. J., D. E. SHAW, and E. G. CARTLEDGE. 1972. Species of Phytophthora and Pythium in Papua

New Guinea. Papua N.G. Agr. J. 23:41-45.

58. TASMANIAN FORESTRY COMMISSION. 1973. Review of eucalypt dieback in forests. Submission to Eucalypt Crown Dieback Seminar, Lakes Entrance, Victoria.

59. VEITCH, H. 1973. Impact of the cinnamon fungus at Wilson's Promontory National Park. Submission to Eucalypt Crown Dieback Seminar, Lakes Entrance, Victoria.

60. WESTE, G. 1973. Information for Dieback Conference. Submission to Eucalypt Crown Dieback Seminar, Lakes Entrance, Victoria.

61. WESTE, G., D. COOKE, and P. TAYLOR. 1972. The invasion of native forest by Phytophthora cinnamomi. II. Post-infection vegetation patterns, regeneration, decline in inoculum, and attempted control. Austral. J. Bot. 21:13-29.

62. WESTE, G., and C. LAW. 1972. The invasion of native forest by Phytophthora cinnamomi. III. Threat to the National Park, Wilson's Promontory, Victoria. Austral. J. Bot. 21:31-51.

63. WESTE, G. M., and P. TAYLOR. 1971. The invasion of native forest by Phytophthora cinnamomi. I. Brisbane Ranges, Victoria. Austral. J. Bot. 19:281-294.

PART III

THE ROOT ENVIRONMENT IN RELATION TO GROWTH AND SURVIVAL OF PLANT PATHOGENS

The Root Environment in Relation to Growth and Survival of Plant Pathogens: An Introduction

GEORGE A. ZENTMYER, *Department of Plant Pathology, University of California, Riverside.*

The papers in this session of the symposium present a comprehensive view of several aspects of the physical environment in relation to soil-borne root pathogens. These approaches to studies of root pathogens are all too frequently overlooked or ignored.

One of the principal points of the paper by Papendick and Campbell is the emphasis on the soil-plant-atmosphere continuum as a dynamic system in which water is continually being absorbed, transported, and evaporated, and in which water moves along gradients in water potential. Also, plant pathogens may exist in water potential equilibrium with various parts of the soil-plant-atmosphere continuum. New techniques described in this paper give means for measuring water potential; these include leaf and soil thermocouple psychrometers and pressure chambers.

The diffusion and concentration of oxygen in the root zone constitute another significant phase of the soil environment in relation to growth and development of soil-borne organisms and eventually to development of root diseases, as shown in the paper by Stolzy and Zentmyer. As the water content of the soil increases, the rate of gas diffusion is reduced, resulting in lower oxygen concentration; a similar situation can result from soil compaction. The main process in soil contributing to soil aeration is diffusion, by which gas exchange between the soil and the atmosphere can result in soil oxygen concentrations similar to oxygen concentration in the atmosphere.

The pH of the rhizosphere is an important physical factor in relation to infection of roots by pathogens and by mycorrhizal fungi. Rhizosphere pH is influenced by the form in which nitrogen is absorbed by roots, as described in the paper by Smiley; ammonium reduces the rhizosphere pH, and nitrate increases it. A significant aspect of the development of root pathogens that has been little recognized is the relationship between ammonium and nitrate nitrogen and the development of diseases caused by different pathogens.

The paper by Lyda and Burnett pointed out that *Phymatotrichum omnivorum* was very tolerant of high levels of CO_2 in liquid culture, growing at 50% CO_2. However, strands did not extend in nonsterilized soil at this concentration of the gas. The information presented on variation in levels of CO_2 at different depths in the soil and at different times of the year is also significant in relation to soil-borne microorganisms.

During one of the discussions of this session, Wilson and Griffin presented information on the effect of a decrease of water potential on respiration of the natural microbial population of the soil. The data show that bacterial respiration and activity predominate over fungal and actinomycete respiration and activity only in soils in which the water potential exceeds approximately —3 bar. Respiratory efficiency of fungi was found to decline as the osmotic potential falls.

Venkata Ram reported a relation between nitrite in the soil and the succession of fungi invading dead roots of tea. Fungi vary considerably in their tolerance to nitrite, and apparently there is a relation between this tolerance and invasion of the tea plant.

Water Potential in the Rhizosphere and Plant and Methods of Measurement and Experimental Control[1]

R. I. PAPENDICK and G. S. CAMPBELL—*Soil Scientist, Agricultural Research Service, U. S. Department of Agriculture, Washington State University, and Assistant Professor of Biophysics and Assistant Soil Scientist, Washington State University, Pullman.*

The water potential in plants and in soil surrounding roots affects both water use and growth (42, 59, 63). Growth and transpiration are generally greatest at high water potentials. Most land plants are periodically subject to water stress that can cause substantial loss in crop quality and yield. Limited growth and low water use efficiency can result directly from an inadequate supply of water or indirectly from an interaction of water with weeds, pathogens, or insects.

Microorganisms living in soil or plants probably have internal water potentials near those of their external surroundings. These potentials can fluctuate widely with time and among the different parts of the soil-plant system. During root infection, the organism is subject to the water potential of the rhizosphere. After penetration, the plant water potential influences growth and survival. Growth of any plant or microorganism is reduced as its potential is lowered below some level; the actual point, however, varies greatly with the species.

Static concepts of soil or plant water, such as field capacity, permanent wilting point, gravitational water, moisture-holding capacity, and others, have limited meaning in defining soil-plant-water relationships. Instead of being static, water in most soil-plant systems moves almost incessantly along an energy gradient from the soil reservoir through the plant and is distributed among the various plant parts or lost according to the physical and physiochemical laws of mass flow, adhesion, cohesion, and evaporation. Static systems are the exception rather than the rule. Plant pathogens and other microorganisms situated at various points along the flow path may thus encounter variable and different water potentials, depending on whether they are dormant in bulk soil, competing with one another in the rhizosphere, penetrating the cortex, growing in the vascular tissue, or embedded in the leaf.

In this paper, we examine water potential relations in the soil-plant system, including the rhizosphere and other habitats commonly occupied by plant pathogens, and discuss methods for controlling the water potential in experimental systems. Other reviews (18-20, 28, 31, 32) may be consulted for details on the energy concept of water as applied in soils and plant pathology.

THE ENERGY CONCEPT APPLIED TO PLANT WATER.—Because of molecular constraints imposed by pressure, solutes, and tissue matrix effects, plant water potential, like soil water potential, may differ considerably from that of pure, free water. Plant water varies from nearly pure to where most or all is part of the structural integrity of cells and tissues and so changed that it can be identified only remotely with water in the pure state. Plant and soil water relations are unified by use of the thermodynamic or energy concept (60, 62), which implies that water is held by forces that lower the water potential compared with pure, free water. The water potential in most soil-plant systems is negative in the traditional case where pure, free water at atmospheric pressure is assigned zero energy.

The water potential of plant cells and tissues (ψ_1) is usually divided into three components:

$$\psi_1 = \psi_m + \psi_o + \psi_p \qquad [1]$$

where ψ_m is the matric potential (negative; from adsorption on colloidal particles or walls), ψ_o is the osmotic potential (negative; from particles in solution), and ψ_p is the pressure potential (positive or negative). In cells, the pressure potential is identified with turgor pressure. In equation 1, ψ_1 is the leaf or plant water potential rather than the total potential, because the latter would include the gravitational component. When elevation differences within a system are less than 1-2 m, changes in the gravitational potential are generally negligible (< 0.1 to 0.2 bar).

The potential concept is readily related to the older mechanical concept which implies that water in unsaturated soil or in the living plant is generally at a negative pressure; i.e., at a pressure less than that for pure, free water at the same location. A relationship previously used in plant work is

$$DPD = OP - TP \qquad [2]$$

[1]Cooperative investigations, ARS, USDA, and Washington State University Agricultural Experiment Station. Scientific Paper No. 4145 of the College of Agriculture, Washington State University.

Trade names are included to provide specific information and do not imply endorsement by the U. S. Department of Agriculture or Washington State University.

in which DPD is the diffusion pressure deficit, OP is the osmotic pressure, and TP is the turgor pressure. Equation 2 can be related to water potential through equation 1 by taking DPD = $-\psi_1$, TP = ψ_p, and OP = $-(\psi_m + \psi_o)$. The weakness in the mechanical concept is that a pressure difference or hydraulic gradient is only one of several possible mechanisms that can cause water flow in a soil-plant system.

Water potential may be expressed as potential energy per unit volume or mass (ergs/cm^3 or joules/kg). Energy per unit volume is dimensionally equivalent to pressure and, hence, allows use of the more traditional unit of bars (10^6 dyne/cm^2 = 1 bar). This is preferred to the purely empirical unit of pressure, the standard atmosphere (1 atmosphere = 1.013 bars).

DYNAMICS OF WATER MOVEMENT IN THE SOIL-PLANT SYSTEM.—Measurement of the water potential within the rhizosphere or at some locations in the plant may be impossible because of instruments being too large. However, it may be possible to infer these water potentials by modeling water transport through the system and predicting the water potential at the location of interest. Flow models described in the literature (21, 29, 53) assume that there is liquid continuity between the bulk soil and evaporating sites in the stomatal chambers of the leaf, and that water flow is from high to low potential, the gradient being the sole driving force.

The mathematical model.—Water movement through the soil and plant is described in terms of a resistance and a driving force. Taking the driving force as the total difference in water potential between the bulk soil and the plant leaf, we write (35)

$$E = -(\psi_1 - \psi_s)/(R_s + R_x + R_1) \qquad [3]$$

where E is the water loss from unit soil area per unit time (cm day^{-1}), R is the resistance (bar day cm^{-1}) to flow along the path from soil (s), root (r), xylem (x), and leaf (l). The water potential difference between any two points in the system is just the *total* water potential difference multiplied by the fraction of the total resistance to flow between the points. Thus, determining the water potential at any point along the transpiration stream involves measuring soil and leaf water potentials and resistances to water flow in each part of the system.

The root presents a major resistance to water flow (4, 9, 11, 22, 59), particularly where water passes through the endodermis. Resistance to flow through small xylem vessels within the leaf also may be high, but not as high as for the root. Resistance to water flow within the main xylem of a healthy plant is negligible compared to other resistances (38, 59). For illustration, we will take $R_x = 0$, $R_1 = 0.2R_p$, and $R_r = 0.8 R_p$, where R_p is the total resistance to flow within the plant. Resistance to water flow may vary with root temperature and aeration, soil water potential, and rate of water uptake (59). However, for our purposes, we will take the resistance to be constant and determine its value at maximum transpiration rate by measurement.

The resistance to water movement in the soil (R_s) can be determined independently as follows: The equation

describing the flux of water near an absorbing root is

$$\frac{q}{A} = -k(\psi)\frac{d\psi}{dr} \qquad [4]$$

where q/A is the flow rate per unit area ($cm^3 cm^{-3} day^{-1}$), $k(\psi)$ is the unsaturated hydraulic conductivity ($cm^2 bar^{-1} day^{-1}$), which is a function of water potential or water content, and $d\psi/dr$ is the water potential gradient (bar cm^{-1}) across the absorbing zone of radius r. Because k decreases very rapidly with a decrease in water potential, q is often more sensitive to water potential changes through k than through the gradient $d\psi/dr$.

To find the resistance to water flow in soil to roots, we will assume steady-state radial flow to the absorbing root (15, 21, 29). The hydraulic conductivity can be described approximately by (33)

$$k(\psi) = k_s\frac{(\psi_e)^n}{\psi} \qquad [5]$$

where k_s is the hydraulic conductivity at saturation, ψ_e is the water potential at which the soil just begins to desaturate (air entry), and n is an empirical constant ranging from 2 for clays to 4 or 5 for sands. For radial flow, the cylindrical area at any distance from the root is $2\pi rl$, where l is the length of absorbing root. If we consider a plant with uniform root distribution of L cm of absorbing root per cm^3 of soil (cm^{-2}) having depth D, then $q/l = -E/DL$ (E in cm/day). The relationships for q/l, A, and $k(\psi)$ can be substituted in equation 4, and an integration performed between the radius at the root surface (r_r) and some distance into the soil which represents the radius of the soil cylinder from which each root is extracting water. If there are L cm of root per cm^3 of soil, then the radius of the soil cylinder from which the root withdraws water will be $(\pi L)^{-1/2}$. The equation for the water potential at the root surface is

$$\psi_r = (\psi_s^{1-n} - EB)^{1/1-n} \qquad [6]$$

where

$$B = (n-1)\frac{\ln(r_r^2\pi L)}{4\pi Dk_sL\psi_e^n} .$$

Note that the argument of the ln term, $r_r^2\pi L$, is just the fraction of the soil volume occupied by roots.

In these equations, the water potentials must be expressed as equivalent tensions (positive numbers). The expression for resistance to flow in the soil, $R_s = (\psi_r - \psi_s)/E$, is somewhat awkward because of the complexity of equation 6, but writing it out will give some insight into the evapotranspiration process. It is

$$R_s = \frac{(\psi_s^{1-n} - EB)^{1/1-n} - \psi_s}{E} . \qquad [7]$$

Equation 7 shows that R_s depends not only on the soil hydraulic conductivity and distance over which the water is flowing, but also on the soil water potential and transpiration rate. Moreover, since n is a positive number greater than 1, the exponent on the quantity within parentheses is negative, so as EB approaches ψ_s^{1-n}, R_s will approach infinity. In other words, for a given soil-plant system described by the parameter B, there will be a soil water potential below which some level of transpiration E cannot be maintained, no matter how low the root water potential. That is, for a given soil water potential, there is a maximum rate at which the soil will supply water to the plant. This has been verified experimentally (45). As this maximum rate is approached, the resistance to water flow in the soil increases rapidly, tending to control the water loss rate at this limiting value.

Application of the model.—Equation 6 can be used to determine the water potential distribution in the rhizosphere. By substituting $1/\pi r^2$ for L in the logarithmic term of equation 6, the soil water potential at various distances r from the root axis can be calculated for a given soil, transpiration rate, and root water potential. Results are graphed in Fig. 1 for three transpiration rates and a soil with properties of a silt loam, and are shown schematically in Fig. 2. Until the soil water potential approaches the potential that limits flow, water potential gradients within the rhizosphere are relatively small, even at relatively high transpiration rates. The soil water potential at which steep gradients develop next to the root surface is lower as the transpiration rate is reduced. As the soil water potential approaches the limiting potential, the gradient becomes steep because the resistance to flow in the dry soil near the root becomes very high (Fig. 1, 2).

Hsieh et al. (36) measured potential gradients of 80 to 120 bars cm^{-1} in the zone 1 to 2 mm from corn plant roots during soil water depletion by transpiration. Root hairs extended about midway into this zone. For example, 24 days after terminating the water supply, the water potentials were −28, −16, and −2.5 bars at distances of 1, 2, and 11 mm from the root surface, respectively. Linear extrapolation gives a water potential of −40 bars at the root surface. These data and Fig. 1 and 2 show that during soil drying by transpiration, the water potential in the first mm adjacent to the root surface can be considerably different from the water potential in the bulk soil.

Equation 7 can be combined with equation 3 to show the water potentials along the flow path of the soil-plant system as water is lost by transpiration, beginning with relatively wet soil. Figure 3A shows diurnal transpiration rates and water potentials at various places along the flow path for a plant having a relatively low root density much like what might exist in the field where root penetration is not restricted. Only the daylight hours are shown; transpiration is taken as zero at night. It was assumed that stomatal closure would reduce transpiration so that the leaf water potential would not drop below −20 bars. These conditions are somewhat arbitrary, but the model will still approximate the behavior of real plants under real conditions. Even at high soil water potentials, the leaf would reach relatively low water potentials during the day with the actual level determined by R_p and E (Fig. 3A). At night, the plant recovers so that its water potential equals the soil water potential. When the soil-plant system starts to limit soil water loss (day 6 in Fig. 3A), transpiration is markedly reduced, and most of the drop in potential is between the bulk soil and root surface. The root surface, xylem, and leaf all approach the same potential. The water potential at the root surface is about equal to the bulk soil water potential until the soil begins to limit water loss. This agrees with Fig. 1. Thus, one would not expect the soil water potential to differ much from the rhizosphere water potential until the bulk soil dried to near the flow-limiting water potential.

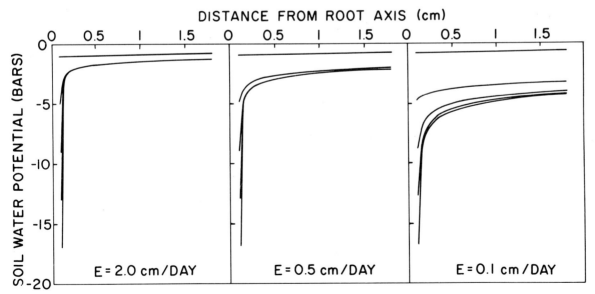

Fig. 1. Steady-state water potential distribution near an absorbing root surface of radius 0.1 cm for three transpiration rates and five root surface potentials (−1, −5, −9, −13, and −17 bars) calculated using equation 6. L = 0.1 cm^{-2} (0.1 cm root/cm^3 soil); values for other parameters are the same as those given in the caption for Fig. 3.

In the field, where water is absorbed at progressively greater depths, the simulation of Fig. 3A is probably reasonable for the zone of maximum absorption, where roots of low density are penetrating relatively moist soil. The data of Papendick et al. (51) for winter wheat indicate that in the drier zone above, which has much higher root density, water absorption slowly continues with time, and the bulk soil water potential approaches that of the leaf water potential. In this case, the leaf water potential sets the lower limit of the bulk soil water potential, excluding the upper 30 cm or so of soil, where influence by evaporative drying is significant. In a pot where root density is greater (50) because of limited soil volume, the plant stresses at a lower soil water potential than in the field (Fig. 3B). The rhizosphere water potential is still about the same as the soil water potential until the plant begins to stress.

In these cases, we have assumed that plant resistance to water flow is constant. Slatyer (59) indicates that resistance decreases with increased transpiration rate. If this feature were taken into account in our model, the result would simply be a change in the shape of the curves. The maxima and minima would not be changed. If, in addition to adding the nonlinear resistance function, we also provided for a small transpiration rate during the night, the amplitude of the leaf water potential fluctuations would be reduced. This is probably the usual field condition, since plant water potentials generally do not reach the level of the soil water potential at night.

Wilt fungi plug conductive tissue and increase the resistance of the xylem to water flow. In one situation, this increase in resistance was most severe in the petiole xylem (27) which, in our model, is included in leaf resistance. To model water flow and water potentials in this diseased plant, we assumed a leaf resistance to flow of 1 bar day cm^{-1} initially which increased at a rate of 2 bar day cm^{-1} per day thereafter. This simulation is shown in Fig. 3C. Total plant resistance at the beginning of the run was 5 bar day cm^{-1}, as with previous simulations. The plant wilted on day 3 even though the soil was relatively wet (−0.3 bar). As the plant wilted, transpiration was reduced and the xylem water potential increased. After day 5, the soil water potential was increased to −0.2 bar, simulating watering. This caused the xylem water potential to increase, but the plant remained wilted throughout much of the day.

Low root density or root volume caused by nutrient deficiency, compact soil, damage by mechanical breakage (frost heaving, transplanting), or disease can presumably lower the water potential of aboveground plant parts and cause wilting even in the presence of high soil water potentials. Kozlowski (42) reported that transpiration by transplanted trees exceeded absorption even when the soil was maintained close to field capacity. Root rot damage to beans that could be relieved by subsoiling (12) is possibly an effect of disease causing increased resistance to water flow through the root, affecting the plant like decreased root density or rooting volume. Techniques for estimating root density in soil are available (49, 56).

DETERMINING WATER POTENTIAL AND ITS COMPONENTS.—Leaf water potential is determined in situ or with freshly excised living tissue. The osmotic component is most readily obtained using sap expressed from killed tissue. Direct measurement of turgor is difficult and hence is usually computed from equation 1 using measured values of ψ_l and ψ_o, with ψ_m neglected or considered as combined with ψ_o. Bulk soil water potential may be measured in situ or on samples in the laboratory.

Methods for determining plant and soil water potentials and their components have been discussed in extensive reviews by Barrs (2) and Wiebe et al. (67). Of the many methods available, thermocouple psychrometry and the pressure bomb appear best suited for most studies.

Thermocouple psychrometry.—Thermocouple psychrometry utilizes the principle that at equilibrium the water potential of the liquid and vapor phases must be equal. Water potential can be inferred from measurements of the relative humidity (rh, %) through the following relation:

$$\psi = (RT/V) \ln(rh/100) \qquad [8]$$

where R is the gas constant, T is the absolute temperature, V is the volume of a mole of water, and ln is the natural

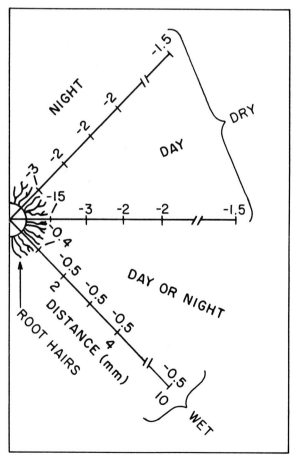

Fig. 2. Diagram showing relationships of soil water potentials in bars at different distances for hypothetical day versus night and wet versus relatively dry soil conditions.

logarithm. At 25°C (298K), the value of RT/V is 1,371 bars. Vapor equilibrium is established between the sample and the air in a small sealed chamber surrounding the sample. The relative humidity of the air in the chamber is then measured with the psychrometer and used to infer water potential. In situ measurement of ψ_l is possible by sealing a small chamber to the leaf surface and measuring the equilibrium relative humidity in the chamber (13, 34). Soil water potential may be measured in situ by enclosing the thermocouple in a small ceramic cup

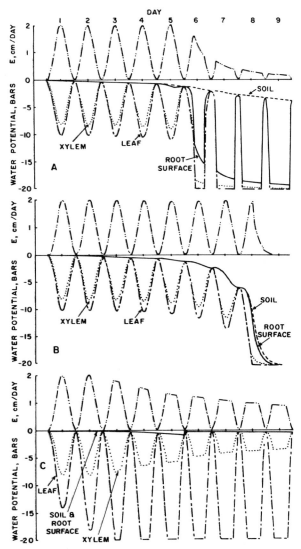

Fig. 3. Simulated daytime transpiration rate and bulk soil, root surface, xylem, and leaf water potentials for *A)* a plant having low root density, $L = 10$ cm^{-2}; *B)* a plant with high root density, $L = 0.1$ cm^{-2}; and *C)* a plant subject to attack by a wilt fungus. Soil parameters used for the calculation (equation 6) were for a sample of Ritzville silt loam, $n = 3.36$, $b = -2.21$, $\psi_e = 0.057$ bar, $k_s = 22$ cm day^{-1}. Plant parameters used were R_p, R_r, and R_l of 5, 4, and 1 bar day cm^{-1}, $L = 0.1$ or 10 cm^{-2}, $D = 30$ cm, and $r_r = 0.1$ cm. Mean daily evapotranspiration rate is 0.5 cm day^{-1}. See text for details on changes in R_l with time in C. The wilting leaf potential was taken as -20 bars.

buried in the soil (55). Psychrometers similar to those used for soil have been sealed in the xylem of trees to measure xylem water potential (66). Sample chamber psychrometers and microvoltmeters for measuring plant osmotic potential and water potential of soil samples are commercially available (Wescor, Inc., Logan, Utah 84321). Techniques and precautions for measuring water potential of soil samples are given by Wiebe et al. (67) and by Campbell and Wilson (14).

Pressure bomb.—The pressure bomb provides a simple, reliable, fast method for determing ψ_l for many plants and is especially adaptable for field use (8, 10, 25, 26, 39, 58). The water potential of the plant cells is taken as equal to the negative of the pressure applied to force sap out the cut end of an excised plant part. The osmotic potential of the xylem sap is sometimes added to the negative pressure reading, but in most cases this small correction is less than 0.5 bar. The osmotic potential of the xylem sap may be determined by applying excess pressure to the sample, collecting the exudate on a filter paper disk, and measuring the water potential with a sample chamber psychrometer. Pressure bomb equipment is commercially available (PMS Equipment Co., Corvallis, Ore.97380).

Osmotic potential.—Osmotic potential is measured using expressed cell sap that is absorbed on filter paper disks and placed in a sample chamber psychrometer. The tissue must be killed; sap expressed from living tissue is subject to filtration by the semipermeable membranes, and thus dilution may cause high values of ψ_o (low solute content). Killing the tissue by exposure to heat, cold, or toxic vapors destroys the semipermeability of the membranes and/or breaks the cells, and allows cell contents to be expressed with only slight pressure. The dilution of vacuolar sap from water in the vascular elements or in the cell wall after killing is probably a negligible effect for most plant species.

Measurements referred to as leaf or plant water potentials have often been made with the psychrometer by using excised leaves or leaf disks. Most indicate reasonable agreement between the psychrometer and the pressure bomb for a number of different plant species. However, one must interpret such readings with caution, since cutting a leaf can influence the water potential of the sample in several ways. Upon excision, the xylem water snaps back from the cut and hangs up in minute pores or pits of the xylem walls, which let water through but block a gas-water interface. The release of xylem tension will cause a general rise in ψ_l as the released water is absorbed by cells. There is also evidence that solutes from broken cells are slowly absorbed by neighboring cells, which would cause a slow decrease in ψ_l with time (3). G. S. Campbell (*unpublished data*) showed that the water potential of fresh excised leaf disks, otherwise undamaged and placed in a psychrometer chamber at constant temperature, decreased rapidly during the first 2 hours, reached a minimum after 3 to 5 hours, and generally increased afterwards. No length of equilibration period could be established which could be used to determine ψ_l. These changes were attributed to normal equilibration processes interacting with leaf metabolism.

Freshly cut bruised leaf disks of winter wheat, equilibrated in a psychrometer chamber, gave water potentials 2 to 3 bars higher (less negative) than those of expressed sap from similar tissue previously frozen over CO_2 ice (52). Sap expressed from fresh tissue read 5 bars or higher than that of sap expressed after freezing the tissue.

Xylem water potential.—The transpiration rate and resistance to flow in the leaf determine the reduction in leaf water potential over that of the xylem. If transpiration were stopped in one leaf of the plant without greatly altering the transpiration rate of the entire plant, the potential of that leaf would become equal to the potential in the xylem. Sealing the leaf in a dark chamber for a few hours and then measuring water potential with a pressure bomb gives reasonable estimates of the xylem water potential (4).

COMPONENTS OF SOIL AND PLANT WATER POTENTIAL.

—In nonsaline unsaturated soil, the matric component dominates the soil water potential. Water moves largely via a matric potential gradient, and, although solutes in saline soil may add to the total potential gradient, the effect on the rate and direction of flow is small (46). This is because solutes generally move with the water. At the root surface, root exudates and salt accumulations could conceivably lower the osmotic component compared with that in bulk soil.

The components of water potential in plant tissues differ with the plant part and relative position along the water flow path. Presumably, water enters the root primarily as liquid, and liquid continuity extends from the bulk soil through the root cortex, endodermis, pericycle, root, stem, leaf xylem, and leaf parenchyma cell walls and stops at the air-water interfaces in the stomatal chambers. Philip (54) and Bonner (7) have proposed that under low soil moisture conditions a vapor gap may surround the roots, and, in this case, the plant absorbs water as vapor.

Absorption of water is most rapid 5 to 10 cm behind the meristematic region of the root tip (59). The resistance seems to be lowest, and thus the rate of flow in the root cortex is highest, in and along the cell walls as illustrated schematically in Fig. 4. Inasmuch as this part of the flow path is exposed to atmospheric pressure, the matric component dominates the water potential. Once across the cortex, the water must cross the endodermis. Here, the Casparian strip appears to block water movement along the cell walls and force the water to move through the cell membranes and protoplasm (Fig. 4). The endodermis appears to be the only point in the transpiration stream where water must cross living membranes and is probably the site of the major flow resistance in the plant. The flow resistance across living membranes depends on metabolic processes. This is one reason plant resistance to water flow depends on temperature and oxygen supply.

Once across the endodermis, bulk water transport is

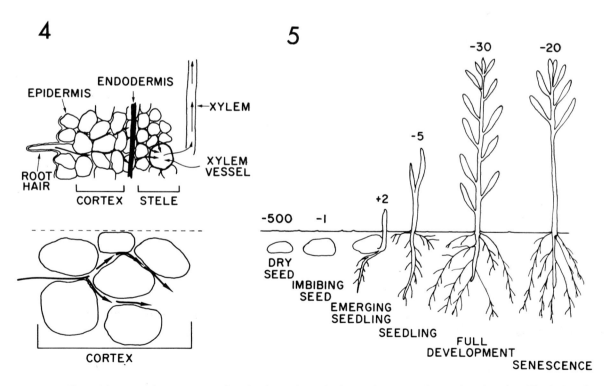

Fig. 4-5. Fig. 4. Diagrammatic transverse section showing major path of water flow across the root into the xylem. The darkened band across the endodermis represents the Casparian strip. Major flow occurs along and between the cell walls except at the endodermis, where water must pass through the cells. Fig. 5. Diagram showing water potentials in bars for different stages in the life cycle of a plant.

through the xylem vessels that terminate at the leaf mesophyll. In actively transpiring plants, a tension (negative potential) develops in the water in the xylem elements. Although it is technically correct to consider this negative potential as a matric component (this being the main one because of the low osmotic potential), it is generally referred to as a negative pressure potential.

Most of the water flows from the xylem terminals through the mesophyll to the evaporating sites in the stomatal chambers outside the cell vacuole; i.e., it flows in the interstices between cells and along cell walls (65). Because this region, like the root cortex, is exposed to atmospheric pressure, the pressure potential would be zero; thus, the water potential in this part of the flow path would be dominated by the matric component.

Although the potential of water in cells adjacent to the flow path is probably very near that for water in the flow path, the components of potential in these separate sites are quite different. In mature, turgid cells, the matric component ψ_m is insignificant compared with the osmotic component ψ_o because the vacuolar sap is essentially a true solution. Within the cells, then, ψ_o plus the pressure component ψ_p must equal ψ_m (or ψ_p for the xylem) outside the cell (plus a small correction for ψ_o). Cells can develop negative turgor pressure, but such pressure probably is small in parenchyma-type cells because cell construction cannot withstand compression and infolding. The relations among the components of water potential in a plant cell are best expressed by a modified Höfler diagram based on terms used in equation 1 and can be found in textbooks on plant-water relations (e.g., see 59, 61).

WATER POTENTIAL VARIATIONS OF PLANTS AND ENVIRONMENTAL EFFECTS.—*Water potentials during the life cycle.*—The water potential of a dry seed may be negative by several hundred bars; it is probably in near equilibrium with the relative humidity of the surrounding atmosphere (Fig. 5). After the seed is planted, it absorbs water until the water potential approaches that of the surrounding medium. As the seedling develops, the water potential of the conducting tissue may become positive up to 1 to 2 bars through active absorption (24), particularly with roots in a high moisture environment. In this case, the shoot may exude nearly pure water while still in the soil, causing the soil microenvironment adjacent to the plant to become quite wet. This occurred when barley shoots emerged through air-dry soil while the plant roots were in soil at high water potential (R. J. Cook and D. M. Griffin, *unpublished data*). Guttation is also common on seedlings after emergence.

Water potentials of most plants are highest when the plants are young and tend to decrease with age (Fig. 5). Early growth usually coincides with high soil moisture and low evaporative demand; both conditions favor high potentials. In addition, young plants have less transpiring surface. As the season progresses, moisture reserves are lowered and water may be available only from deeper in the soil, where the rooting density is less. Often coupled with this are increases in the evaporative demand and the transpirational capacity, all of which lower the plant water potential. With the beginning of senescence, the

water potential of the plant tissue begins to increase again and continues to do so until reversed by death and desiccation (complete equilibration with the atmosphere). Presumably, the increase in water potential at senescence results from a decrease in solute content of the cells due to translocation to other plant parts.

Seasonal changes.—An example of seasonal changes in plant water potential is that of fall-planted wheat in low rainfall (24-32 cm annual precipitation) areas of the northwestern United States (18, 52). After emergence and during winter, when most precipitation is received, midday plant water potentials range between −10 and −15 bars with 3 to 5 bars turgor pressure. During the early spring, after active growth begins, potentials range between −12 and −18 bars, still with 3 to 5 bars turgor. As the season progresses, leaf and pressure potential decrease, and by early June daytime turgor is very low (< 1 bar), although plants often do not show much wilting. Leaf potentials decrease to −35 to −40 bars or lower as plants near maturity. In the high precipitation area, the trends are similar; however, low turgor occurs later in the season. Cary (16) and Cary and Wright (17) have shown similar trends in seasonal changes of leaf water potential for several crops under both irrigated and dryland conditions.

Diurnal changes.—Diurnal variation of water potential can be wide (high at night, low during daytime) in many plants. Scholander et al. (58) reported diurnal fluctuations of 10 to 20 bars for several vascular species. Campbell and Campbell (13) showed that the leaf water potential for dryland wheat ranged from a low of −28 bars at midday to −12 bars during early morning. Less understood is how the osmotic component fluctuates. Our experience with winter wheat indicates that ψ_o may vary 8 to 12 bars (highest potentials at night or early morning) for water-stressed wheat (daytime leaf water potential of −34 to −40 bars) and 3 to 5 bars when wheat is less stressed (daytime leaf water potential of −15 to −20 bars).

Environmental effects.—The low water potentials of stressed plants in the field are often not achieved with plants grown in the greenhouse or growth chamber (17). The low potentials of field-grown plants could not be achieved in the greenhouse without permanent wilting of the plant (R. J. Cook and R. I. Papendick, *unpublished data*). The wheat in the greenhouse was grown in containers 45 cm deep. After vernalization and completion of tillering, water was applied through tubes leading to the container base in limited quantity such that established plants were stressed just short of permanent wilting. Leaf water potentials never decreased below −25 bars. In contrast, leaf water potentials of the same wheat variety grown in a low rainfall area reached lows of −35 to −40 bars, and plants exhibited only mild wilting during the daytime. Wheat grown in a higher rainfall area developed minimal potentials of −25 to −30 bars with occasional symptoms of daytime wilting.

Obviously, the water relations of these stressed greenhouse plants differed considerably from those of the field. If, hypothetically, a greenhouse plant or a plant from the high rainfall area were transplanted to the low rainfall area, neither would presumably have a sufficiently low osmotic potential to survive under that low water condition.

Leaf water potentials of −35 to −40 bars late in the growth period were achieved in the greenhouse with wheat plants grown in 150-cm-long soil cores extracted undisturbed from the field (data not shown). The soil was taken when relatively moist, the plants were established as with the previous experiment, and during growth only 2 cm of water was added to the bottom of the columns at about heading stage. Examination after maturity showed root growth to the end of the columns, with root density generally decreasing with depth. Water absorption was probably limited throughout much of the growth period by low root density at the deeper depths, where water was more readily available. However, the difference in plant water relations between the plants grown in the shallow containers and in the tall columns has not been fully resolved.

A wide range in plant water potentials is expected on the basis of the widely different environments in which plants grow, either through adaptation or cultivation. Drought-tolerant plants apparently continue metabolic function at lower osmotic potentials than species that cannot survive the dry environment. Conversely, plants that are adapted to low water potentials do poorly in wetter areas, probably because of competition by species better adapted to wet soil. Plant water potentials of certain cultivated crops may be raised 10 to 15 bars when grown under irrigation. However, irrigation of plants adapted to dry conditions does not necessarily raise their water potential to that of plants normally grown under high soil moisture (16).

CONTROL OF WATER POTENTIAL IN THE SOIL-PLANT SYSTEM.

—Water potential of the rhizosphere or plant must be controlled by controlling the water potential of the root zone medium and factors that influence plant resistance to water flow, or by controlling atmospheric factors that affect water use rates and photosynthesis.

The root zone water potential may be controlled by adjusting the matric potential of the soil through adjustment of water content. Alternatively, the roots may be grown in culture solutions or in soil with control of water potential achieved by adjusting the osmotic component using soluble salts or organic substances.

Matric control.—It is difficult to grow plants under conditions of constant low matric potential. When a column of dry soil is watered at the surface, a zone of constant water content near saturation develops at the application point and extends close to the wetting front. The gradation from very wet to dry soil is abrupt. Continued addition of water wets the soil to a greater depth, but the wetting pattern remains unchanged. When water application ceases, further movement into dry soil below is still rapid, though decreasing, for some time (several days), after which movement becomes very slow. Equilibrium within the column is approached too slowly to provide constant root zone water potential for growing plants.

Nevertheless, methods have been devised to control the soil water potential by adjusting the matric component. One is to use porous ceramic cylinders, cones, or plates buried in the soil or used as container walls and connected to a water reservoir that can be raised or lowered to achieve the desired potential (43). The soil slab adjacent to the ceramic surface must be kept thin (about 1 to 2 cm) where plants are transpiring. With large plants, the supply rate with this soil thickness often is still insufficient, causing the soil away from the ceramic surface to dry and the roots to become massed around the ceramic surface. The lowest potential achievable with this technique is about −0.8 bar.

Another method is to use a soil chamber separated from a solution chamber by a semipermeable cellulose (Visking) membrane (23, 68). The whole system can be jacketed by a water chamber for temperature control. The matric potential is controlled by the osmotic concentration of a solution of polyethylene glycol (carbowax). At equilibrium, the matric soil water potential is equal to the osmotic water potential of the solution. Equipment setup involves considerable time but relatively simple materials. Again, because of the high resistance to water flow in dry soil, the soil slabs must be 1 cm or less thick. The system has been used to maintain a constant water potential down to −2.5 bars with growing plants (1). However, microbial decomposition of the membrane limits the experimental period. Zur (69) found that even with autoclaved soil, the membrane could not withstand microbial attack for longer than 12 days. Usually plants are first established in the soil slab. Then the temporary walls of the slab are replaced by the membrane at the beginning of the test.

The root hair screen technique developed by Hsieh et al. (36) provides perhaps the best method yet devised for close control of the matric potential in the root hair zone throughout the growth range of most plants. The plants are grown on a fine screen placed over soil in such a way that only the root hairs extend through the screen into the soil beneath. Water is supplied and matric potential control is achieved either through a hanging water column or by atomizing water or nutrient solution directly on the soil. Water potential of soil in the root hair zone may be measured directly with a psychrometer or inferred from water content measurements using gamma-ray attenuation.

A commonly used method of imposing differential stress on plants is to allow the plants to dry the soil in a container to some predetermined mean water content and then to add water to raise the matric potential to near −0.1 bar (64). Well-developed plants grown in usual greenhouse containers usually have dense root systems, and thus the water potential of the bulk soil will be reasonably uniform throughout at a given time as extraction proceeds, provided the soil is wetted uniformly at the outset. However, the method is of limited use in many studies because soil and plant water potential fluctuates over a range of values. Also, there is the possibility that water potential may vary over different parts of the root zone.

Osmotic control.—Osmotic control of the root system water potential is achieved with solution cultures or by adding inorganic salts to soils or other water-adsorbing media. Stirring solution cultures or frequent leaching of soil with osmotic solutions assures a reasonably uniform potential throughout the root system. Using soluble salts does not appear to be a reliable method for controlling water potential over extended periods for many plants because of specific ion effects and alterations of osmotic relations within the plant (44). Several researchers (5, 6, 40) have shown that for different plant species a decrease in the osmotic potential of the root medium through use of inorganic salts causes a similar decrease in the osmotic potential of the plant. The plant adjustment is rapid, and because the difference in potential between the root medium and the plant parts is maintained, turgor potential is unchanged. Previously, it was thought that growth inhibition caused by lowered soil water potential was the same whether the lowering was due to the osmotic (soluble salts) or matric component. More recent work and reinterpretations suggest that a given reduction in water potential through salts in solution inhibits growth less than the same reduction through the matric component. Slatyer (59) suggests that where the turgor potential remains near the original value as the osmotic potential of the root medium is lowered, part of the detrimental effect on plant growth is probably due to the decrease in osmotic potential per se and part is due to direct osmotic effects on internal water deficits.

Osmotic control of solution cultures across most of the growth range has also been achieved through use of organic substances that are thought not to permeate roots (37, 47, 57). Polyethylene glycol or mannitol have been used most extensively as solutes. Carbowax 6000, which did not penetrate cell membranes (47), presumably would duplicate the effects of low soil moisture more closely than would low molecular weight carbowaxes, mannitol, or other solutes that are more permeating. Care is needed to avoid root injury, which might enhance solute uptake and thus cause toxic effects (37). Where plants are to be subjected to stress, the osmotic potential of the solution culture should be lowered gradually to avoid damaging the plant (57). The use of a nonpermeating solute in solution cultures should eliminate specific ion effects and affect metabolic function much like a change in matric potential.

Control through regulating root temperature.—Control of root zone temperature offers a means of controlling plant water potential temporarily because root temperature influences plant resistance to water flow (11). Glenn (30) showed that lowering the root zone temperature below some critical value markedly increased plant resistance to water flow and decreased leaf water potential. Decreasing root temperature from 25°C to 5°C caused the leaf water potential of beans to decrease from −9 to −17 bars, and that of corn from −26 to −45 bars. The large increase in resistance occurred around 10°C for both species. Thus, stress could be induced by lowering soil temperature and exposing the tops to high evaporative demand. Plants are often stressed in this way unintentionally in the greenhouse when cold water is applied to pots in the presence of high evaporative demand. Adaptation of the plant to the cold soil treatment and alteration of the root metabolism by continued cooling make this method unreliable for imposing stress for more than a few hours. Apparently, the permeability of different plant species is affected differently by temperature (11).

Control through regulating evaporative demand.—To an extent, plant and rhizosphere water potentials can be controlled in the growth chamber or greenhouse by adjusting the evaporative demand. Figure 1 indicates that low root surface water potentials can be achieved in progressively wetter soil by increasing the prevailing transpiration rate. The transpiration level can be increased or maintained (until soil resistance limits water flow) by lowering the relative humidity of the air and increasing air circulation to reduce the resistance of the leaf-air boundary layer. Millar et al. (48) found that the potential drop between leaves and soil of field onions growing in relatively moist soil (−0.05 to −0.2 bars) decreased (became more negative) as the transpiration rate increased; however, the relation was nonlinear. Transpiration increased markedly when the difference between leaf and soil water potential exceeded about −5 bars, and the same tendency held with other soils and environmental conditions. This type of relation causes a discrepancy in the present water transport models, which assume that transpiration rate is proportional to the water potential difference between leaves and soil.

Control through regulating light intensity.—Altering the light intensity may be another means for controlling plant water potential. Potato plants in a growth chamber began wilting at water potentials of about −6 bars; plants in the greenhouse, where the lighting intensity was about double that in the growth chamber, developed potentials of about −10 bars before wilting (M. D. Campbell, *unpublished data*). In the field, stress was not evident until water potentials were −12 bars. Moreover, reducing the light intensity 50% with shade cloth increased daytime leaf osmotic potentials of winter wheat 2 to 4 bars compared with unshaded wheat (G. S. Campbell et al., *unpublished data*). Possibly, the lowered light intensity reduced photosynthesis, thereby decreasing the solute concentration within the plant cell.

Effects of nitrogen fertility.—Plant osmotic potentials are generally decreased by increased N fertility, even in the presence of adequate soil water (52). High N may increase photosynthetic activity, thereby increasing the solute concentration in the plant cells. Also, with high N, plants deplete soil water to lower potentials than with low N. With limited soil water, leaf water potentials with high N fertility were as much as 18 bars lower than with low N fertility (52). The lower potential with high N has been attributed to a greater rooting density, which would increase absorption of soil water (41), but it is also possible that the lower soil moisture level resulted from reduced plant osmotic potential.

Control of water potential in the field.—Control of plant water potential in the field is generally difficult because of soil variability and weather uncertainty. In most cases, greatest control is possible through altering the soil water content. Methods to accomplish this with particular reference to disease control have been discussed (20). If soil water content cannot be altered by irrigation, plant water potential can be controlled somewhat through managements that may affect water use rates: plant density, shading, use of species or varieties having different rooting characteristics, early versus late planting, and rate of fertilizer nitrogen application. For example, in the eastern Washington dryland area, winter wheat grown in rows 30 cm apart developed leaf water potentials 2 to 3 bars lower than wheat in rows 90 cm apart (52). Similarly, early fall plantings of wheat tended to stress earlier and more severely than late plantings (52).

LITERATURE CITED

1. BABALOLA, O., L. BOERSMA, and C. T. YOUNGBERG. 1968. Photosynthesis and transpiration of Monterey pine seedlings as a function of soil water suction and soil temperature. Plant Physiol. 43:515-521.

2. BARRS, H. D. 1968. Determination of water deficits in plant tissues, p. 235-368. *In* T. T. Kozlowski (ed.), Water deficits and plant growth. Vol. I. Academic Press, New York and London.

3. BARRS, H. D., and P. J. KRAMER. 1969. Water potential increase in sliced leaf tissue as a cause of error in vapor phase determinations of water potential. Plant Physiol. 44:959-964.

4. BEGG, JOHN E., and NEIL C. TURNER. 1970. Water potential gradients in field tobacco. Plant Physiol. 46:343-346.

5. BERNSTEIN, L. 1961. Osmotic adjustment of plants to saline media. I. Steady state. Amer. J. Bot. 48:909-918.

6. BERNSTEIN, L. 1963. Osmotic adjustment of plants to saline media. II. Dynamic phase. Amer. J. Bot. 50:360-370.

7. BONNER, J. 1959. Water transport. Science 129:447-450.

8. BOYER, J. S. 1967. Leaf water potential measured with a pressure chamber. Plant Physiol. 42:133-137.

9. BOYER, J. S. 1971. Resistances to water transport in soybean, bean, and sunflower. Crop Sci. 11:403-407.

10. BOYER, J. S., and S. R. GHORASKY. 1971. Rapid field measurement of leaf water potential in soybean. Agron. J. 63:344-345.

11. BROUWER, R. 1965. Water movement across the root. Symp. Soc. Exp. Biol. 19:131-149.

12. BURKE, D. W., D. E. MILLER, L. D. HOLMES, and A. W. BARKER. 1972. Counteracting bean root rot by loosening the soil. Phytopathology 62:306-309.

13. CAMPBELL, G. S., and M. D. CAMPBELL. 1973. Evaluation of a thermocouple hygrometer for measuring leaf water potential in situ. Agron. J. 66:24-27.

14. CAMPBELL, G. S., and A. M. WILSON. 1972. Water potential measurements on soil samples, p. 142-149. *In* R. W. Brown and B. P. Van Haveren (ed.), Psychrometry in water relations research. Utah Agr. Exp. Stn.

15. CAMPBELL, M. D. 1972. The lower limit of soil water potential for potato growth (Solanum tuberosum). Ph.D. thesis, Wash. State Univ., Pullman. 49 p.

16. CARY, J. W. 1971. Energy levels of water in a community of plants as influenced by soil moisture. Ecology 52:710-715.

17. CARY, J. W., and J. L. WRIGHT. 1971. Response of plant water potential to the irrigated environment of southern Idaho. Agron. J. 63:691-695.

18. COOK, R. J. 1973. Influence of low plant and soil water potentials on diseases caused by soilborne fungi. Phytopathology 63:451-458.

19. COOK, R. J., and R. I. PAPENDICK. 1971. Effect of soil water on microbial growth, antagonism and nutrient availability in relation to soil-borne fungal diseases of plants, p. 81-88. *In* T. A. Toussoun et al. (ed), Root diseases and soil-borne pathogens. Univ. Calif. Press, Berkeley and Los Angeles.

20. COOK, R. J., and R. I. PAPENDICK. 1972. Influence of water potential of soils and plants on root disease. Annu. Rev. Phytopathol. 10:349-374.

21. COWAN, I. R. 1965. Transport of water in the soil-plant-atmosphere system. J. Appl. Ecol. 2:221-239.

22. COWAN, I. R., and F. L. MILTHORPE. 1968. Plant factors influencing the water status of plant tissues, p. 173-193. *In* T. T. Kozlowski (ed.), Water deficits and plant growth. Vol. I. Academic Press, New York and London.

23. COX, L. M., and L. BOERSMA. 1967. Transpiration as a function of soil temperature and soil water stress. Plant Physiol. 42:550-556.

24. CRAFTS, A. S. 1968. Water deficits and physiological processes, p. 85-133. *In* T. T. Kozlowski (ed.), Water deficits and plant growth. Vol. II. Academic Press, New York and London.

25. DEROO, H. C. 1969. Leaf water potentials of sorghum and corn estimated with a pressure chamber. Agron. J. 61:969-970.

26. DEROO, H. C. 1970. Leaf water potentials of tobacco, estimated with the pressure bomb. Tobacco Sci. XIV:105-106.

27. DUNIWAY, J. M. 1971. Resistance to water movement in tomato plants infected with Fusarium. Nature (London) 230:252-253.

28. DUNIWAY, J. M. 1973. Pathogen-induced changes in host water relations. Phytopathology 63:458-466.

29. GARDNER, W. R. 1960. Dynamic aspects of water availability to plants. Soil Sci. 89:63-73.

30. GLENN, L. F. 1972. The limiting resistance to water uptake by plants: soil water potential and temperature effects. Ph.D. thesis, Wash. State Univ., Pullman. 41 p.

31. GRIFFIN, D. M. 1969. Soil water in the ecology of fungi. Annu. Rev. Phytopathol. 7:289-310.

32. GRIFFIN, D. M. 1972. Ecology of soil fungi. Syracuse Univ. Press, New York. 193 p.

33. HILLEL, D., and W. R. GARDNER. 1969. Steady infiltration into crust-topped profiles. Soil Sci. 108:137-142.

34. HOFFMAN, G. J., and S. L. RAWLINS. 1972. Silver-foil psychrometer for measuring leaf water potential in situ. Science 177:802-804.

35. HONERT, T. H. VAN DEN. 1948. Water transport in plants as a catenary process. Disc. Faraday Soc. 3:146-153.

36. HSIEH, J. J. C., W. H. GARDNER, and G. S. CAMPBELL. 1972. Experimental control of soil content in the vicinity of root hairs. Proc. Soil Sci. Soc. Amer. 36:418-421.

37. JANES, B. E. 1961. Use of polyethylene glycol as a solvent to increase the osmotic pressure of nutrient solution in studies on the physiology of water in plants. Plant Physiol. 36 (suppl.):XXIV.

38. JENSEN, R. D., S. A. TAYLOR, and H. H. WIEBE. 1961. Negative transport and resistance to water flow through plants. Plant Physiol. 36:633-638.

39. KAUFMAN, M. R. 1968. Evaluation of the pressure chamber technique for estimating plant water potential of forest tree species. Forest Sci. 14:369-374.

40. KIRKHAM, MARY BETH, W. R. GARDNER, and G. C. GERLOFF. 1969. Leaf water potential of differentially salinized plants. Plant Physiol. 44:1378-1382.

41. KMOCH, H. G., R. E. RAMIG, R. L. FOX, and F. E. KOEHLER. 1957. Root development of winter wheat as influenced by soil moisture and nitrogen fertilization. Agron. J. 49:20-25.

42. KOZLOWSKI, T. T. 1968. Importance of water to plants, p. 1-21. *In* T. T. Kozlowski (ed.), Water deficits and plant growth. Vol. I. Academic Press, New York and London.

43. KRAMER, PAUL J. 1949. Plant and soil water relationships. McGraw-Hill Book Co., Inc., New York. 347 p.

44. LAGERWERFF, J. V., and H. E. EAGLE. 1961. Osmotic and specific effects of salts on beans. Plant Physiol. 36:472-477.

45. LANG, A. R. G., and W. R. GARDNER. 1960. Limitation to water flux from soils to plants. Agron. J. 62:693-695.

46. LETEY, J., W. D. KEMPER, and L. NOONAN. 1969. The effect of osmotic pressure gradients on water movement in unsaturated soil. Proc. Soil Sci. Soc. Amer. 33:15-18.

47. MICHEL, BURLYN E. 1971. Further comparisons between carbowax 6000 and mannitol as suppressants of cucumber hypocotyl elongation. Plant Physiol. 48:513-516.

48. MILLAR, A. A., W. R. GARDNER, and S. M. GOLTZ. 1971. Internal water status and water transport in seed onion plants. Agron. J. 63:779-784.

49. NEWMAN, E. I. 1965. A method of estimating the total length of root in a sample. J. Appl. Ecol. 2:139-145.

50. NEWMAN, E. I., and ROSALIE E. ANDREWS. 1973. Uptake of phosphorous and potassium in relation to root growth and root density. Plant Soil 38:49-69.

51. PAPENDICK, R. I., V. L. COCHRAN, and W. M. WOODY. 1971. Soil water potential and water content profiles with wheat under low spring and summer rainfall. Agron. J. 63:731-734.

52. PAPENDICK, R. I., and R. J. COOK. 1974. Plant water stress and development of root rot in wheat subjected to different cultural practices. Phytopathology 64:358-363.

53. PHILIP, J. R. 1957. The physical principles of soil water movement during the irrigation cycle. Proc. Int. Congr. Irrig. Drain. 8:125-154.

54. PHILIP, J. R. 1958. The osmotic cell, solute diffusibility and the plant water economy. Plant Physiol. 33:264-271.

55. RAWLINS, S. L., and F. N. DALTON. 1967. Psychrometric measurement of soil water potential without precise temperature control. Proc. Soil Sci. Soc. Amer. 31:297-301.

56. REICOSKY, D. C., R. J. MILLINGTON, and D. B. PETERS. 1970. A comparison of three methods for estimating root length. Agron. J. 62:451-453.

57. RUF, ROBERT H., RICHARD E. ECKERT, JR., and RICHARD O. GIFFORD. 1963. Osmotic adjustment of cell sap to increases in root medium osmotic stress. Soil Sci. 96:326-330.

58. SCHOLANDER, P. F., H. T. HAMMEL, E. D. BRADSTREET, and E. A. HEMMINGSEN. 1965. Sap pressure in vascular plants. Science 148:339-346.

59. SLATYER, R. O. 1967. Plant-water relationships. Academic Press, New York and London. 366 p.

60. TAYLOR, S. A. 1968. Terminology in plant and soil water relations, p. 49-72. *In* T. T. Kozlowski (ed.), Water deficits and plant growth. Vol. I. Academic Press, New York and London.

61. TAYLOR, S. A., and G. L. ASHCROFT. 1972. Physical edaphology. W. H. Freeman and Co., San Francisco. 533 p.

62. TAYLOR, S. A., and R. O. SLATYER. 1962. Proposals for a unified terminology in studies of plant-soil-water relations. UNESCO Arid Zone Res. 16:339-349.

63. VAADIA, Y., F. C. RANEY, and R. M. HAGEN. 1961. Plant water deficits and physiological processes. Annu. Rev. Plant Physiol. 12:265-292.

64. WADLEIGH, C. H., and A. D. AYERS. 1945. Growth and biochemical composition of bean plants as conditioned by soil moisture tension and salt concentration. Plant Physiol. 23:485-495.

65. WEATHERLEY, P. E. 1965. The state and movement of water in the leaf. Symp. Soc. Exp. Biol. 19:157-184.

66. WIEBE, H. H., R. W. BROWN, T. W. DANIEL, and E. CAMPBELL. 1970. Water potential measurements in trees. Bioscience 20:225-226.

67. WIEBE, H. H., G. S. CAMPBELL, W. H. GARDNER, S. L. RAWLINS, J. W. CARY, and R. W. BROWN. 1971. Measurement of plant and soil water status. Utah Agr. Exp. Stn. Bull. 484. 71 p.

68. ZUR, B. 1966. Osmotic control of the matric soil-water potential. I. Soil-water system. Soil Sci. 102:394-398.

69. ZUR, B. 1966. Osmotic control of the matric soil-water potential. II. Soil-plant system. Soil Sci. 103:30-38.

Dynamics and Measurement of Oxygen Diffusion and Concentration in the Root Zone and Other Microsites[1]

L. H. STOLZY, G. A. ZENTMYER, and M. H. ROULIER—*Professor, Department of Soil Science and Agricultural Engineering, Professor of Plant Pathology, and Postgraduate Research Soil Physicist, University of California, Riverside.*

Oxygen is an atmospheric element essential to most forms of life. Fourteen years after Priestley discovered oxygen in 1774, Ingenhouz reported that the oxygen concentration in seed pods was about the same as that in air. Since then, oxygen has been a multidisciplinary subject.

Oxygen is the most abundant atom in the earth's crust (53.77%), mainly because it reacts with other elements in the crust such as silicon and iron. Oxygen makes up 90% of the volume of the earth's crust, and 0.01% of all oxygen is molecular oxygen in the atmosphere (12). A large exchange of oxygen between the hydrosphere and atmosphere takes place due to the exchange of gaseous oxygen in the atmosphere with dissolved oxygen in the ocean. Photosynthesis in the biosphere results in an atmospheric turnover of oxygen about every 5,400 years (12).

Because of the great volume and mobility of the air medium, the oxygen concentration in the atmosphere remains at about 20.93%, which provides the earth crust with a uniform and adequate oxygen supply. The process whereby gases consumed or produced in the earth crust are exchanged for gases in the aerial atmosphere is known as soil aeration.

SOIL AERATION AND MICROSITES.— Measurement of the gas composition of the soil atmosphere has been the most popular index used to describe soil aeration. A knowledge of the gas composition of the gaseous constituents in soil (O_2 and CO_2) is insufficient for understanding the effect of soil aeration on plant growth, although the composition of gas samples removed from the soil is being used as a guide for research in solution cultures. We cannot assume that the gas phase oxygen concentration in soil can be compared to data obtained by bubbling gases of various oxygen concentrations through solution culture to determine whether the soil oxygen concentrations are limiting for root growth. Nonetheless, roots do differ from organisms living in the soil, in that roots are connected to a plant top that lives in the atmosphere above. Soil-borne organisms carry out their life cycle in the soil, and the various physiological processes (e.g., respiration, reproduction, hatching, and growth stages)

are greatly influenced only by soil environmental factors. The concentration of oxygen in its gas phase in soil could be in equilibrium with the soil solution surrounding the organism.

Many other indices have been used to describe soil aeration: (a) gas porosity, (b) diffusion in the gas phase, (c) saturation or waterlogging, (d) air permeability, (e) oxidation-reduction potential, and (f) diffusion through the gas-liquid-solid medium surrounding a cylindrical platinum electrode. Other chemical and biological methods are considered in the articles on soil aeration by Baver (3), Black (4), Domsch (10), Russell (26), Stolzy (30), and Taylor and Ashcroft (34). Of course, the best index of soil aeration conditions is determined by the type of study in which the data will be used.

Diffusion in the gas phase through the larger continuous pore spaces in soils is the main process contributing to soil aeration. Diffusion is a direct consequence of the random thermal motion of molecules. At 25°C, oxygen molecules travel at a velocity of about 1,600 miles/hour with a collision frequency of 4,430/second and a mean free path of 905 Å (31). The various soil gases have different molecular size and weight and will diffuse at different rates. As a result of this random motion, more gas molecules tend to move from points of high to points of low concentration than from points of low to points of high concentration. This net movement of molecular species from points of high to points of low concentration is generally directly proportional to the concentration gradient, the cross-sectional area available for diffusion, and the time. Because soil gases tend to contain more CO_2 and less O_2 than the atmosphere, CO_2 diffuses out of the soil and O_2 diffuses into the soil. Oxygen exchange between the soil and atmosphere is rapid under many field conditions, giving a soil oxygen concentration not greatly different from that in the atmosphere. As soil water content increases, however, the rate of gas diffusion is reduced, and lower oxygen concentration may occur. A similar effect may occur with the increase in soil compaction associated with reduced gas-filled porosity.

The theory of gaseous diffusion in soil to include the presence of aggregates was advanced by Currie (9). Previous theories on diffusion were based on the assumption that soils were homogeneous with respect to random distribution of pores; this may be true for soils

[1]Supported in part by NSF-RANN Grant No. GI 34733X.

high in silt and sand content but not for aggregated soils. Currie pointed out that pores between the aggregates (noncapillary) will drain first, and the more complex aggregate pores (capillary) will be full of water [Stolzy and Van Gundy (33)]. Microenvironments within soil may have very low oxygen concentration. Greenwood (13) indicated that water-saturated aggregates can be anaerobic, and root tips penetrating such aggregates would encounter low oxygen concentrations. The soil aggregate is probably the most important microenvironment for microorganisms within the soil (33). It is the site of intimate contact between soil particles, water, air, and microorganisms. The aggregate, as the solid phase of the soil, controls the quantity and distribution of water and air, both of which are active in the physiology and metabolism of soil microorganisms.

Greenwood and Goodman (14) used a microelectrode to measure distribution of oxygen in saturated spherical soil aggregates under laboratory conditions. A method has not been divised that could directly measure oxygen in microsites under field conditions. However, the fact that *Clostridium*, a strict anaerobe, has been recovered from well-aerated sandy loam soils in the field indicates that anaerobic microsites exist under natural conditions.

A field method that gives a clear indication of anaerobic microsites in well-aerated soil is the measurement of gas production as an index of denitrification (25). The two principal products of denitrification are nitrogen (N_2) and nitrous oxide (N_2O). Because a large part of the N lost during denitrification is in the form of N_2, it would be most useful to measure N_2 concentration changes in the soil atmosphere. However, it is difficult to measure small deviations from the high background level (about 80% by volume) unless an N source that is tagged by enrichment or depletion of the stable isotope ^{15}N is used. Studies involving the use of ^{15}N-labeled material are generally limited to lysimeters. The other main product of denitrification, N_2O, accounts for a minor part of the evolved N, but, with the appropriate instruments, small changes in concentration can be measured because the ambient level in the atmosphere is so low (about 0.3 ppm by volume). Nitrous oxide has been identified in the atmosphere of various field soils (1, 2, 6, 7, 25), where it is usually an intermediate in denitrification. Inasmuch as N_2O is produced in soil only during denitrification, its presence in the soil atmosphere at concentrations above ambient is proof that denitrification has taken place in anaerobic microsites in what is considered a well-aerated soil. Analysis of oxygen concentrations in gas samples taken from a Ramona sandy loam field soil indicated a well-aerated soil (Fig. 1). However, the amount of N_2O in the profile was several times higher than ambient, indicating a large number of anaerobic microsites. Addition of water to the soil increased the number of saturated aggregates, causing an increase in the production of N_2O (Fig. 2). The arrows indicate the times of irrigations or rainfalls; these times correlated well with increased N_2O concentrations. Studies on N_2O concentrations in the soil atmosphere are useful in identifying the occurrence of and studying the effect of anaerobic conditions in the soil profile on the ecology of microorganisms.

DIFFUSION THROUGH GAS-LIQUID-SOLID MEDIA.—In many instances, the evaluation of soil aeration has been based on the diffusion rate of oxygen in the gas phase. Such measurements fail to include the effect of soil solution, which may surround the root and cause an impedance for gas exchange between root and gas-filled pore spaces. A method for measuring the rate of oxygen diffusing through soil solution to a platinum wire electrode inserted into the ground is described by Letey and Stolzy (18). When an electric potential is applied between a platinum microelectrode inserted in the soil solution and a reference electrode, oxygen is reduced at the platinum surface. An electric current flows between the two electrodes and is proportional to the rate of oxygen reduction. The assumption is that the platinum wire is in a soil environment similar to that of a root or microorganism, and so the higher the oxygen diffusion rate (ODR), the higher will be the supply of oxygen to the respiration site. Critical ODR values in soils for the roots of many plant species are compared in different reviews (32). A review by Stolzy and Van Gundy (33) compares critical ODR for microflora and microfauna.

RESPIRATION RATES AND SOIL TEMPERATURE.—Soil temperature strongly influences the rate of biological sources and sinks for gases in soils. Profile variation in soil temperature is well known, and a proliferating root system would encounter different temperatures. Several factors are affected by temperature change: respiration rate, diffusion

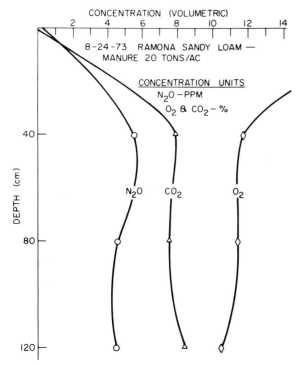

Fig. 1. Nitrous oxide, oxygen, and carbon dioxide in the soil atmosphere at different depths of a Ramona sandy loam on which 20 tons/acre (45 tons/hectare) of manure was applied.

coefficients of oxygen in air and in water, and the oxygen solubility coefficient (Fig. 3) (21). Cannon (8) attributed the higher oxygen concentrations in the soil atmosphere needed to maintain normal root growth at higher temperatures to the decreasing solubility of oxygen in the soil solution with temperature increase. The solubility, however, is only one of the factors that determine the rate of oxygen supply to the root or microorganism surface. With increasing temperature, the solubility of oxygen decreases and the diffusion coefficient through both gas and liquid increases. Diffusion of oxygen through water increases in the range of 3 to 4% per degree Centigrade, and the solubility decreases approximately 1.6% per degree, indicating a net increase in the rate of oxygen supply with increased temperature (19). For a given oxygen concentration in a soil pore and a given respiration rate, the oxygen concentration at the root surface will be higher under higher temperatures.

PLANT AERATION.—Roots of field-grown plants receive oxygen either by diffusion through the soil and root wall (soil aeration) or by diffusion from the atmosphere via gas spaces between the plant cells (plant aeration) (Fig. 4). Plant aeration is the dominant pathway for bog plants and paddy rice, but for most field crops soil aeration is dominant.

The occurrence of air space tissues in plants was the subject of two reviews by Sifton (28, 29). The small voids between adjacent cells are called intercellular spaces, and these are distinct from large regularly shaped voids (chambers) and large irregularly shaped voids (lacunae). Two mechanisms have been proposed for gas space formation in roots. McPherson (23) demonstrated that a lack of oxygen in inner root cells of maize results in deterioration and death of the protoplasm in groups of cells, followed by loss of cell turgidity and collapse. Large voids formed by the death and decay of existing cells have a lysigenous origin and have been observed within 40 μ of the root tip zone. Schramm (27) suggested that differential growth between outer root cells (which are better aerated) and inner cells causes tearing apart of cells and gas space formation in cortex tissues. These types of voids are described as schizogenous.

The influence of cultural practices on gas space development was noted by Norris (24) when he observed that spaces in corn roots grown in solution are more numerous than when grown in soil. Yu et al. (37) grew several plant species in soil and examined the effect of flooding treatments on root porosity. They found that roots developed in a flooded soil had a higher root porosity than roots growing in drained soil. Large air spaces were found in barley root cortex when the plants were grown in a nonaerated solution (5). Luxmoore and Stolzy (20) did not observe any difference between 0.75 and 8 ppm solution oxygen concentration on the porosity (volume of gas spaces per unit volume of root) of maize and rice roots. Both treatments induced high porosity. Varade et al. (35), growing wheat in solution culture, found that high light intensity and temperatures induced high root porosity for only certain varieties of wheat.

A concentration gradient of oxygen in gas spaces of various root tissues exists from the stem base to the root tip. Evans and Ebert (11) followed the diffusion of $^{15}O_2$

down the primary root of broad bean seedlings with Geiger tube detectors. Using closed plastic tubes as a model, they established that diffusion was the mechanism involved. Jensen et al. (15) showed that $^{18}O_2$ moved through corn roots. The amount of oxygen transported across membranes was a linear function of the number of corn roots that penetrated the membrane. The above experiments provide evidence for the continuity of gas spaces that occurs between cells.

MODELS: OXYGEN DIFFUSION IN SOIL-PLANT SYSTEMS.—A root is often approximated by a cylinder in model studies. Lemon (17) considered steady state radial oxygen diffusion to a cylindrical root model

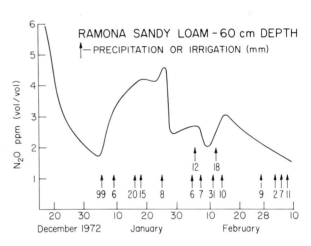

Fig. 2. Nitrous oxide concentration at 60 cm depth in a Ramona sandy loam soil compared with times of rainfall and irrigation [Roulier et al. (25)].

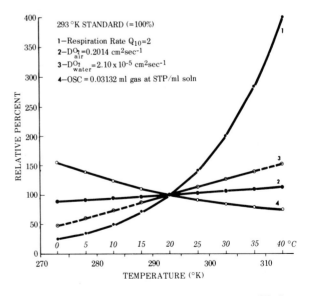

Fig. 3. Relative changes in respiration rate, oxygen diffusion coefficient in air and in water, and oxygen solubility coefficient with change in temperature [Luxmoore and Stolzy (21)].

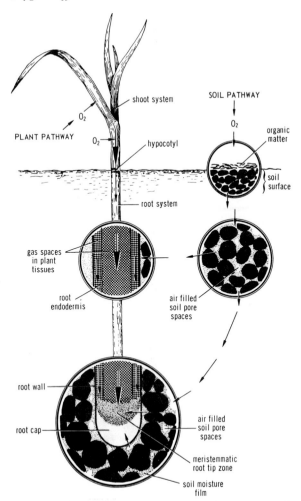

Fig. 4. Soil and plant aeration pathways [Luxmoore et al. (22)].

account the respiratory sink of the tissue as a function of oxygen concentration and position down the root. Such a model was developed and forms a basis for the assessment of root oxygen relations (22). From this model, it was predicted that percentage of plant aeration increases as the root length decreases and as root radius increases. Increases in water film thickness around the root result in a decrease in the total respiration rate of the root, and this sink is supplied with an increase in the amount of oxygen supplied by plant aeration.

The above model has been applied to the tip zone (1 cm) of a root (21). A need exists to also apply it to the study of microorganisms in soils. There are significant differences in the geometry and physics of gas exchange for a root tip and for microorganisms. The region adjacent to the root tip (rhizosphere) has a large microbial population that could be considered an integral part of a root tip model. Adaptation of the above model should be utilized in a consideration of root tips and microorganisms.

LITERATURE CITED

1. ALBRECHT, B., C. JUNGE, and H. ZAKOSEK. 1970. Der N₂O-Gehalt der Bodenluft in drei Bodenprofilen. Z. Pflanz. Bod. 15:205-211.
2. ARNOLD, P. W. 1954. Losses of nitrous oxide from soil. J. Soil Sci. 5:116-128.
3. BAVER, L. D. 1956. Soil Physics, 3rd ed. John Wiley & Sons, New York. 489 p.
4. BLACK, C. A. 1968. Soil-plant relationships. John Wiley & Sons, New York. 332 p.
5. BRYANT, A. E. 1934. Comparison of anatomical and histological differences between roots of barley grown in aerated and nonaerated culture solutions. Plant Physiol. 9:389-391.
6. BURFORD, J. R., and R. J. MILLINGTON. 1968. Nitrous oxide in the atmosphere of a red-brown earth. Trans. 9th Int. Congr. Soil Sci. (Adelaide) II:505-511.
7. BURFORD, J. R., and R. C. STEFFANSON. 1973. Measurement of losses of nitrogen from soils. Soil Biol. Biochem. 5:133-141.
8. CANNON, W. A. 1925. Physiological features of roots with especial reference to the relation of roots to the aeration of soil. Carnegie Inst. Wash. Pub. 368:1-168.
9. CURRIE, J. A. 1961. Gaseous diffusion in the aeration of aggregated soils. Soil Sci. 92:40-45.
10. DOMSCH, K. H. 1962. Bodenatmung, Sammelbericht über Methoden und Ergebnisse, Zentr. Bakteriol. Parasitnek. 116:33-78.
11. EVANS, N. T. S., and M. EBERT. 1960. Radioactive oxygen in the study of gas transport down the root of Vicia faba. J. Exp. Bot. 11:246-257.
12. GILBERT, D. L. 1964. Atmosphere and evolution, p. 641-643. *In* F. Dickens and E. Neil (ed.), Oxygen in the animal organism. Macmillan Co., New York.
13. GREENWOOD, D. J. 1967. The effect of oxygen concentration on the decomposition of organic materials in soil. Plant Soil 14:360-376.
14. GREENWOOD, D. J., and D. GOODMAN. 1967. Direct measurements of the distribution of oxygen in soil aggregates and in columns of fine soil crumbs. J. Soil Sci. 18:182-186.
15. JENSEN, C. R., J. LETEY, and L. H. STOLZY. 1964. Labeled oxygen: transport through growing corn roots. Science 144:550-552.
16. JENSEN, C. R., L. H. STOLZY, and J. LETEY. 1967. Tracer studies of oxygen diffusion through roots of barley, corn and rice. Soil Sci. 103:23-29.

surrounded by a water film of constant thickness. His model did not account for any longitudinal diffusion within the root and so may only apply near the root tip, where all oxygen is supplied by the soil.

Using a model, Woolley (36) derived an equation for the length of a root that may be supported entirely by plant aeration. His model did not allow for any radial diffusion to the root from the soil, and it assumed a uniform root respiration with distance along the root. This model may be a useful approximation for plants growing in soils with a high water content.

Jensen et al. (16), in their analysis of steady state oxygen diffusion through roots, assumed no respiratory loss of oxygen and proposed a relationship for the concentration of oxygen within a cylindrical root model at a distance along the root length. They predicted that for a given root length, rice roots will have a greater relative concentration of oxygen than will maize or barley roots.

All models of oxygen diffusion in the soil-plant system make assumptions that oversimplify the conditions of field plants. A more useful model would include both longitudinal and radial diffusion and would take into

17. LEMON, E. R. 1962. Soil aeration and plant root relations. I. Theory. Agron. J. 54:167-170.

18. LETEY, J., and L. H. STOLZY. 1964. Measurement of oxygen diffusion rates with the platinum microelectrode. I. Theory and equipment. Hilgardia 35:545-554.

19. LETEY, J., L. H. STOLZY, G. B. BLANK, and O. R. LUNT. 1961. Effect of temperature on oxygen-diffusion rates and subsequent shoot growth, root growth, and mineral content of two plant species. Soil Sci. 92:314-321.

20. LUXMOORE, R. J., and L. H. STOLZY. 1969. Root porosity and growth responses of rice and maize to oxygen supply. Agron. J. 61:202-204.

21. LUXMOORE, R. J., and L. H. STOLZY. 1972. Oxygen diffusion in the soil-plant system. V. Oxygen concentration and temperature effects on oxygen relations predicted for maize roots. Agron. J. 64:720-725.

22. LUXMOORE, R. J., L. H. STOLZY, and J. LETEY. 1970. Oxygen diffusion in the soil-plant system. I. A model. Agron. J. 62:317-332.

23. MC PHERSON, D. C. 1939. Cortical air spaces in the roots of Zea mays L. New Phytol. 38:190-202.

24. NORRIS, F. DE LA M. 1913. Production of air passages in the root of Zea mays by variation of the culture media. Proc. Bristol Nat. Soc. 4:134-136.

25. ROULIER, M. H., L. H. STOLZY, D. D. FOCHT, and N. R. FETTER. 1974. Nitrous oxide as an index of denitrification in field soils. Proc. Soil Sci. Soc. Amer. (In press).

26. RUSSELL, M. B. 1952. Soil aeration and plant growth, p. 253-301. *In* B. T. Shaw (ed.), Soil physical conditions and plant growth. Academic Press, New York and London.

27. SCHRAMM, R. J., JR. 1960. Anatomical and physiological development of roots in relation to aeration of the substrate. Ph.D. thesis, Duke Univ. (Libr. Congr. Card No. Mic. 60-6001). Univ. Microfilms, Ann Arbor, Mich. 204 p.

28. SIFTON, H. B. 1945. Air space tissues in plants. Bot. Rev. 11:108-143.

29. SIFTON, H. B. 1957. Air space tissues in plants. II. Bot. Rev. 23:303-312.

30. STOLZY, L. H. 1971. Soil aeration and gas exchange in relation to grasses, p. 247-257. *In* V. B. Younger and C. M. McKell (ed.), The biology utilization of grasses. Academic Press, New York and London.

31. STOLZY, L. H. 1973. Soil atmosphere, p. 335-361. *In* E. W. Carson (ed.), The plant root and its environment. The Univ. Press, Charlottesville, Va.

32. STOLZY, L. H., and J. LETEY. 1964. Characterizing soil oxygen conditions with a platinum microelectrode. Adv. Agron. 16:249-279.

33. STOLZY, L. H., and S. D. VAN GUNDY. 1968. The soil as an environment for microflora and microfauna. Phytopathology 58:889-899.

34. TAYLOR, S. A., and G. L. ASHCROFT. 1972. Physical edaphology. W. H. Freeman Co., San Francisco. p. 352-391.

35. VARADE, S. B., L. H. STOLZY, and J. LETEY. 1970. Influence of temperature, light intensity and aeration on growth and root porosity of wheat Triticum aestivum. Agron. J. 62:505-507.

36. WOOLLEY, J. T. 1965. Drainage requirements of plants. Conf. Proc., Drainage for efficient crop production. Amer. Soc. Agr. Eng., St. Joseph, Mich. p. 2-5.

37. YU, P. T., L. H. STOLZY, and J. LETEY. 1969. Survival of plants under prolonged flooded conditions. Agron. J. 61:844-847.

Forms of Nitrogen and the pH in the Root Zone and Their Importance to Root Infections

RICHARD W. SMILEY—*Turf Pathologist, Cornell University, Ithaca, New York.*

D. M. Huber and associates (39, 40) recently emphasized that the specific form of N, rather than N per se, is a major factor influencing disease severity. Some diseases are most severe when ammonium nitrogen (NH_4-N) is the primary form of inorganic N in the root zone; others are most severe when nitrate-nitrogen (NO_3-N) predominates. This could be attributed in part to metabolic differences within NH_4- versus NO_3-fed plants (55, 80); however, absorption of NH_4-N by roots also reduces the rhizosphere soil pH, and absorption of NO_3-N tends to increase it (62, 75). The hypothesis presented here is an extension of Huber's observations and shows that diseases favored more by soil alkalinity are suppressed by NH_4-N and enhanced by NO_3-N, and that the reverse holds for root diseases favored by soil acidity. The discussion is restricted to the parasitic activity of root pathogens and excludes the effects of N on their saprophytic survival.

FORM OF NITROGEN AND THE IONIC EQUILIBRIUM OF PLANTS.—The principal factor affecting the cation-anion uptake balance of plants is the predominant form in which N is absorbed (18, 48). An electrochemical equilibrium must be maintained in plants and in the medium in which they are grown, and, therefore, an excessive uptake of cations (as with NH_4-N) or anions (as with NO_3-N) is in part balanced by an exchange of charges (as H^+, OH^-, or HCO_3^-) with the growth medium. The resultant changes in rhizosphere pH may be large since N is required by plants in much higher concentrations than are other nutrients that are absorbed in the ionic form (23).

Assimilation of NO_3-N into organic forms results in an alkaline effect within the plant. The excess of negative charges must be

$$NO_3^- + 8H^+ + 8e^- \longrightarrow NH_3 + 2H_2O + OH^-$$

neutralized, either by a transfer to stable metabolites (particularly organic anions) within the plant or by a transfer into the nutrient solution as OH^- and HCO_3^- (20). If organic anions accumulate within the plant, they must be accompanied by mineral cation uptake for the ionic equilibrium to be maintained. Some plants (e.g., peas, beans, radishes, apples) assimilate much of the absorbed NO_3-N in the roots, whereas others (e.g., oats, corn, beets, clover) transport most of the unchanged NO_3^-

ion (and urea) to the shoot before it is assimilated (63). If the NO_3^- is largely assimilated in the root, then excretion of HCO_3^- and OH^- from the root to the nutrient medium is much greater than if the major site of reduction occurs in aerial plant parts. If assimilation occurs largely in the shoot, organic anion accumulation and cation uptake become important in maintaining the ionic equilibrium of the plant.

Conversely, plants feeding predominantly on NH_4-N absorb a much higher proportion of cations than anions. Acidification results from the assimilation of NH_4-N, and the protons (H^+) are to a large extent

$$NH_4^+ \longrightarrow NH_3 + H^+$$

returned to the nutrient medium because NH_4-N is assimilated largely, if not entirely, in the root. Equilibration may also be accomplished within the root by production of organic anions; however, this tendency is reduced because the acidity that results from the assimilatory process tends to depress the dissociation of organic acids. A near stoichiometry of cation (K^+, Na^+, Ca^{++}, and NH_4^+) uptake and H^+ release by roots of intact maize plants occurs when these ions are supplied individually in the low concentrations typical of soil solutions (8).

ROOT ENVIRONMENT pH.—*In Nutrient Solutions.*—It is difficult to restrict pH changes in nutrient solutions that have unbalanced supplies of NH_4-N and NO_3-N (74). Absorption of N by tomato roots repeatedly modified the pH to 4.0 for NH_4-N, to 4.5 for urea-N, and to 6.2 for NO_3-N in circulating nutrient solutions that were regularly adjusted to pH 5.5 (48). The pH of the capillary layer around apple tree roots in sand cultures changed rapidly to match the root surface pH (4.0-4.5 for NH_4-N, 5.6 for NO_3-N, and 5.0 for no N), regardless of the pH of constantly renewed solutions flowing outside the capillary film (6). However, roots were highly buffered by organic acids and their internal pH was practically independent of the external acidity (60, 74).

In Soil.—It was formerly believed that the rhizosphere zone in soil is more acidic than the surrounding soil because root tissues and surfaces are acidic, because CO_2 is released by roots and the rhizosphere microflora, and because acid-producing bacteria are stimulated by roots.

This concept must be dispelled. Corn rhizospheres tend to be nearer neutrality than the surrounding soil (86), and Nye (61) recently indicated that roots are more likely to increase than to decrease the rhizosphere pH. Nye's contention has been supported by measurements of rhizosphere pH (62, 67, 75) in soils where NO_3-N was the predominant form of N, the usual situation in most agricultural soils.

I have grown dicotylendonous and monocotylendonous plants in the field and in pots in a glasshouse. In these studies, the most important single factor influencing the rhizosphere pH in a given plant-soil system was the predominant form of N absorbed by roots (75, 77). The influences of plant age, soil type, soil amendment by acidification or liming, soil fumigation, organic amendment, source of N, and nitrification inhibitors on the pH of nonrhizosphere (bulk) soil and wheat rhizospheres were also observed. The rhizosphere pH of wheat seedlings was generally lower than the bulk soil pH in soils fertilized with $NH_4H_2PO_4$ and $(NH_4)_2SO_4$, higher in soils fertilized with $Ca(NO_3)_2$, and relatively unchanged in nonfertilized soils and those fertilized with NH_4NO_3. The rhizosphere pH was inversely correlated with the NH_4-N concentration in both acid and alkaline soils. Nitrification of NH_4-N amplified the differences in bulk pH and minimized differences in rhizosphere pH of NH_4-N versus NO_3-N fertilized soils; the reverse was true when nitrification was suppressed. Nitrification was suppressed longer in an acid soil than in an alkaline soil, and thus larger differences in rhizosphere pH also persisted longer in the acid soil. Measurable changes in both the rhizosphere and bulk soil pH were eliminated by adding large amounts of lime to acid soils in the glasshouse and in the field, and differences in rhizosphere pH were amplified when alkaline soils were acidified. Where N had been applied to fumigated soil (e.g., where nitrification was suppressed), the bulk soil pH remained relatively unchanged, but large differences in rhizosphere pH occurred just as in the nonfumigated soil. Differences in rhizosphere pH became greater as the N application rates were increased.

FORM OF NITROGEN, pH, AND PATHOGENESIS OF SOIL-BORNE PLANT PARASITES.—The influence of pH on soil-borne plant-pathogenic fungi and the root diseases that they cause is well known. It is equally well known that generalizations about the optimal pH range for severe disease development are difficult to make because of the exceptions. *Ophiobolus graminis* and *Verticillium albo-atrum* are most destructive to plants in alkaline soils (17, 35), but they may also cause severe diseases in acid soils (70, 77, 96). The reverse applies to diseases caused by *Sclerotium rolfsii* and *Fusarium oxysporum*. These discrepancies are possibly the result of pH measurements on nonrhizosphere soil rather than on the rhizosphere soil, which is most important to the parasite during pathogenesis, particularly in soils treated with N fertilizers. Much of the confusion concerning effects of N and pH on soil-borne diseases appears to have resulted from comparisons of (a) N and pH effects in nutrient culture with those in soil, (b) studies where different forms of N were used, (c) studies where the extent of nitrification

of NH_4-fertilizers was not assessed before or during pathogenesis (e.g., where plants were probably absorbing considerably different proportions of NH_4-N and NO_3-N), and (d) pH values that were based on measurements of bulk soil and that were probably irrelevant to pathogenesis.

In a survey of the literature on the forms of N in disease, a correlation is evident if certain qualifications are first accepted. Little if any emphasis can be given to (a) data obtained in partially or totally sterile soil or in nutrient solution cultures, (b) instances where disease was obviously also limited by host resistance, moisture, or temperature, (c) instances where form of N was not reported, (d) instances where form of N absorbed by roots during pathogenesis was not reasonably predictable (e.g., based upon fertilizer placement and timing relative to seeding, potential for completion of nitrification prior to infection, etc.), and (e) instances where the effect of N and/or pH was presumed to be primarily on the saprophytic rather than on the parasitic activities of the pathogen.

Diseases caused by *Phymatotrichum omnivorum*, *Thielaviopsis basicola*, *O. graminis*, *V. albo-atrum*, *Streptomyces scabies*, and possibly *Phytophthora* spp. are favored more by soil alkalinity than acidity; these same diseases are suppressed by NH_4-N and/or enhanced by NO_3-N (Table 1). In contrast, diseases caused by *Sclerotium rolfsii* and *Fusarium* spp. are favored more by soil acidity than alkalinity and are suppressed by NO_3-N and/or enhanced by NH_4-N. Diseases caused by *Pythium* spp. or *Rhizoctonia solani* are favored by environmental factors unfavorable for growth of the host. *Plasmodiophora brassicae*, *Armillaria mellea*, *Poria weirii*, and *Fomes annosus* cause diseases that respond to changes in soil pH, but their response to forms of N is poorly established. No attempt will be made to relate specific mechanisms to each N-form:pH correlation, since the relationships for each particular disease undoubtedly result from at least several of the numerous mechanisms that have been proposed to explain how N influences the host-pathogen interaction. These mechanisms were discussed in a recent review (41) of nitrogen forms and plant diseases, including those caused by fungi, bacteria, viruses, and nematodes.

Diseases Suppressed by an Acid Environment and by NH_4-Fertilizers.—*Ophiobolus graminis.*—Take-all of wheat and Ophiobolus-patch of turf have been suppressed by soil applications of NH_4-N fertilizers (25, 33, 39, 77, 78), but fertilizers supplying NO_3-N failed to suppress them (31, 39, 77). Ammonium-N is ineffective for control in acid soils that are devoid of all NO_3-N (38), presumably due to adverse effects on root metabolism. This may explain why I (*unpublished data*) succeeded in suppressing take-all with additions to soil of particulate NH_4-N fertilizers pretreated with a nitrification inhibitor, but not when the fertilizer and inhibitor were supplied as a solution mixed homogeneously with N-deficient soils. The severity of take-all and Ophiobolus-patch was also increased when the soil pH was increased and was decreased by soil acidification (77, 78).

The suppressive influence of NH_4-N and lack of control with NO_3-N correlated with the rhizosphere pH (77). The

TABLE 1. Influence of the N-form and soil pH on development of diseases caused by root parasites

Parasite	Crop	Suppression by	Authority
Phymatotrichum omnivorum	Cotton	low pH	19
		NH_4-N	46
		not NO_3-N	81
Thielaviopsis basicola	Tobacco	low pH	3, 57
		NH_4-N	82
		not NO_3-N	29
Ophiobolus graminis	Cereals, turf	low pH	17
		NH_4-N, low pH	78
		NH_4-N	25, 33
		NH_4-N, not NO_3-N	39, 77
		not NO_3-N	95
Verticillium dahliae and	Tomato	not NO_3-N	69
V. albo-atrum		NH_4-N	97
		low pH	45
	Eggplant, cotton,	low pH	19, 35
	potato	NH_4-N, not NO_3-N	40
Streptomyces scabies	Potato	NH_4-N	64
		NH_4-N, not NO_3-N	40
		not NO_3-N	49
		low pH	52
Sclerotium rolfsii	Assorted	NO_3-N	36, 50, 56
		high pH	5, 44
Fusarium solani f. phaseoli	Bean	NO_3-N, not NH_4-N	54, 94
		not NH_4-N	79
Fusarium roseum f. cerealis	Wheat	high pH	9
'Culmorum'		not NH_4-N	30
Fusarium oxysporum f.	Tomato	high pH	45, 73
lycopersici			
Fusarium oxysporum f.	Cotton	high pH	19, 84
vasinfectum		NO_3-N, not NH_4-N	59
Fusarium oxysporum f.	Chrysanthemum	NO_3-N, not NH_4-N	98
chrysanthemi		high pH	98

suppression apparently resulted from direct inhibition of the pathogen at rhizosphere pH values less than 5.0 and from indirect (biological) control at higher pH values.

TABLE 2. Ectotrophic growth (mm/21 days) of *Ophiobolus graminis* upon roots of intact wheat plants in soil[a]

	Fertilizers[b]	
Soil	$(NH_4)_2SO_4$	$Ca(NO_3)_2$
Nonfumigated	3.7 a[c]	16.7 a
Fumigated (methyl bromide)	13.9 b	15.6 a
Fumigated + nonfumigated (1%, w/w)	4.6 a	18.6 a

[a] *Ophiobolus graminis*-infested potato-dextrose agar (5 mm diam) placed under nonsterile wheat seed.

[b] Fertilizers, as prills or crystals, were treated with 2% (w/w-based on N) 2-chloro-6-(trichloromethyl) pyridine (N-Serve 24) and then mixed uniformly with soil (0.15 g N/kg soil).

[c] Average growth on 45 seminal roots (3 replications, 5 plants/pot, 3 roots/plant). Common letters denote insignificant differences at the 99% confidence level.

Ammonium-N reduced the rhizosphere pH to values less favorable to the pathogen and thus reduced its inoculum potential and increased its vulnerability to antagonism from microorganisms. The magnitude of change in rhizosphere pH appeared to exert a greater influence on take-all than did the pH value per se (76, 77). The influence of NH_4-N on take-all was greatest on alkaline soils that had low concentrations of NO_3-N and that had been frequently cropped with cereals or grasses. There may be more antagonists (primarily species of *Pseudomonas* and *Streptomyces*) along roots in NH_4-N, compared to NO_3-N, fertilized soils (Smiley and Rovira, *unpublished*). The antagonism of these isolates toward *O. graminis* in culture was greatest when the agar was at pH 5.5 compared to 7.0. The form-of-N effect on take-all, therefore, appears to have both a biological (Table 2) and a chemical basis, and each is governed largely by rhizosphere pH.

Thielaviopsis basicola.—Black root rot of tobacco is suppressed by soil applications of NH_4-N (43) provided that the acidity is not increased to the extent that it injures the tobacco roots (58). The use of ammoniacal fertilizers was only recommended in fields where control of root rot became necessary (82). The effect of NH_4-N is apparently more than the acidification of the soil, since partial acidification of soil with some inorganic acids failed to suppress root rot (21). Extreme soil acidity suppressed black root rot (57) but did not suppress growth of *T. basicola* on the roots (3) or in culture (7). The pathogen became increasingly virulent as the acidity was increased in soils that were sterilized before inoculum was added (43). Therefore, ammonium in nonsterile soil acts to suppress black root rot, and this effect is possibly related to the acidity of the rhizosphere rather than that of the bulk soil. The suppressive effect of low pH may result from increased microbial competition or antagonism rather than from an exclusively direct effect on the pathogen.

Streptomyces scabies.—Soil acidity, either natural or sulfur-induced, was suggested more than 50 years ago to control common scab of potatoes (52). Scab has also been suppressed by applying NH_4-N fertilizers to soil and increased by applying NO_3-N (40, 49, 64). The effect of NH_4-N is not likely to be through acidification of the bulk soil mass, however, since the benefits of NH_4-N were increased when nitrification inhibitors were also used. Urea-formaldehyde solutions also helped control scab when they were placed near the seed piece (93), but not when they were placed away from it (13) or were broadcast and disked into the soil, except at very high application rates (72).

Verticillium spp.—Wilt of cotton and eggplant is generally associated with alkaline soils or alkaline sites within a field (24, 35). Verticillium wilt of potatoes has been suppressed more by NH_4-N than by NO_3-N (40). Enhancement of this effect by treatment of soil with a nitrification inhibitor or with the nematocidal soil fumigant 1,3-Dichloropropene (Telone) occurred without a reduction in pathogen populations. Telone and several other soil fumigants inhibited nitrification and increased the NH_4-N concentration in soil (28). Potatoes absorbed and assimilated NH_4-N preferentially to NO_3-N (80).

Phymatotrichum omnivorum.—Cotton root rot occurs in the Blackland soils of Texas and in other calcareous and alkaline soils. Field surveys indicated a linear relationship, even within the same field, between the pH and root rot incidence and severity (22). Little or no disease spread down the row in acid soils (83). High application rates for NH_4-N fertilizers suppressed root rot (46), and it has been recommended that NO_3-N be avoided (81). Applications of ammonium sulphate and monoammonium phosphate were equally effective in suppressing root rot (46, 81), but phosphorus alone, i.e., in nonammoniacal fertilizers, aggravated the disease (46). Disease suppression by NH_4-N became more pronounced in noncalcareous, neutral to acid soils compared to the calcareous soils that are highly buffered against changes in soil pH (1).

Phytophthora spp.—The incidence and severity of tobacco black shank, caused by *P. parasitica*, have been suppressed in acid soils (47). Lillian Fraser (N.S.W. Department of Agriculture, Australia, *personal communication*, 1972) also observed that root decay on citrus, caused by *P. citrophthora*, occurred mostly in neutral to alkaline soils. She subsequently found that infection was very slight at pH values less than 4.5 and high in soils above pH 5.4. Inoculation techniques were effective only when NO_3-N fertilizers were used. *Phytophthora* spp. are a component of the citrus-replant problem in southern California. Citrus growth in this area is better on acid soils (14), and the influence of pH can be eliminated by soil fumigation.

Diseases Suppressed by an Alkaline Environment and by NO_3-Fertilizers.—*Sclerotium rolfsii.*—This pathogen incites diseases in acid sandy soils of tropical and temperate regions, and has been suppressed in alkaline soils (44). Suppression of southern Sclerotium rot of sugar beets was directly correlated with the rate of applied N, with NO_3-N being most conducive to control (50). Nitrate-N fertilizers also suppressed the incidence of southern stem rot of peanuts and the severity of southern blight of tomatoes (56). Liming reduced the disease in peppers and sweet potatoes (5), but exceptions occurred, in that high pH did not always control the diseases in sugar beets in Arizona.

Fusarium spp.—Vascular wilts and cortical root rots caused by *formae speciales* of *F. oxysporum, F. roseum, F. nivale,* and *F. solani* are generally most severe in acid soils (27). Fertilizer studies with bentgrass turf (32) revealed a negative correlation between soil pH and the number of *Fusarium nivale*-caused patches. Low pH increased the rate of *F. oxysporum* growth along roots (91), and Rishbeth (68) indicated that NH_4-N fertilizers led to rapid onset of wilt. Wilt of chrysanthemum was reduced by liming soils and by supplying N predominantly as NO_3-N rather than as NH_4-N (98). Cotton wilt was also greater in NH_4-N- than in NO_3-N-treated soil (2). Fusarium wilt of maize was reduced by NO_3-N (59). Ammonium-N has been associated with increased severity of wheat and bean root rots caused by *Fusarium* (30, 94). Germination of conidia and chlamydospores and early penetration and pathogenesis were favored more by NH_4-N or reduced N in soil than by NO_3-N (16, 87). Secondary infections from runner hyphae of *F. solani* f. sp. *phaseoli* on bean roots were more numerous in soils fertilized with NH_4-N than in those fertilized with NO_3-N (94), and the detrimental effects increased where NH_4-N had been treated with nitrification inhibitors (53). The effect was shown only if N and the inoculum were intimately mixed in soil (94). Root nutrition was apparently not a factor, and this may explain why Burke and Nelson (12) were unable to show differences in root rot where NH_4-N or NO_3-N were banded to the side of bean seedlings shortly after emergence.

Diseases of Uncertain Affinity with Respect to N-Form and pH.—*Pythium spp.*—Damping-off of conifer seedlings and root rot of poinsettia are least severe in acidic soils (7, 34, 37, 42, 90). Conifers and poinsettia are acid-tolerant (7, 10). Damping-off of conifer seedlings is increased as the level of NO_3-N increases (37). This contrasts with sugar beet [an alka-tolerant plant (74)] root rot, caused by a *Pythium* spp.-*Aphanomyces levi* complex, which is suppressed by liming and is increased in severity by NH_4-N (4) and by acid soils (34). Damping-off of table beets is also suppressed by NO_3-N but not by NH_4-N (99).

The principal factor influencing damping-off is the growth rate (vigor) of the host (34), acid-tolerant plants having their greatest resistance in acid soils and the reverse holding for alka-tolerant plants. The effect of N-forms on damping-off may be through changes in rhizosphere pH that alter the susceptibility of roots to attack by *Pythium*. However, many examples could also be cited to illustrate that these trends are not without exception.

Rhizoctonia solani.—A review of the literature concerning effects of soil pH and N fertilizers on diseases caused by *R. solani* led to the conclusion that there is little, if any, direct influence of pH on the inoculum potential, but that pH could have an indirect influence on disease potential (6). It now appears that N and pH affect these diseases by altering the disease susceptibility of plants in the same manner that occurs with *Pythium*.

The Ectotrophic Growth Habit.—The precise manner by which soil conditions affect pathogenesis is determined largely by the parasitic growth habit of the fungus (26). The rhizosphere environment exerts its maximum effect upon specialized root-infecting fungi (*Ophiobolus, Phymatotrichum, Sclerotium,* and *Thielaviopsis*), which are characterized by the ectotrophic growth habit, and exerts a considerably smaller effect upon the less-specialized fungi (*Aphanomyces, Phytophthora, Pythium,* and *Rhizoctonia*), which are sheltered within the root once penetration has occurred. Those pathogens that have the ability to penetrate with a minimum of ectotrophic growth (*Verticillium*) are influenced less by adverse environmental conditions than those that produce a prepenetration thallus or mycelial mat (*Ophiobolus* and *F. solani*) (71). I have observed that suppression of take-all of wheat by NH_4-fertilizers is not attributable to a significantly reduced frequency of primary infections by *O. graminis*, but rather to a restriction of ectotrophic growth away from the point of infection, compared to that in soil having mostly NO_3-N. The behavior of the pathogen prior to, during, and after penetration is thus of major importance in considering the effects of rhizosphere environments on soil-borne diseases. The ectotrophic growth habit of fungi, the environmental factors that influence it, and its importance to root colonization and infection have been thoroughly discussed (11, 27). The significance of ectotrophic growth lies in its potential for continuous and indefinite spread over the host root system. Therefore, it is not effectively or consistently checked by host resistance, but is extremely susceptible to adverse environmental conditions and to intense competition, antibiosis, and hyperparasitism in the rhizosphere.

Fungal symbionts of pine mycorrhizal associations colonize roots most effectively in slightly acid soils (85). The concentration of NO_3-N in acid media had no significant influence on colonization of glass fibers by *Rhizopogon luteolus*, but, compared to low NO_3-N, high levels of NO_3-N in acid soil caused a highly significant reduction in colonization of pine roots (85). Mycorrhizal associations are suppressed in alkaline soils with substantial NO_3-N concentrations, but not in those deficient in NO_3-N (66). Thus, the suppressive influence of NO_3-N on mycorrhizal fungi also appears to be mediated by an increase in rhizosphere pH.

CONCLUSIONS.—This survey supports the hypothesis that the form-of-N effect on diseases caused by soil-borne plant pathogens is often also a pH effect, with NH_4-N acting to reduce the rhizosphere pH and with NO_3-N acting to increase it. The absorption by roots of predominantly NH_4-N versus NO_3-N induces rhizosphere pH differences up to 2.5 units. In the critical pH range, a difference of a few tenths of a pH unit may profoundly affect the growth of a fungus (51) and thus alter its inoculum potential. If the inoculum potential is sufficiently reduced, pathogenicity may be limited directly; however, disease control could also be achieved with slight reductions in the pathogen's growth if that reduction places the pathogen at a disadvantage relative to the saprophytic rhizosphere microflora. Inoculation of the soil with antagonists is generally ineffective as a biological control method, unless the environmental conditions are also modified to favor the antagonist (92). The importance of rhizosphere pH to parasitic activities of root pathogens has been stressed (71), and it has been suggested that some unsuccessful attempts to control disease by altering the bulk soil pH may have occurred because of the neutralizing effect at the root surface. The use of N fertilizers to modify the root surface pH during the early phases of plant growth alleviates these difficulties and thus avoids the necessity for suppressing the parasite in the bulk soil. This approach would be most important for diseases that remain to be controlled by fungicides or by host resistance. Unfortunately, many of the diseases (e.g., those caused by *F. solani* and by species of *Aphanomyces, Pythium,* and *Rhizoctonia*) that are not controlled genetically (88) are also variable in their response to the form of N applied to soil and in their response to soil pH. Resistance to nonvascular root pathogens is mostly multigenic when it exists, and therefore quite sensitive to environmental conditions (89). Manipulation of the rhizosphere pH could be an important means for complementing multigenic resistance to pathogenic species of *F. roseum*. In contrast, this system would be of little use for favoring monogenic resistance to vascular pathogens (*F. oxysporum, V. albo-atrum,* and *V. dahliae*), since this form of resistance is relatively insensitive to the environment.

Although the correlation of N-forms and pH effects on root disease is instructive, it must not be assumed that the change in rhizosphere pH is necessarily the active mechanism for the form-of-N effect on a specific disease. The influence of NH_4-N versus NO_3-N has many other effects on host roots and their rhizospheres. An adverse pH may be operative in one case and not in another. The rhizosphere pH, for instance, is unlikely to be the controlling factor for vascular diseases caused by *Verticillium* spp. and other pathogens that do not rely heavily upon ectotrophic growth during pathogenesis. Additionally, any one mechanism is not likely to be effective over a large range in pH (15). In my experience, take-all of wheat can be positively suppressed if the rhizosphere pH is reduced to very low values. However, at higher pH values, the suppression depends upon the presence of an antagonistic microflora and thus varies with each soil. Since the magnitude of pH change depends upon the amount of each absorbed N-form, this method of disease control is likely to be ineffective for dryland crops in regions of low rainfall where fertilizers are necessarily limited to low application rates.

Technically, it is not difficult to maintain a high $NH_4:NO_3$ ratio in soil; the primary problem is economic. Economic and public pressure has led to new approaches for increasing the efficiency of fertilizer-N (28, 65). Nitrification can now be suppressed sufficiently long enough so the form of N available for uptake by roots in the upper soil profile can be manipulated with reasonable accuracy during the early stages of plant growth. This is especially important for diseases of plants that have short to moderate growth periods. Technological advances should allow us to manage the nitrification process in many commercial fields, just as it has been done in experimental agriculture.

ACKNOWLEDGEMENTS.—The author wishes to thank R. J. Cook, USDA Agricultural Research Service, and his previous employers for their cooperation and financial assistance: The Plant Pathology Department, Washington State University at Pullman; The Division of Soils, Commonwealth Scientific and Industrial Research Organization at Adelaide, South Australia; and The North Atlantic Treaty Organization (postdoctoral fellowship at Adelaide, South Australia).

LITERATURE CITED

1. ADAMS, J. E., R. C. WILSON, L. E. HESSLER, and D. R. ERGLE. 1939. Chemistry and growth of cotton in relation to soil fertility and root-rot. Proc. Soil Sci. Soc. Amer. 4:329-332.
2. ALBERT, W. B. 1946. The effects of certain nutrient treatments upon the resistance of cotton to Fusarium vasinfectum. Phytopathology 36:703-716.
3. ANDERSON, P. J., A. V. OSMUN, and W. L. DORAN. 1926. Soil reaction and black root-rot of tobacco. Mass. Agr. Exp. Stn. Bull. 229:117-136.
4. ARRENHIUS, O. 1923. (Experiments in the control of beet root rot). Rev. Appl. Mycol. 3:74-75. 1924. (Abstr.).
5. AYCOCK, R. 1966. Stem rot and other diseases caused by Sclerotium rolfsii. N. C. Agr. Exp. Stn. Tech. Bull. 176. 202 p.
6. BAKER, R., and C. A. MARTINSON. 1970. Epidemiology of diseases caused by Rhizoctonia solani, p. 172-188. *In* J. R. Parmeter, Jr. (ed.), Rhizoctonia solani, biology and pathology. Univ. Calif. Press, Berkeley and Los Angeles.
7. BATEMAN, D. F. 1962. Relation of soil pH to development of poinsettia root-rots. Phytopathology 52:559-566.
8. BECKING, J. H. 1956. On the mechanism of ammonium ion uptake by maize roots. Acta. Bot. Neerl. 5:1-79.
9. BENNETT, F. T. 1939. Fusarium disease of cereals. Agr. Progr. 16:64-69.
10. BENZIAN, B. 1970. Nutrition of young conifers and soil fumigation, p. 222-225. *In* T. A. Toussoun et al. (ed.), Root diseases and soil-borne pathogens. Univ. Calif. Press, Berkeley and Los Angeles.
11. BOWEN, G. D., and C. THEODOROU. 1973. Growth of ectomycorrhizal fungi around seeds and roots, p. 107-150. *In* G. C. Marks and T. T. Kozlowski (ed.), Ectomycorrhizae: their ecology and physiology. Academic Press, New York and London.
12. BURKE, D. W., and C. E. NELSON. 1968. NH_4^+ verus NO_3^- fertilization of dry field beans on Fusarium-infested land. Wash. Agr. Exp. Stn. Circ. 490. 9 p.
13. CETAS, R. C., and R. L. SAWYER. 1962. Evaluation of Uracide for the control of common scab of potatoes on Long Island. Amer. Potato J. 39:456-459.
14. CHAPMAN, H. D. 1965. Chemical factors of the soil as they affect micro-organisms, p. 120-141. *In* K. F. Baker and W. C. Snyder (ed.), Ecology of soil-borne plant pathogens. Univ. Calif. Press, Berkeley and Los Angeles.
15. COCHRANE, V. W. 1959. Physiology of fungi. John Wiley and Sons, New York. 525 p.
16. COOK, R. J., and M. N. SCHROTH. 1965. Carbon and nitrogen compounds and germination of chlamydospores of Fusarium solani f. phaseoli. Phytopathology 55:254-256.
17. DAVIS, R. J. 1925. Studies on Ophiobolus graminis Sacc. and the take-all disease of wheat. J. Agr. Res. 31:801-825.
18. DEKOCK, P. C., and E. A. KIRKBY. 1969. Uptake by plants of various forms of nitrogen and effects on plant composition. Ministry Agr. Fish. Food Tech. Bull. 15:7-14.
19. DICKSON, J. G. 1956. Diseases of field crops. McGraw-Hill Book Co., Inc., New York. 517 p.
20. DIJKSHOORN, W. 1962. Metabolic regulation of the alkaline effect of nitrate utilization in plants. Nature (London) 194:165-167.
21. DORAN, W. L. 1931. Increasing soil acidity as a means of controlling black root-rot of tobacco. Mass. Agr. Exp. Stn. Bull. 276:117-146.
22. EZEKIEL, W. N., J. J. TAUBENHAUS, and E. C. CARLYLE. 1930. Soil reaction effects on Phymatotrichum root-rot. Phytopathology 20:803-815.
23. FRIED, M., and H. BROESHART. 1967. The soil-plant system in relation to inorganic nutrition. Academic Press, New York and London. 358 p.
24. GARRETT, S D. 1947. Report on investigations of Verticillium wilt. Empire Cotton Growing Rev. 24:101-102.
25. GARRETT, S. D. 1948. Soil conditions and take-all disease of wheat. IX. Interaction between host plant nutrition, disease escape and disease resistance. Ann. Appl. Biol. 35:14-17.
26. GARRETT, S. D. 1956. Biology of root-infecting fungi. Cambridge Univ. Press, Cambridge. 293 p.
27. GARRETT, S. D. 1970. Pathogenic root-infecting fungi. Cambridge Univ. Press, Cambridge. 294 p.
28. GASSER, J. K. R. 1970. Nitrification inhibitors—their occurrence, production and effects of their use on crop yields and composition. Soils Fert. 33:547-554.
29. GILBERT, W. W. 1909. The root-rot of tobacco caused by Thielavia basicola. U.S. Dept. Agr. Bur. Plant Ind. Bull. 158. 55 p.
30. GLYNNE, M. D. 1951. Effects of cultural treatments on wheat and on incidence of eye spot, lodging, take-all and weeds. Field experiment, 1945-1948. Ann. Appl. Biol. 38:665-688.
31. GLYNNE, M. D., and D. B. SLOPE. 1959. Effects of previous wheat crop, seed rate and nitrogen on eyespot, take-all, weeds and yields of two varieties of winter wheat.

Field experiment, 1954-1956. Ann. Appl. Biol. 47:187-199.

32. GOSS, R. L., and C. J. GOULD. 1968. Some interrelationships between fertility levels and Fusarium patch disease of turfgrasses. J. Sports Turf Res. Inst. 44:19-26.

33. GOULD, C. J., R. L. GOSS, and V. L. MILLER. 1966. Effect of fungicides and other materials on control of Ophiobolus patch disease on bent grass. J. Sports Turf Res. Inst. 42:41-48.

34. GRIFFIN, D. M. 1958. Influence of pH on the incidence of damping-off. Trans. Br. Mycol. Soc. 41:438-490.

35. HAENSELER, C. M. 1928. Effect of soil reaction on Verticillium wilt of egg-plant. Annu. Rept. N. J. Agr. Exp. Stn. 1927-1928:267-273.

36. HARRISON, A. L. 1961. Control of Sclerotium rolfsii with chemicals. Phytopathology 51:124-128.

37. HARTLEY, C. 1921. Damping-off in forest nurseries. U.S. Dept. Agr. Bull. 934. 99 p.

38. HORNBY, D., and C. A. I. GORING. 1972. Effects of ammonium and nitrate nutrition on take-all disease of wheat in pots. Ann. Appl. Biol. 72:225-231.

39. HUBER, D. M., C. G. PAINTER, H. C. McKAY, and D. L. PETERSON. 1968. Effect of nitrogen fertilization on take-all of winter wheat. Phytopathology 58:1470-1472.

40. HUBER, D. M., and R. D. WATSON. 1970. Effect of organic amendment on soil-borne plant pathogens. Phytopathology 60:22-26.

41. HUBER, D. M., and R. D. WATSON. 1974. Nitrogen forms and plant disease. Annu. Rev. Phytopathol. (In press).

42. JACKSON, L. W. R. 1940. Effects of H-ion and Al-ion concentrations on damping-off of conifers and certain causative fungi. Phytopathology 30:563-578.

43. JOHNSON, J., and R. E. HARTMAN. 1919. Influence of soil environment on the root-rot of tobacco. J. Agr. Res. 17:41-86.

44. JOHNSON, S. P. 1953. Some factors in the control of the southern blight organism, Sclerotium rolfsii. Phytopathology 43:363-368.

45. JONES, J. P., and A. J. OVERMAN. 1971. Control of Fusarium wilt of tomato with lime and soil fumigants. Phytopathology 61:1415-1417.

46. JORDAN, H. V., H. A. NELSON, and J. E. ADAMS. 1939. Relation of fertilizers, crop residues and tillage to yields of cotton and incidence of root-rot. Proc. Soil Sci. Soc. Amer. 4:325-328.

47. KINCAID, R. R., and N. GAMMON, JR. 1954. Incidence of tobacco black shank directly related to soil pH. Plant Dis. Reptr. 38:852-853.

48. KIRKBY, E. A. 1969. Ion uptake and ionic balance in plants in relation to the form of nitrogen nutrition, p. 215-235. *In* I. H. Rorison (ed.), Ecological aspects of the mineral nutrition of plants. Blackwell Sci. Pub., Oxford.

49. LAPWOOD, D. M., and P. W. DYSON. 1966. An effect of nitrogen on the formation of potato tubers and the incidence of common scab (Streptomyces scabies). Plant Pathol. 15:9-14.

50. LEACH, L. D., and A. E. DAVEY. 1942. Reducing southern Sclerotium rot of sugar beet with nitrogenous fertilizers. J. Agr. Res. 64:1-18.

51. LILLY, V. G., and H. L. BARNETT. 1947. The influence of pH and certain growth factors on mycelial growth and perithecial formation by Sordaria fimicola. Amer. J. Bot. 34:131-138.

52. MARTIN, W. H. 1920. The relation of sulphur to soil acidity and to the control of potato scab. Soil Sci. 9:393-409.

53. MAUER, C. L. 1962. Effect of carbon substrates and carbon to nitrogen ratio on bean root-rot. M. S. thesis, Colo. State Univ., Fort Collins. 50 p.

54. MAUER, C. L., and R. BAKER. 1965. Ecology of plant

pathogens in soil. II. Influence of glucose, cellulose, and inorganic nitrogen amendments on development of bean root-rot. Phytopathology 55:69-72.

55. MCKEE, H. S. 1962. Nitrogen metabolism in plants. The Clarendon Press, Oxford. 728 p.

56. MOHR, H. C., and G. M. WATKINS. 1959. The nature of resistance to southern blight in tomato and the influence of nutrition on its expression. Proc. Amer. Soc. Hort. Sci. 74:484-493.

57. MORGAN, M. F., and P. J. ANDERSON. 1927. Relation of soil reaction to black root-rot and good tobacco. Conn. Agr. Exp. Stn. Tobacco Bull. 8:47-49.

58. MORGAN, M. F., P. J. ANDERSON, and H. DORSEY. 1929. Soil reaction and liming as factors in tobacco production in Connecticut. Conn. Agr. Exp. Stn. Bull. 306:733-806.

59. MOSTAFA, M. A., and M. K. MOAWAD. 1965. Studies on Fusarium wilt disease of maize in United Arab Republic. I. Pathogenicity of Fusarium and Gibberella, singly and in association under different nitrogeneous manurial treatments. J. Bot. (United Arab Republic) 8:55-73.

60. NIGHTINGALE, G. T. 1934. Ammonium and nitrate nutrition of dormant delicious apple trees at 48° F. Bot. Gaz. XCV:437-452.

61. NYE, P. H. 1968. Processes in the root environment. J. Soil Sci. 19:205-215.

62. OZANNE, P. G., and S. A. BARBER. 1970. Rhizosphere differences between species in relation to the uptake of phosphate. p. 1 (e) 1-4. Austral. Plant Nutr. Conf., Mount Gambier, South Australia.

63. PATE, J. S. 1973. Uptake, assimilation and transport of nitrogen compounds by plants. Soil Biol. Biochem. 5:109-119.

64. POTTER, H. S., M. G. NORRIS, and C. E. LYONS. 1971. Potato scab control studies in Michigan using N-Serve nitrogen stabilizer for nitrification inhibition. Down To Earth 27(3):23-24.

65. PRASAD, R., G. B. RAJALE, and B. A. LAKHDIVE. 1971. Nitrification retarders and slow-release nitrogen fertilizers. Adv. Agron. 23:337-383.

66. RICHARDS, B. N. 1965. Mycorrhizal development of loblolly pine seedlings in relation to soil reaction and the supply of nitrate. Plant Soil 22:187-199.

67. RILEY, D., and S. A. BARBER. 1969. Bicarbonate accumulation and pH changes at the soybean [Glycine max (L.) Merr.] root-soil interface. Proc. Soil Sci. Soc. Amer. 33:905-908.

68. RISHBETH, J. 1957. Fusarium wilt of bananas in Jamaica. II. Some aspects of host-parasite relationships. Ann. Bot. 21:215-245.

69. ROBERTS, F. M. 1943. Factors influencing infection of the tomato by Verticillium albo-atrum. Ann. Appl. Biol. 30:327-331.

70. ROSEN, H. R., and J. A. ELLIOTT. 1923. Pathogenicity of Ophiobolus cariceti in its relationship to weakened plants. J. Agr. Res. 25:351-358.

71. SCHROTH, M. N., and D. C. HILDEBRAND. 1964. Influence of plant exudates on root-infecting fungi. Annu. Rev. Phytopathol. 2:101-132.

72. SCHULTZ, T. H., K. C. BERGER, H. M. DARLING, and M. H. FLEISCHFRESSER. 1961. Urea formaldehyde concentrate-85 for scab control in potatoes. Amer. Potato J. 38:85-88.

73. SHERWOOD, E. C. 1923. Hydrogen-ion concentration as related to the Fusarium wilt of tomato seedlings. Amer. J. Bot. 10:537-552.

74. SMALL, J. 1954. Modern aspects of pH. Balliere, Tindall and Cox, London. 237 p.

75. SMILEY, R. W. 1974. Rhizosphere pH as influenced by

plants, soils, and nitrogen fertilizers. Proc. Soil Sci. Soc. Amer. (In press).

76. SMILEY, R. W. 1974. Take-all of wheat as influenced by organic amendments and nitrogen fertilizers. Phytopathology. (In press.)

*77. SMILEY, R. W., and R. J. COOK. 1973. Relationship between take-all of wheat and rhizosphere pH in soils fertilized with ammonium versus nitrate-nitrogen. Phytopathology 63:882-890.

78. SMITH, J. D. 1956. Fungi and turf diseases. 6. Ophiobolus patch disease. J. Sports Turf Res. Inst. 32:180-202.

79. SNYDER, W. D., M. N. SCHROTH, and R. CHRISTOU. 1959. Effect of plant residues on root-rot of bean. Phytopathology 49:755-756.

80. STREET, H. E., and D. E. G. SHEAT. 1958. The absorption and availability of nitrate and ammonia, p. 150-165. *In* W. Ruhland (ed.), Encyclopedia plant physiology. Vol. VIII. Nitrogen metabolism. Springer-Verlag, Berlin.

81. STREETS, R. B. 1937. Phymatotrichum (cotton or Texas) root-rot in Arizona. Ariz. Agr. Exp. Stn. Tech. Bull. 71:299-410.

82. SWANBACK, T. R., and P. J. ANDERSON. 1947. Fertilizing Connecticut tobacco. Conn. Agr. Exp. Stn. Bull. 503. 52 p.

83. TAUBENHAUS, J. J., and W. N. EZEKIEL. 1930. Recent studies on Phymatotrichum root-rot. Amer. J. Bot. 17:554-571.

84. TAUBENHAUS, J. J., W. N. EZEKIEL, and D. T. KILLOUGH. 1928. Relation of cotton root-rot and Fusarium wilt to the acidity and alkalinity of the soil. Texas Agr. Exp. Stn. Bull. 389. 19 p.

85. THEODOROU, C., and G. D. BOWEN. 1969. The influence of pH and nitrate on mycorrhizal associations of Pinus radiata D. Don. Austral. J. Bot. 17:59-67.

86. THOM, C., and H. HUMFIELD. 1932. Notes on the associations of micro-organisms and roots. Soil Sci. 34:29-36.

87. TOUSSOUN, T. A., S. M. NASH, and W. C. SNYDER. 1960. The effect of nitrogen sources and glucose on the pathogenesis of Fusarium solani f. phaseoli. Phytopathology 50:137-140.

88. WALKER, J. C. 1965. Host resistance as it relates to root pathogens and soil microorganisms, p. 314-320. *In* K. F. Baker and W. C. Snyder (ed.), Ecology of soil-borne plant pathogens. Univ. Calif. Press, Berkeley and Los Angeles.

89. WALKER, J. C. 1969. Plant Pathology. McGraw-Hill Book Co., Inc., New York. 819 p.

90. WARCUP, J. H. 1952. Effect of partial sterilization by steam or Formalin and damping-off of Sitka spruce. Trans. Br. Mycol. Soc. 35:248-262.

91. WARDLAW, C. W. 1941. The banana in Central America. IV. Panama disease. Nature (London) 147:380-381.

92. WEINDLING, R., and H. S. FAWCETT. 1936. Experiments in the control of Rhizoctonia damping-off of citrus seedlings. Hilgardia 10:1-16.

93. WEINHOLD, A. R., T. BOWMAN, and J. BISHOP. 1964. Urea-formaldehyde for the control of common scab of potato. Amer. Potato J. 41:319-321.

94. WEINKE, K. E. 1962. The influence of nitrogen on the root disease of bean caused by Fusarium solani f. phaseoli. Ph.D. thesis, Univ. Calif., Berkeley.

95. WESTE, G., and L. B. THROWER. 1971. The effect of added nitrate on the growth of Ophiobolus graminis. Plant Soil 35:161-172.

96. WILHELM, S. 1950. Verticillium wilt in acid soils. Phytopathology 40:776-777.

97. WILHELM, S. 1951. Effect of various soil amendments on the inoculum potential of the Verticillium wilt fungus. Phytopathology 41:684-690.

98. WOLTZ, S. S., and A. W. ENGELHARD. 1973. Fusarium wilt of chrysanthemum: effect of nitrogen source and lime on disease development. Phytopathology 63:115-157.

99. YALE, J. W., JR., and E. K. VAUGHAN. 1962. Effects of mineral fertilizers on damping-off of table beets. Phytopathology 52:1285-1287.

The Role of Carbon Dioxide in Growth and Survival of Phymatotrichum omnivorum

STUART D. LYDA and EARL BURNETT—*Department of Plant Sciences, Texas A&M University, College Station, Texas, and Blackland Conservation Research Center, Temple, Texas.*

Phymatotrichum root rot is largely confined to the alkaline, calcareous soils of the southwestern United States and northern Mexico. In Texas the disease is prevalent in the central blacklands, coastal bend, and Lower Rio Grande Valley. These soils are predominantly montmorillonitic clays with extensive capacity for expanding and shrinking. Soil pH is near 8.2, the content of organic matter is approximately 4.1%, the exchange capacity is approximately 64 me/100 g soil (7), and the moisture-holding capacity is 45%. Most of the soil is subtended by a white or yellowish limestone layer (13).

Sclerotia of *P. omnivorum* have been found in the soil profile to depths of 244 cm (15). We have found nearly 95% of the sclerotium population situated in a zone 30 to 90 cm deep (*unpublished*). The fungus grows from these deep-seated propagules as strands ramifying through the soil. Upon contact with a plant root, the strands ascend its outside surface, and eventually the plant succumbs as the root is girdled near the soil surface.

Deep tillage of the soil was advocated as a control measure for Phymatotrichum root rot of cotton when Shear and Miles (16) stated that "lack of proper aeration of the soil is one of the most important factors favoring the development of the root-rot fungus. Deeper plowing than usually practiced in ordinary cultivation methods improves the aeration of the soil and was therefore tried." They achieved favorable disease control with deep plowing in the fall, but not with deep plowing in the spring. Extensive disruption of the soil profile by rototilling or ditching ameliorated the disease for several years following the treatment. Improved soil aeration was proposed as one explanation for the beneficial responses noted.

Under laboratory conditions, CO_2 accumulated in soil supporting growth of *P. omnivorum*, and sclerotial initiation was coincident with an elevated CO_2 content of the soil gaseous environment. We proposed that CO_2 or the dissolved bicarbonate ion influenced sclerotium formation by *P. omnivorum* (10).

We have studied the influence of gases on the growth and survival of this organism and have analyzed changes in soil gases to depths of 180 cm for 3 years under field conditions. Changes in CO_2 concentrations have striking effects on the growth and development of *P. omnivorum* and upon disease development. This paper will present some of those responses.

INFLUENCE OF CO_2 ON GROWTH OF P. OMNIVORUM IN VITRO.—A comprehensive review of the effects of gaseous environments on the growth and metabolism of fungi was presented by Tabak and Cooke (17). Although most filamentous fungi are sensitive to high CO_2 concentrations, some can tolerate high levels. Burges and Fenton (1) reported that vertical distribution of fungi in the soil profile was influenced by their tolerance to CO_2. Three well-known strand or rhizomorph-forming, soil-borne fungi (*Armillaria mellea, Ophiobolus graminis,* and *Phymatotrichum omnivorum*) tolerate high CO_2 pressures. Raabe and Gold (14) found that *A. mellea* made good hyphal and rhizomorph growth on potato-dextrose agar in environments of 30% CO_2, but that no growth occurred above 60% CO_2. The rhizomorphs became flattened and plaque-like at and above 30% CO_2.

Fellows (3) observed that *O. graminis* grew on solid or liquid potato-dextrose medium in reduced O_2 and elevated CO_2 environments. Growth diminution was not noticed until the O_2 content dropped below 6% or the CO_2 content was greater than 18%. He concluded that "the gases ordinarily found in soils, whether free or in solution, are not present in sufficient quantity to affect the growth of *Ophiobolus graminis* very materially."

Neal and Wester (12) reported that *P. omnivorum* tolerated high concentrations of CO_2. It grew on neutral carrot agar with an ambient environment composed of 75% CO_2; however, the growth rate was reduced to about 77% of the control. No growth occurred at 100% CO_2, but growth began after the agar plates were placed in aerobic conditions. The fungus grew as well in 25% CO_2 as it did in the air control. They postulated that deep plowing or subsoiling may provide the necessary aeration for the germination of sclerotia.

Gunasekaran (5) made comparative growth studies of *P. omnivorum* under four levels of CO_2 (0.03, 0.5, 5.0, and 50%) and four concentrations of glycogen (0.5, 1.0, 2.0, and 4.0%). The growth rate on agar medium was similar with 0.03 to 5.0% CO_2, but was reduced by 50% CO_2. In liquid media, growth was 68% of the control for the 50% CO_2 level after 30 days. The dry weight of mycelium produced was proportional to the glycogen concentration. Gunasekaran also observed that more glucose was metabolized through the Embden-Meyerhof-Parnas pathway when *P. omnivorum* was grown in chambers purged with 5% CO_2 rather than with air (6).

We have found that mycelium starved for 72 hours in sterile, distilled water assimilated $H^{14}CO_3$ into organic acids, amino acids, sugars, and starch when challenged for 1 hour (Table 1). The site of incorporation has not been determined.

GROWTH OF P. OMNIVORUM AND ROOT ROT DEVELOPMENT IN NONSTERILE SOIL AS AFFECTED BY CO₂ CONCENTRATIONS.

—Growth rates of *Phymatotrichum* mycelial strands were determined by placing sclerotia in glass tubes (5.1 × 122 cm) filled with nonsterile Houston black clay (HBC). The horizontal soil columns were flushed with different gas mixtures (air, 0.5, 5.0, and 50% CO_2), with the balance of each CO_2 mixture being air, at a flow rate of 12 ml/min. No visible growth occurred in soils flushed with air or 50% CO_2, but the fungus grew through the soil at a rate of 1.6 cm/day in soils flushed with 0.5 and 5.0% CO_2 (Fig. 1A).

We do not understand why the fungus grows in the presence of 50% CO_2 on agar or liquid media but not in soil. Presumably, it is due to the composition of the soil solution under this high CO_2 pressure.

Whitney and Gardner (19) studied the effect of CO_2 on soil reaction (pH) in 23 calcareous soils. They found that pH is almost a straight line function of the log of the CO_2 pressure from 0.0003 to 1 atmosphere (pH = b log CO_2 + k, where b is the regression coefficient and k a constant). If one uses the data presented by Bradfield (2) for the solubility of $CaCO_3$ in water at 25°C and various CO_2 pressures and applies it to the equation of Whitney and Gardner, the influence of CO_2 pressures on soil pH can be calculated (Table 2).

Bradfield (2) stated that a linear relationship existed between calcium in solution and the cube root of the CO_2 pressure. Equations derived from data in Table 2 were used to calculate the influence of CO_2 pressure on the pH and calcium concentration in soil solutions of HBC (Table 3).

At CO_2 pressures of 0.5 atmosphere, the soil hydrogen ion concentration increases from 6.02×10^{-9} to 1.5×10^{-7} mole/liter under atmospheric conditions. The calcium content of the soil solution would theoretically increase from 0.35×10^{-3} at atmospheric conditions to 7.05×10^{-3} mole/liter under 0.5 atmosphere CO_2 pressure.

Although we have not determined why the fungus does not grow in soil purged with 50% CO_2, it is obvious that many factors could be involved. The drop in pH from 8.22 to 6.82 not only affects the bicarbonate:carbonic acid ratio, but also affects the concentration of these species, since $CaCO_3$ is further dissolved and the concentration of calcium increases nearly 20 fold at pH 6.82.

Similar tests were conducted to determine whether the soil gaseous environment influences the rate of Phymatotrichum root rot development. Eighteen acrylic cylinders (5.7 × 36 cm) were equipped with a gas port at each base. A layer of glass wool was inserted at the bottom of each cylinder and covered with nonsterile HBC (50 g). Sclerotia (1 g) of *P. omnivorum* were spread on the soil layer, and the cylinders were filled with nonsterile HBC and adjusted to field capacity. Three cotton seeds (Watson GL-16) were planted in each cylinder. Sixteen soil cylinders were connected to manifolds attached to cylinders of compressed air or 5% CO_2. Fine-metering valves were adjusted for a flow rate of 15 ml/min. After 12 days, cotton seedlings emerged in all

cylinders. The surface of each cylinder was then sealed with silicone rubber except for a small opening contiguous with the plant. One plant was left in each cylinder. Two cylinders containing sclerotia and plants were used as controls and were not purged with any gas.

Daily measurements were made of plant growth, and observations were made on root and strand growth. All plants grew faster in 5% CO_2 environments than in air (Fig. 1B). The roots were large and white until they became severely damaged by *P. omnivorum*. Strands of the fungus were much larger and reached the surface sooner in the 5% CO_2 treatment than in soil purged with air. There was fungus growth in these air-purged soils, but the fungus did not grow in cylinders without cotton plants. We attributed this to an accumulation of CO_2 in these soils from root and microbial respiration. The rate of soil purging with air was evidently insufficient to remove all the liberated CO_2.

Cotton plants in the two cylinders not flushed with air or CO_2 both died of Phymatotrichum root rot 33 days after planting. In soils flushed with 5% CO_2, the first dead plant was observed 38 days after planting. Three of eight plants were dead in the CO_2 treatment before the first plant in the air treatment succumbed (44 days). All plants eventually succumbed to Phymatotrichum root rot.

Several reports have indicated that elevated CO_2 levels increase root growth of plants. A stimulatory effect of aeration by 5% CO_2 on excised wheat roots was noted by Talbot and Street (18). Roots incubated in air for 14 days were 94 mm long, compared with 121 mm for those incubated in 5% CO_2. Lateral roots were several times more numerous in the 5% CO_2 treatment. These same phenomena were observed with pea roots grown in water culture under various CO_2 pressures (4), and with cotton roots grown in nutrient solution under different CO_2 concentrations (8).

Lundegardh (9) found that 2-8% CO_2 favored the development of root rot of wheat caused by *Gibberella saubinetii* and *Fusarium* spp. The number of infected plants was higher in the elevated CO_2 environments, with the exception of soils infested with *Fusarium culmorum*, in which 50% of the plants died in both treatments (air and CO_2).

TABLE 1. Distribution of ^{14}C radioactivity in mycelial extracts of *Phymatotrichum omnivorum* incubated in a synthetic medium[a] (pH 8.0) containing 250 μCi $NaH^{14}CO_3$, and sampled at 1, 6, 12, and 24 hours

Fraction	Hours			
	1	6	12	24
	cpm/g dry wt			
Water extract				
Organic acids	60,888	291,756	208,642	83,628
Amino acids	18,478	46,467	62,809	16,853
Sugars	3,442	31,846	132,483	7,303
Ether extract	9,783	35,928	41,594	24,898
Starch	130,000	676,000	1,870,000	2,373,000

[a]Medium composition in g/liter distilled water: glucose (4.0), NH_4NO_3 (1.18), K_2HPO_4 (2.98), KH_2PO_4 (0.02), KCl (0.15), $MgSO_4$ (7 H_2O) (0.75). Trace elements: Zn^{+2}, Fe^{+3}, Cu^{+2}, Mo^{+5}, Mn^{+2} (2.5 mg/liter).

SEASONAL FLUCTUATIONS OF CO_2 IN HOUSTON BLACK CLAY.—Field studies were conducted for 3 years at the Blackland Conservation Research Center at Temple, Texas, to monitor seasonal fluctuations in soil gases. Twelve cylinders (2.44×0.96 m) were forced into the soil by a static load of steel plates. Three gas-sampling tubes were forced into the soil beginning 15 cm from the top of each cylinder and continuing every 15 cm for 180 cm. This provided 36 gas-sampling tubes/cylinder, or a total of 432 tubes for the 12

cylinders. A gas sample (1 ml) was removed from each tube and analyzed by gas chromatography for CO_2, N_2, and O_2. The analyses were made at monthly intervals for over 3 years (March 1969 to September 1972).

Similar trends were observed each year for soil CO_2 fluctuations (Fig. 1D). Carbon dioxide reached its highest level at the 30- and 60-cm depths in June and then declined as soil moisture was depleted. The maximum CO_2 concentration at the 90-cm depth was reached in July, and that for the lower sampling depths in August or

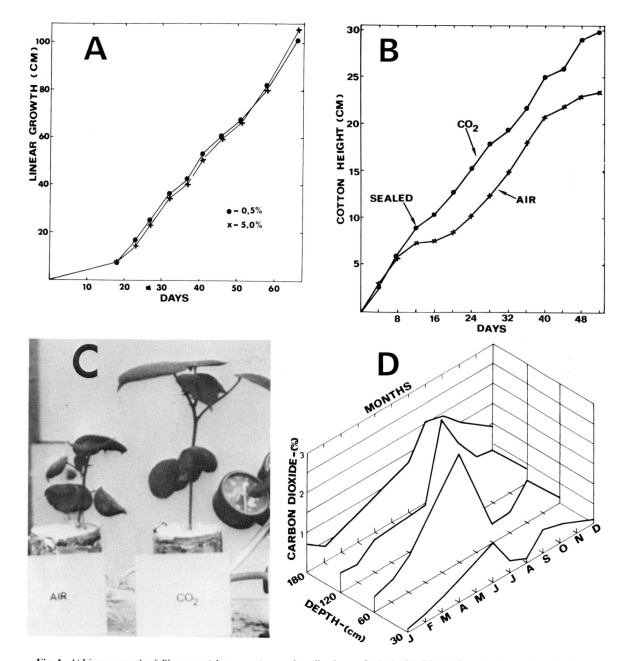

Fig. 1. *A)* Linear growth of *Phymatotrichum omnivorum* in soil columns flushed with CO_2. *B)* Growth of cotton in soil columns flushed with 5% CO_2 or air. *C)* Differential response of cotton to two soil gaseous environments, air and 5% CO_2, 16 days after planting. *D)* Monthly fluctuations of CO_2 at various soil depths in Houston black clay.

September. A rise in CO_2 was noted after rainfalls of sufficient quantity to close the large shrinkage cracks. The blacklands area is a dryland region, and soil moisture depends upon natural rainfall (75-90 cm/yr).

Under field conditions, the maximum level of CO_2 ranged from 3 to 5%, which also was found to be beneficial for growth of *P. omnivorum* through nonsterile soil columns under laboratory conditions. An increased partial pressure of CO_2 would lower the pH and increase the bicarbonate:carbonate ratio in the soil solution. No free carbonate was detectable in the soil solution of HBC using a titrimetric method of analysis and phenolphthalein-methyl orange as end-point indicators.

A soil environment with elevated CO_2 is believed to be conducive for growth of *P. omnivorum*, perhaps through induction of morphological changes (strands) and the suppression of other microbes. Disease expression in cotton was coincident with periods of maximum CO_2 concentrations in the upper 60 cm of soil.

INFLUENCE OF CO_2 ON SURVIVAL OF P. OMNIVORUM.—Phymatotrichum root rot does not occur in sodic soils. The pH of these soils is higher than in calcareous soils, which decreases the bicarbonate:carbonate ratio with increased partial pressures of CO_2.

To study the influence of these changes on sclerotium production, the calcium of HBC was exchanged with sodium. Various proportions of sodium-HBC were added to HBC to give soil mixtures ranging from 100% HBC to 100% sodium-HBC in 10% increments on a w/w basis. Sclerotia were produced in sterilized soil (250 g) covered with sorghum seed (25 g) in Erlenmeyer flasks (500 ml). Water was added to bring the soil to a saturated condition. The flasks were autoclaved at 121°C for 1

hour, and, after cooling, the soil was inoculated with one sorghum seed colonized with mycelium of *P. omnivorum*. Flasks were incubated at 28°C for 8 weeks, after which the sclerotia were recovered from the soil by wet-sieving through a U.S. Standard Sieve Series No. 18 (1-mm opening). There was a 98% reduction in sclerotial yield in the two extremes of soil. Sclerotial oven-dry weights averaged 6.52 g/250 g HBC and 0.12 g in the sodium-saturated HBC. The yield decline was exponential (Fig. 2A). Individual dry weights of sclerotia also decreased from 1.64 mg/sclerotium in HBC to 0.30 mg/sclerotium in 100% sodium-HBC (Fig. 2B).

Another test was made to verify whether the sclerotium-producing capacity of HBC could be restored in 100% sodium-HBC by removal of sodium and resaturation with calcium (Table 4). More sclerotia were produced in HBC resaturated with calcium than in regular HBC; however, the hydrogen ion concentration also increased 1.55 times (from pH 8.32 to 8.03). The bicarbonate content increased slightly in the calcium-resaturated HBC.

In another test, a forest soil (pH 4.77) from Overton, Texas, was used to determine whether *P. omnivorum* would grow and produce sclerotia. The same method as that used for producing sclerotium inoculum in HBC was employed. The fungus grew profusely on the sterilized sorghum seed at the soil surface, but failed to produce strands or sclerotia after 8 weeks. At this pH, CO_2 would be predominantly in the H_2CO_3 form.

DISCUSSION.—Pressures of CO_2 greater than 0.0003 atmosphere affect growth and survival of *P. omnivorum* and the development of Phymatotrichum root rot of cotton. The fungus withstands high pressures of CO_2 (0.5 to 0.75 atmosphere) when cultured in liquid or agar

TABLE 2. The solubility of $CaCO_3$ in water at 25°C and various pressures of CO_2 [after Bradfield (2, p. 9)] and the calculated pH of the soil solution obtained at various CO_2 pressures

$^PCO_2$	log $^PCO_2$	$(^PCO_2)^{1/3}$	$(Ca^{+2}) \times 10^3$	pH
atm	atm	atm	mole/liter	
0.00031	−3.5086	0.0677	0.52	8.30
0.00334	−2.4762	0.1494	1.17	7.62
0.0160	−1.7958	0.2513	2.01	7.18
0.0432	−1.3645	0.3509	2.87	6.90
0.1116	−0.9523	0.4814	4.03	6.64
0.9684	−0.0139	0.9893	8.91	6.05

TABLE 3. Influence of CO_2 pressures on pH of Houston black clay and its predicted influence on the solubility of $CaCO_3$

$^PCO_2$	$(^PCO_2)^{1/3}$	pH		$(Ca^{+2}) \times 10^3$
		Calculated	Observed	
atm	atm			mole/liter
0.0003	0.0669	8.30[a]	8.22	0.35[b]
0.005	0.1710	8.16		1.31
0.05	0.3684	6.87	7.42	3.13
0.5	0.7937	6.22	6.82	7.05

[a]pH = −0.6444 log $(^PCO_2)$ + 6.0291.
[b]$(Ca^{+2}) \times 10^3 = 9.215(^PCO_2)^{1/3} − 0.265$.

media, but mycelial dry weight decreases with CO_2 concentrations greater than 0.25 atmosphere. The fungus grows in soil as mycelial strands, and the rate of strand extension is enhanced by CO_2 pressures of 0.005 to 0.05

atmosphere. The mycelial strands are more durable than individual hyphae, and in the presence of suitable substrate, they form sclerotia.

Sclerotia are found deep in the soil (30-244 cm), where environmental fluctuations are minimized and microbial activity is subdued compared to the 0-30-cm zone. Carbon dioxide increases with soil depth to pressures ranging between 0.01 and 0.03 atmosphere in HBC, depending upon moisture content and sampling site. A high moisture content favors CO_2 accumulation in the soil by sealing cracks and forming a barrier that retards equilibration with the aboveground atmosphere.

Why is Phymatotrichum root rot found in localized spots within a field, and why is it found in some fields within a given region but not in adjacent fields? Perhaps localized environmental conditions within a particular soil type favor production and survival of large sclerotium populations. Gas samples taken from soil at different depths in the field showed large variations from one site to another at similar depths and sampling sites. Apparently, CO_2 accumulates within localized pockets in the soil profile. Such pockets might serve as localized niches for growth and survival of *P. omnivorum*. The fungus probably liberates CO_2 in the soil when colonizing a substrate, which further contributes to a CO_2 environment favorable for sclerotium production. Also, in the vicinity where sclerotia are formed, it might restrict the growth of other soil microbes that are more sensitive to CO_2.

Macauley and Griffin (11) postulated that the HCO_3^- ion plays a part in regulating mycelium production in *Chaetomium* sp., *Cochliobolus spicifer*, *Cochliobolus sativus*, *Curvularia* sp., *Fusarium acuminatum*, *Fusarium oxysporum*, *Gibberella zea*, and *Trichoderma* sp. Mycelium dry weight production was most noticeably affected at high pH values (pH 6 to 7), when the CO_2 pressure was 0.1 to 0.15 atmosphere. More mycelium was found at pH 4 or 5, which supported their hypothesis that HCO_3^- was affecting the growth of some soil fungi.

Phymatotrichum root rot occurs primarily in soils with a pH of 8.0 or slightly higher where dissolved CO_2 is mainly in the form of HCO_3^-. Sclerotia were not produced in soil of pH 4.77, where the predominant form of dissolved CO_2 is in the form of H_2CO_3, although visible hyphal growth developed. The number and size of sclerotia formed at pH 9.0 or greater were reduced where much of the dissolved CO_2 occurred as HCO_3^- and CO_3^{2-}.

We believe that the HCO_3^- ion plays a regulatory role in the formation of mycelial strands and sclerotia. The nature of this role is unknown, but it could be through an effect on gluconeogenesis and glycogen synthesis. The concentration of HCO_3^- in soil solution, and the HCO_3^-

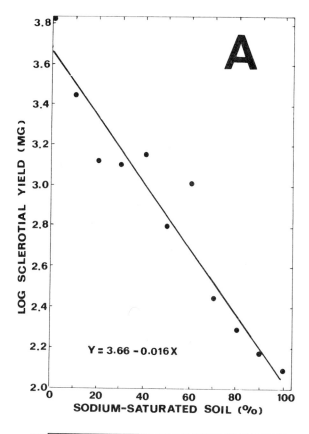

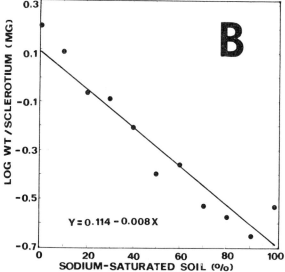

Fig. 2. *A)* Sclerotial yield of *Phymatotrichum omnivorum* in Houston black clay under various levels of sodium saturation. *B)* Dry weight/sclerotium as affected by degree of sodium saturation.

TABLE 4. Sclerotium production by *Phymatotrichum omnivorum* in Houston black clay (HBC), in HBC saturated with sodium chloride, and in sodium-HBC resaturated with calcium chloride after removal of sodium

Soil	Sclerotium production/ 250 g soil	pH
HBC	6.52	8.22
HBC → sodium-HBC	0.12	9.81
Sodium-HBC → calcium-HBC	7.51	8.03

:H_2CO_3 or HCO_3^-:CO_3^{-2} ratios may be critical. Careful manipulation of soils to alter the HCO_3^- concentration or the ratios between its associated and dissociated species may afford a means of reducing the sclerotium population to mitigate Phymatotrichum root rot.

ACKNOWLEDGMENT.—Some of the research presented in this paper was supported in part by Contract Agreement 12-14-100-9339(41) from the USDA Soil and Water Conservation Research Division, and from Cooperative Agreement 70-178 with Cotton Incorporated. The manuscript is a contribution of the Texas Agricultural Experiment Station and the Soil and Water Conservation Research Division, ARS, USDA.

LITERATURE CITED

1. BURGES,A., and E. FENTON. 1953. The effect of carbon dioxide on the growth of certain soil fungi. Trans. Br. Mycol. Soc. 36:104-108.

2. BRADFIELD, R. 1941. Calcium in the soil. I. Physico-chemical reactions. Proc. Soil Sci. Soc. Amer. 6:8-15.

3. FELLOWS, H. 1928. The influence of oxygen and carbon dioxide on the growth of Ophiobolus graminis in pure culture. J. Agr. Res. 37:349-355.

4. GEISLER, G. 1963. Morphogenetic influence of (CO_2 + HCO_3^-) on roots. Plant Physiol. 38:77-80.

5. GUNASEKARAN, M. 1970. Physiological studies on Phymatotrichum omnivorum (Shear) Duggar (with special reference to glycogen metabolism). Ph.D. dissertation, Tex. A&M Univ., College Station. 112 p.

6. GUNASEKARAN, M. 1972. Physiological studies on Phymatotrichum omnivorum. I. Pathways of glucose catabolism. Arch. Mikrobiol. 83:328-331.

7. KUNZE, G. W., and E. H. TEMPLIN. 1956. Houston black clay, the type Grumusol. II. Mineralogical and chemical characterization. Proc. Soil Sci. Soc. Amer. 1956:91-96.

8. LEONARD, O. A., and J. A. PINCKARD. 1946. Effect of various oxygen and carbon dioxide concentrations on cotton root development. Plant Physiol. 21:18-36.

9. LUNDEGARDH, V. H. 1923. Die Bedeutung des Kohlensauregehalts und der Wasserstoffionkonzentration des Bodens für die Entstenhung der Fusariosen. Bot. Notiser 1923:25-52.

10. LYDA, S. D., and E. BURNETT. 1971. Changes in carbon dioxide levels during sclerotial formation by Phymatotrichum omnivorum. Phytopathology 61:858-861.

11. MACAULEY, B. J., and D. M. GRIFFIN. 1969. Effect of carbon dioxide and the bicarbonate ion on the growth of some soil fungi. Trans. Br. Mycol. Soc. 53:223-228.

12. NEAL, D. C., and R. E. WESTER. 1932. Effects of anaerobic conditions on the growth of the cotton-root-rot fungus, Phymatotrichum omnivorum. Phytopathology 22:917-920.

13. PAMMEL, L. H. 1889. Cotton root-rot. Tex. Agr. Exp. Stn. Annu. Rept. 2:61-86.

14. RAABE, R. D., and A. H. GOLD. 1967. Effects of different levels of carbon dioxide on the growth of Armillaria mellea in culture. Phytopathology 57:101.

15. ROGERS, C. H. 1942. Cotton root-rot studies with special reference to sclerotia, cover crops, rotations, tillage, seeding rates, soil fungicides, and effects on seed quality. Tex. Agr. Exp. Stn. Bull. 614. 45 p.

16. SHEAR, C. L., and G. F. MILES. 1907. The control of Texas root-rot of cotton. USDA Plant Ind. Bull. 261. p. 39-42.

17. TABAK, H. H., and W. B. COOKE. 1968. The effects of gaseous environments on the growth and metabolism of fungi. Bot. Rev. 34:126-252.

18. TALBOT, B., and H. E. STREET. 1968. Studies on the growth in culture of excised wheat roots. VI. Influence of carbon dioxide on growth and branching. Physiol. Plant. 21:800-805.

19. WHITNEY, R. S., and R. GARDNER. 1943. The effect of carbon dioxide on soil reaction. Soil Sci. 55:127-141.

The Influence of Food Base on Rhizomorph Growth and Pathogenicity of Armillaria mellea *Isolates*

D. B. REDFERN—*Forestry Commission, Northern Research Station, Roslin, Midlothian, Scotland.*

Armillaria mellea (Vahl ex Fr.) Kummer is a common cause of death in young conifer plantations established on sites formerly occupied either by broad-leaved forest or by a mixture of coniferous and broad-leaved species. Killing attacks are almost invariably associated with stumps of broad-leaved trees. Peace (7) stated that *A. mellea* is essentially a fungus of areas with a hardwood history, and Day (2), in 1929, wrote that "in natural or semi-natural forest, it [*A. mellea*] seems to be confined to broad-leaved forest, pure or mixed with conifers. In pure natural conifer forest it seems at least to be rare, but in pure conifer plantations established on the site of old broad-leaved forest it is extremely common at least during the first rotation."

Stumps of conifers left after thinning are readily invaded by *A. mellea* (5), but they are generally thought to be of little or no significance in sustaining attacks. Peace (7) believed that the fungus acts in a purely saprophytic role in this situation and that there is no increase in parasitic activity in the stand. He concluded that where conifers completely replace hardwoods, *A. mellea* damage is absent or much reduced in the second conifer rotation. Greig (4) also found evidence from a study of conifer crops on former hardwood sites in Britain indicating that "the vigour of *A. mellea* appears to decline" as the hardwood stumps disintegrate. The bulk of field observations suggest that hardwood stumps are superior to conifer stumps as food bases for *A. mellea*.

An alternative hypothesis was put forward by Childs and Zeller (1) when they observed that *A. mellea* attacks were serious in apple orchards planted on land cleared of oak (*Quercus garyana* Dougl.) but rare on former Douglas fir [*Pseudotsuga menziesii* (Mirb.) Franco] land, although both site types were heavily infested with rhizomorphs. The authors attributed this difference to strains of varying pathogenicity, one on oak and the other on fir; however, it could also be associated with food base differences.

Although the overwhelming majority of killing attacks are associated with hardwood stumps, severe attacks can occur where the previous crop consisted entirely of coniferous species. In the United States, Weisse and Riffle (14) recorded killing in *Pinus ponderosa* following a previous crop of the same species, and Swift (13), in Rhodesia, reported killing in *Pinus elliottii* planted as a second rotation of conifers after 35-year-old *Cupressus lusitanica* had been felled on a formerly indigenous hardwood forest site. Three similar attacks investigated in the present study are described later. These examples are rare exceptions, but they do raise the question of what *A. mellea* activity can be expected in the future in successive conifer rotations on sites where killing is now occurring in the first crop of conifers after hardwoods. Can killing be expected to cease or will it continue in later rotations?

In experiments to determine the value of the food base provided for *A. mellea* by the roots of various tree species, Redfern (10) found that segments of conifer roots inoculated with *A. mellea* produced approximately the same total length and dry weight of rhizomorphs as segments of hardwood roots, but the latter produced significantly more rhizomorph systems and growing tips than the former. By contrast, Morrison (6), who conducted similar laboratory experiments using a greater number of isolates, found that, with the exception of one isolate, greater dry weights of rhizomorphs were produced from hardwood root segments than from conifer segments. When stumps were inoculated in the field, Morrison found that the number of rhizomorph systems as well as the total length of rhizomorphs per stump was greater for hardwood stumps than for conifer stumps. In a similar study, which also included measurement, in situ of rhizomorph production by naturally infected stumps, Rishbeth (12) concluded that pine is inferior to oak as a substrate for *A. mellea* in terms of the number and weight of rhizomorphs produced.

However, the success of a specialized root-infecting fungus [sensu, Garrett (3, p. 21)] largely depends on its parasitic ability. And, although with *A. mellea* there is evidence to show that isolates differ greatly in both pathogenicity (8, 15) and in ability to form rhizomorphs in vivo (6, 12), there is no evidence that these characteristics are related. For this reason, experiments were performed to assess the influence of food base on pathogenicity as well as on rhizomorph production.

EXPERIMENT 1.—Norway spruce [*Picea abies* (L.) Karst.] and Douglas fir, which were chosen as examples of susceptible and resistant hosts, respectively (9), were inoculated in a greenhouse with one isolate of *A. mellea* growing on root segments of Red spruce (*Picea rubens* Sarg.) and Red maple (*Acer rubrum* L.). Sixteen plots, 60 × 60 cm and containing soil to a depth of 20 cm, were formed by subdividing a greenhouse bench. Three randomized blocks of four treatments occupied 12 plots in the center of the bench; i.e., there were two hosts and two food bases in factorial combination, and two plots at either end acted as one control plot for each food base/species combination. Forty-nine trees (15-25 cm tall) were planted in each plot at 7.5-cm spacing. Freshly excised spruce and maple root segments (2.5-3.0 cm diameter, 6.0 cm long) were infected with *A. mellea* in the manner described below for experiment 2. Eighteen segments of one species were buried in a regular pattern about 1-2 cm deep in each plot, and a 2.5-cm layer of sphagnum peat was spread on top of the soil. Deaths of trees killed by *A. mellea* were recorded, and after 8 months, rhizomorphs and root systems were washed from the soil. Roots were then examined for infections, and rhizomorphs were oven-dried and weighed.

Food base had no significant effect on either the percentage of trees infected or on rhizomorph production, but host species had a significant ($P < 0.05$) effect on both (Table 1). Norway spruce was more susceptible than Douglas fir, but, surprisingly, it also induced the formation of a greater dry weight of rhizomorphs. The most probable explanation is that dead Norway spruce acted as foci for extra rhizomorph growth from the inocula by enabling rhizomorphs to grow into the peat around their root collars; peat is known to greatly stimulate rhizomorph growth (11). It is unlikely that an effect of this magnitude could be accounted for by the additional food base provided by dead trees (a mean of 8.5/plot) because the volume of their root systems would have been insignificant compared to the mean volume of inoculum segments per plot (831 cm³).

In the experiment just described, differences between the two food bases were great enough (although not quite significant) to suggest they might be real and could be detected in a larger experiment. In a second experiment to test for differences, Sitka spruce plants [*Picea sitchensis* (Bong.) Carr.] were inoculated with four isolates of *A. mellea* growing on root segments of Scots pine and sycamore.

EXPERIMENT 2.—In April 1971, 25 Sitka spruce plants, 10-20 cm tall, were planted in each of 96 plastic buckets (30 cm diameter, 25 cm deep) that were provided with drainage holes and filled with a sandy loam soil of pH 6.7. Buckets were sunk in soil outdoors in 12 groups of eight; i.e., in a factorial combination of four isolates of *A. mellea* and two food bases replicated 12 times.

Inoculum preparation was a two-stage process. First, segments of sycamore (*Acer pseudoplatanus* L.) stem (3 cm long, 3-7 cm diameter) were placed in 2-liter wide-necked flasks (20/flask), and a mixture of 400 g sand and 12 g maize meal was poured on top, followed by 100 ml water. After autoclaving, flasks were inoculated with one of four *A. mellea* isolates by pipetting 10 ml of a suspension formed by macerating 3-week-old cultures in sterile water. Flasks were incubated at room temperature in the dark for 3 months. In the second stage, these stem segments were used to infect 10-cm lengths of freshly excised and living roots of Scots pine (*Pinus sylvestris* L.) and sycamore, 2.5-5.5 cm in diameter, by attaching one

TABLE 1. Rhizomorph production and pathogenicity of *Armillaria mellea* to *Picea abies* and *Pseudotsuga menziesii* when growing on root segments of *Picea rubens* and *Acer rubrum*

Food base	Trees infected, %		Dry weight of rhizomorphs, g	
	P. abies	*P. menziesii*	*P. abies*	*P. menziesii*
P. rubens	12.2	3.4	2.23	1.49
A. rubrum	22.4	10.2	2.49	1.68
Mean	17.3	6.8	2.36	1.59

TABLE 2. Rhizomorph production and pathogenicity (to *Picea sitchensis*) of four isolates (1-4) of *Armillaria mellea* growing on root segments of *Acer pseudoplatanus* (S) and *Pinus sylvestris* (P)

Isolate of *A. mellea* and substrate	Infected, %			Rhizomorph dry weight, g
	Dead	Living	Dead + living	
S1	46.5 b[a]	30.9 a	77.4 a	2.06 c
P1	19.9 c	21.9 a	41.9 b	0.76 e
S2	1.7 e	5.3 b	7.0 c	2.03 c
P2	1.0 e	7.0 b	8.0 c	2.74 b
S3	2.0 e	5.0 b	7.0 c	4.13 a
P3	8.0 d	10.0 b	17.9 c	1.21 d
S4	61.3 a	21.3 a	82.7 a	1.18 d
P4	49.2 b	27.1 a	76.3 a	0.68 e
			LSD 0.05 = 0.40	

[a]In each column, figures indicated by dissimilar letters are significantly different at the 5% level. Least significant differences are not given for the percentages of trees dead, living, and dead + living because the analysis was performed on data transformed by angles.

segment of appropriate diameter firmly to the end of each root length with two rubber bands. At this time, the five segments destined for each treatment plot were selected in such a way from the range of diameters available that the total volume in each plot would be equal. Joined segments were incubated in moist nonsterile sand. When rhizomorphs emerged from the noninoculated ends (in 8 weeks), infection was judged to be complete. The joined segments were then separated, and the newly infected root segments were ready for use. They were prepared in this manner because rhizomorph production from living root segments infected by *A. mellea* is superior to that from autoclaved ones colonized under sterile conditions (10).

The previously grouped five inocula of each isolate/food base combination were allocated at random to one of the eight plots in each block and in July 1971 were buried 1-2 cm below the soil surface in a standard configuration. A 2.5-cm layer of sphagnum peat was added to the soil surface to maintain soil moisture. Buckets were watered during rare periods of hot dry weather. Plants that exhibited various foliage symptoms and that were also girdled at the root collar by mycelial fans of *A. mellea* were recorded as killed by *A. mellea*. Resinosis of stem bases was a common but not invariable symptom.

Eighteen months after inoculation, rhizomorphs, tree roots, and inoculum segments in each plot were washed free of soil. Roots of living trees were examined for rhizomorph contacts and infections. Rhizomorphs were collected, freed of any adhering soil particles with a camel's hair brush, oven-dried at 95° C, and weighed. Inoculum segments were scrubbed, and the volume of each was determined by weighing in water. The degree to which inocula were colonized by *A. mellea* was determined visually after splitting. This assessment was checked by observing the formation of characteristic light brown-pink mycelial tufts on 40 split inocula from one replicate that had been incubated in a moist chamber at room temperature for 7-10 days. Finally, inoculum segments were oven-dried at 105° C and weighed.

Variation in pathogenicity.—Differences in pathogenicity among isolates were striking (Table 2). When growing on sycamore, isolates 1 and 4 were highly pathogenic, whereas isolates 2 and 3 were almost nonpathogenic. Disease escape due to rhizomorph growth habit was not a factor for isolates 2 and 3 because rhizomorph contacts that had not resulted in infection were found on up to half the trees in some plots. These relative differences were maintained with pine as a food base, except that isolate 3 killed more trees than did isolate 2. Food base effects were also apparent; isolates 1 and 4 killed significantly fewer trees with pine as a food base than with sycamore. For isolate 3, the reverse was true. Food base had no significant effect on isolate 2.

Variation in rhizomorph production.—On sycamore, isolate 3 produced more rhizomorphs than the other three isolates, whereas on pine, isolate 2 was outstanding and isolate 3 was in second place. Comparing food bases, rhizomorph production was significantly less for isolates 1, 3, and 4 on pine than it was on sycamore, but for isolate 2, pine was better than sycamore.

In previous experiments comparing rhizomorph production from root segments of equal volume (10), rhizomorph weights were corrected for differences in initial density of the roots used as food bases. In this experiment, differences were so small that correction was considered unnecessary: the mean volume of root segments per plot was 553 cm³ for sycamore and 558 cm³ for pine, and the initial densities of roots of the two species were 0.41 g cm⁻³ and 0.43 g cm⁻³, respectively.

Relationship between rhizomorph production and pathogenicity.—Regression analysis showed that for the four isolates together there was a highly significant negative correlation between percentage of trees killed and dry weight of rhizomorphs produced for both substrates: r = −0.564 for sycamore and −0.488 for pine, P < 0.001 in both cases. At the same time, strong reservations must be maintained about such a relationship because only four isolates were used and they fell into two distinct groups: isolates 2 and 3 were almost nonpathogenic and, except for isolate 2 growing on sycamore roots, produced a significantly greater dry weight of rhizomorphs than the highly pathogenic isolates 1 and 4.

Considering each isolate/food base combination individually, there was a significant positive correlation between percentage of trees killed and weight of rhizomorphs produced for isolate 1 on both substrates (r = 0.584 for sycamore, P < 0.05, and 0.831 for pine, P < 0.001). There were no significant correlations for other isolates. This relationship may be connected with the group killing that occurred with both pathogenic isolates (1 and 4) but which was much more marked with isolate 1. Group killing accounted for a sharp increase in the rate of killing by isolate 1 in the 2nd year after inoculation (no killing occurred until week 40), whereas killing by isolate 4, which began 15 weeks after inoculation, progressed more slowly. At the end of the experiment, dichotomously branched rhizomorph systems, similar to those already described from peat by Redfern (11), were found within groups of dead trees in isolate 1 plots. Therefore, as mentioned above for experiment 1, it is possible that increased rhizomorph growth and killing in some plots was caused by rhizomorphs entering the peat, either directly or as new ones formed from the root collars of dead trees, and rapidly killing adjacent trees. There are several reasons why it is unlikely that this relationship was due simply to the fact that variation between inocula and variation in soil conditions resulted in differences in rhizomorph production between plots and consequential differences in killing. The reasons are: there were five inocula per plot, soil was carefully sieved and mixed, and between-plot variation in rhizomorph production for the other three isolates on both food bases was much smaller than for isolate 1.

Loss in density of root inoculum segments.—Isolates 1-4 reduced the density (g cm⁻³) of sycamore roots to 0.27 (34.5%), 0.25 (39.3%), 0.25 (38.8%), and 0.27 (35.7%), respectively. Figures in parentheses are density reductions expressed as a percentage of initial density. The same figures for pine were: 0.32 (26.7%), 0.28 (36.1%), 0.31 (28.3%), and 0.28 (34.6%). Differences were probably due entirely to *A. mellea*, since replacement by other fungi was insignificant. Density reductions reflect rhizomorph production to some extent, but clearly (Table 2) some isolates are more efficient than others; e.g., isolate 3 versus 2 on sycamore.

Isolate characteristics.—Isolates 1 and 2 were selected as "conifer isolates," and, although differing greatly in pathogenicity and ability to produce rhizomorphs, both came from the same plantation, a 40-year-old crop of Scots pine in South Scotland in which killing was associated with pine stumps. Isolate 1 fruited during the experiment but only on sycamore root segments. Isolate 3 came from an isolated group of mixed conifers and broad-leaved trees in open moorland in Scotland, and isolate 4, selected as a "hardwood isolate," came from an ash stump (*Fraxinus* sp.) at the Cambridge Botanic Garden, England.

Morrison (6) described two main types of rhizomorph growth habit in soil that produced two types of rhizomorph system: Type I, formed by monopodial branching, and Type II, formed by dichotomous branching. Type II was divided into two subgroups (a and b); group a systems were more fragile and less frequently branched than those of group b.

In this experiment, some difficulty was experienced in separating rhizomorph systems that were reasonably intact from the tangle of root systems. However, isolates 2 and 3 clearly corresponded to Morrison's Type I, whereas 1 and 4 were of Type II. Rhizomorphs of isolate 4 were notably fragile, and this isolate probably corresponded to Type IIa. The pathogenicity results reported above confirm Morrison's findings that Type II isolates are more pathogenic than those of Type I.

FIELD OBSERVATIONS.—In East Scotland, a survey of one of several groups of trees killed by *A. mellea* in a 34-year-old crop of Scots pine revealed that in an area 25 × 25 m, 17 stumps of the previous crop were conifers and only two were hardwoods (*Quercus* sp.). Only the heartwood of these stumps remained, and the stumps appeared to be exhausted as food bases. The area also included 45 trees of the present crop killed by *A. mellea* and 44 infected thinning stumps. Similar results were obtained in another Scots pine crop, approximately 40 years old, in South Scotland (mentioned above). Except for rare birch stumps (*Betula* sp.), the previous crop had consisted entirely of large conifers. Trees were currently being killed, and thinning stumps were heavily infected. Finally, in an example from Wales, 34% mortality was recorded by 1971 in experimental plots of Sitka spruce planted in 1960 on a site with two previous conifer rotations, the first of mixed conifers and the second of Sitka spruce planted in 1929.

Hardwood stumps, although rare, were not completely absent from any of the three sites just described. Living hardwood trees could also be found on or near all sites. Although infection may have originated in these hardwood trees or stumps, they were so rare it is unlikely they made any significant contribution to the nutriton of *A. mellea* in the current rotation. Even in the case quoted by Swift (13) and mentioned above, it is unlikely that 35-year-old hardwood stumps would constitute a significant food base compared with freshly invaded conifer stumps.

DISCUSSION.—The great variation in pathogenicity shown by isolates of *A. mellea*, particularly between isolates taken from within 100 m of each other on the same site (isolates 1 and 2), may be as important as food base effects when we consider the possibility of continued *A. mellea* killing in succeeding conifer rotations on infested sites.

If *A. mellea* exists as a mosaic of clones in natural broad-leaved forests, establishment of susceptible conifer plantations might create a situation in which pathogenic clones could become dominant, whereas they would formerly have been kept in check by the very high resistance of broad-leaved trees. Any broad-leaved trees that became moribund would be colonized by the clone forming the rhizomorph network in contact with its roots. Conversion of broad-leaved forests to conifer plantations may therefore favor pathogenic strains as well as present the fungus with a different food base.

The relationship between rhizomorph growth habit and pathogenicity discovered by Morrison (6) and confirmed by the results described above provides a useful method (either directly, or indirectly through some other correlation) of screening isolates in a study of clone distribution.

Morrison (6) and Rishbeth (12) both concluded that the association of hardwood food bases with killing by *A. mellea* can be attributed to two factors: the superiority of hardwood roots over conifer roots as food bases for rhizomorph production, and the fact that stumps of those broad-leaved trees in which resistance to infection is maintained by regrowth after cutting (resulting in a variable delay in colonization) are less quickly exhausted as food bases than are conifer stumps that die rapidly after felling. Results reported here broadly support the first conclusion and amplify it to the extent that for some isolates sycamore roots were superior to pine in terms of pathogenicity as well as rhizomorph production.

However, in addition to those two factors, a third, site history, may be involved in the association of *A. mellea* killing with hardwood stumps alone. An increase in world demand for coniferous timber and changes in the economics of timber production have resulted in conversion of broad-leaved forests to conifer plantations during the last 50 years. Attention has become focused on killing attacks in young plantations of susceptible conifers and, therefore, on the hardwood food bases with which this killing is necessarily associated. There is certainly good evidence to show that, stump for stump, hardwood stumps are indeed superior to conifer stumps as food bases. But how significant is the difference? How much less effective are conifer stumps?

Examples quoted previously indicate that killing attacks can develop in association with conifer stumps and that they may be sustained by thinning stumps and perhaps to some extent by tree-to-tree spread as indicated by expanding groups of dead trees. In the experiments described above, although sycamore roots provided a somewhat superior food base to pine roots for rhizomorph production and pathogenicity, pine was still quite effective, and for one isolate it was superior to sycamore. Even the fact that conifer stumps are effective as food bases for a shorter time than hardwood stumps may be counterbalanced in a conifer crop by the short rotations generally adopted (40-50 years) and the 5-year thinning interval that will ensure a steady supply of new food bases. Thus, although there is little doubt that hardwood stumps provide the best food base for *A. mellea*, conifer stumps are probably much more effective than has hitherto been realized. There is certainly sufficient evidence to indicate that *A. mellea* killing may continue even after successive rotations of conifers, rather than decline following disintegration of the hardwood stumps in which infection originated.

As a corollary, mixed crops of broad-leaved and coniferous species may well provide maximum opportunity for *A. mellea* to perpetuate itself. Inasmuch as conifer stumps are more quickly colonized than hardwood stumps, felling and thinning would provide a mixture of stumps available for colonization over a more extended period of time (ensuring a more uniform supply of food bases) than if the crop consisted entirely of one species. The effects of substrate on different *A. mellea* isolates suggest that mixed crops might also maintain a more diverse population of *A. mellea* clones than a monoculture of either a broad-leaved or coniferous species.

ACKNOWLEDGEMENTS.—Experiment 1 described in this paper was carried out at the Forest Research Laboratory, Fredericton, New Brunswick, Canada. I am grateful to J. Rishbeth, Botany School, Cambridge, England, for providing one isolate of *A. mellea* used in this work. I also wish to express my gratitude to D. H. Stewart for statistical analyses, to S. Harris for technical assistance, and to S. C. Gregory for helpful discussions.

LITERATURE CITED

1. CHILDS, L., and S. M. ZELLER. 1929. Observations on Armillaria root rot of orchard trees. Phytopathology 19:869-873.
2. DAY, W. R. 1929. Environment and disease. A discussion on the parasitism of Armillaria mellea (Vahl) Fr. Forestry 3:94-103.
3. GARRETT, S. D. 1970. Pathogenic root-infecting fungi. Cambridge Univ. Press, London. 294 p.
4. GREIG, B. J. W. 1962. Fomes annosus (Fr.) Cke and other root-rotting fungi on conifers on ex-hardwood sites. Forestry 35:164-182.
5. LOW, J. D., and R. J. GLADMAN. 1962. Present day research on Fomes annosus in Britain by the Forestry Commission, p. 56-65. IUFRO Sect. 24, conf. and study tour on Fomes annosus, Scotland, June 1960.
6. MORRISON, D. J. 1972. Studies on the biology of Armillaria mellea. Ph.D. thesis, Univ. Cambridge. 168 p.
7. PEACE, T. R. 1962. Pathology of trees and shrubs. Oxford Univ. Press. 753 p.
8. RAABE, R. D. 1967. Variation in pathogenicity and virulence in Armillaria mellea. Phytopathology 57:73-75.
9. REDFERN, D. B. 1966. Root infection by Armillaria mellea. Ph.D. thesis, Univ. Cambridge. 136 p.
10. REDFERN, D. B. 1970. The ecology of Armillaria mellea: rhizomorph growth through soil, p. 147-149. In T. A. Toussoun et al. (ed.), Root diseases and soil-borne pathogens. Univ. Calif. Press, Berkeley and Los Angeles.
11. REDFERN, D. B. 1973. Growth and behaviour of Armillaria mellea rhizomorphs in soil. Trans. Br. Mycol. Soc. 61:569-581.
12. RISHBETH, J. 1972. The production of rhizomorphs by Armillaria mellea from stumps. Eur. J. Forest Pathol. 2:193-205.
13. SWIFT, M. J. 1972. The ecology of Armillaria mellea Vahl (ex Fries) in the indigenous and exotic woodlands of Rhodesia. Forestry 45:67-86.
14. WEISSE, M. J., and J. W. RIFFLE. 1971. Armillaria root rot in a Ponderosa pine plantation in New Mexico. Plant Dis. Reptr. 55:823-824.
15. WILBUR, W., D. E. MUNNECKE, and E. F. DARLEY. 1972. Seasonal development of Armillaria root rot of peach as influenced by fungal isolates. Phytopathology 62:567-570.

PART IV

CROP RESIDUES AND AMENDMENTS IN RELATION TO SURVIVAL AND CONTROL OF ROOT-INFECTING FUNGI

Crop Residues and Amendments in Relation to Survival and Control of Root-Infecting Fungi: An Introduction

G. C. PAPAVIZAS—*Research Microbiologist, Soilborne Diseases Laboratory, Agricultural Research Service, U.S. Department of Agriculture, Beltsville, Maryland.*

An impressive amount of experimental work has been devoted to crop residues and soil amendments in relation to soil-borne plant pathogens. This work has suggested that the direct or indirect use of amendments may give rise to some sort of biological or cultural control of root-infecting fungi. This body of research has been concerned with several different aspects of the interrelationship between soil amendments and plant pathogens: the effects of amendments on the saprophytic activities of the pathogens in soil; the mechanisms of biological control and the effects of amendments on the inoculum density of the pathogens; stimulatory and suppressive substances, both volatile and nonvolatile, associated with amendment decomposition; the replenishment of food bases in soil by amendments; the carbon-to-nitrogen balance of amendments and its implications in root disease suppression; and the influence of amendments on soil fungistasis. Considerable knowledge has been accumulated in these areas and more, I am certain, will come in the future.

Despite this high caliber research of worldwide scope and significance, plant pathologists and soil microbiologists are keenly aware of the shortcomings inherent in our research with amendments and the complexity of the problems stemming from the use of soil amendments. Research difficulties, of course, are not new to plant pathologists performing research on root-infecting fungi or to those working in other areas of plant pathology. Because of frustrations, discouragement, and failures, many scientists have not continued their investigations with amendments. The lack of papers on the subject during the past several years has made this attitude apparent.

One major difficulty with our amendment research may be our haste in applying research findings before the research has been adequately completed. Also, if some of this research had been done in the field rather than in the laboratory or greenhouse, results might have been more useful. It is in the field that we encounter the most difficulties with amendment additions. This drawback is particularly unfortunate because farmers are more interested in field results than in basic research accomplishments. Some researchers believe amendments and crop residues usually fail to provide some degree of control in the field, even though they may be effective in the greenhouse. Erratic results are often observed in the field, even when an amendment has proved effective in the laboratory and greenhouse. Moreover, amendments are effective at such high rates that normally 5-10 tons of dry materials/acre would be needed to obtain some degree of control. In mechanized farming, these rates would be prohibitive or economically unfeasible. Also, amendments and crop residues may occasionally be toxic to plants, especially under wet, cool conditions. Despite these shortcomings, amendments and crop residues can and should be one of our best weapons for alleviating soil-borne plant diseases without seriously polluting the soil environment with pesticides.

Before amendments can be effectively applied in a disease situation, however, plant pathologists and soil microbiologists must continue research in this area. Special consideration should be given to minimal amounts of an amendment needed to suppress root disease fungi, to the mechanisms and spectra of their action, and to the precise effects of amendments on various pathogens. With additional research, methods can be devised to control root-infecting fungi with minimal amounts of amendments or to alleviate root disease problems with amendments used as an integrated part of an economical, feasible disease control program. We hope this session of the symposium will provide impetus for researchers to learn more about the significance and function of crop residues and soil amendments.

Systems and Mechanisms of Residue Possession by Pioneer Fungal Colonists

G. W. BRUEHL—*Department of Plant Pathology, Washington State University, Pullman.*

Residue will be considered primarily as potential food (energy and raw material), rather than as a physical or spatial support. Residue is often considered dead tissue, but this is not necessarily so. When the tree is cut, the stump and roots are residue. When the leaves are stripped from sugarcane, or the young top is felled, these are residues. When a papaya or pineapple planting is destroyed, the mass of succulent green leaves is residue. The roots of all the above plants are alive at the time the planting is destroyed. In contrast, the vegetative structures of many annual crops (wheat, oats, barley, corn) are usually dead or practically so at the time of harvest. The vitality of the discarded plant parts influences the ability of microorganisms to invade and possess them. In general, the greater the vitality of the tissue, the greater is the exclusion of saprophytes.

Possessing the residue means controlling the food within it, but only that part of the food which the organism in question is enzymatically capable of utilizing. Once the residue has no more food potential for that organism, that residue essentially ceases to exist. Thus, in a natural sequence of residue destruction, the sequence of organisms invading the residue may represent little direct competition to each other.

SYSTEMS OF FOOD (RESIDUE) MANAGEMENT.

Pathogenic fungi are diverse; they provide for the future in many ways. Each of several life patterns has certain weaknesses and strengths in respect to survival. Most soil-borne pathogens cause endemic rather than epidemic or sporadic diseases. This implies some level of stability or longevity of inoculum. Before discussing our subject, let us consider *Puccinia graminis* as it develops in the northern Great Plains of North America.

For survival, a pack of wolves depends upon its ability to discover and attack a moving prey. It has little means for storing food or preserving it for the future. Urediospores of *P. graminis* are plant wolves. The spores are produced in abundance. They are disseminated aerially. Survival depends upon chance encounters with a suscept under proper conditions. There is no secure storage of food for any period of time, and populations fluctuate widely. Pathogens of this type have practically no possession of residues or control of food reserves. Avoiding this vulnerability is a major problem faced by the less mobile soil-borne organisms. Typical soil fungi are less adapted to ride the winds; they are more limited spatially, and, as Park (38) states, they play "the waiting game". Time partially substitutes for space. Whereas motility, chemotaxis, and growth through soil are significant for some fungi, usually the root or residue comes to the propagule rather than the fungus going to the potential food.

A few systems of food conservation (residue possession in the sense of this paper) will be contrasted. In the broad sense, energy can be under passive possession (stored within resting structures) (36), under active possession (undigested substrate controlled by fungal hyphae with the competitors excluded) (36), or under a combination of the two.

Passive possession.—Delicate host tissues are most effectively possessed by fungi that invade thoroughly, digest extensively, store the surplus food in resting structures, and then abandon the fragile, exhausted host remains to scavengers. *Pythium* spp. that attack succulent roots and *Typhula* spp. that rot wheat leaves exemplify this type of system. These fungi produce delicate hyphae that die quickly; they make little or no attempt at longevity by virtue of active or dormant hyphae within the host residue. They play the waiting game with long-lived oospores or sclerotia. Host remains are not fouled by potent antibiotics. Indeed, destruction of remnant tissues by scavengers probably benefits the pathogen, as it may aid in dispersing the oospores or sclerotia. The exploited tissue is not worth defending anyway.

Some will argue that passive possession as described is not possession of residues. These organisms are pioneer colonists, being there first; they possess the residue more surely by hoarding within resting structures than by persisting as active hyphae and contesting the remains with all challengers. The hit-and-run (exploit and hoard) system contributes greatly to the success of many widely distributed soil-borne pathogens (*Pythium, Phytophthora, Typhula*).

Active possession.—The organism invades the substrate, usually as a parasite rather than as a saprophyte, becomes established in some tissue in which it persists, and is metabolically active within the dead host remains. Active possession relies upon slow utilization of the substrate and upon persistent defense of it against competitors. The active possessor does not retreat into a dormant structure.

Cephalosporium gramineum is a xylem parasite of winter wheat. It invades the host systemically, produces no known specialized resting structures, and depends for possession upon its prior colonization of invaded tissues and upon continued production of small quantities of a wide-spectrum antifungal antibiotic. Its hyphae are tough and persistent. The substrate is utilized sparingly. When the substrate is gone, the fungus dies.

Unlike *Fusarium oxysporum* and *Verticillium dahliae*,

C. gramineum produces no true chlamydospores or microsclerotia. It persists in straw in moist, natural soil by remaining metabolically active and by constantly producing low levels of a broad-spectrum, primarily antifungal, antibiotic (8). When the straw no longer serves as food, the fungus dies. It is controlled by short rotations. *Ophiobolus graminis, Cercosporella herpotrichoides,* and *Fusarium graminearum* have some characteristics of the active possessor category.

Combination possession.—The parenchyma-blighting, chlamydospore-forming fusaria (*F. oxysporum, F. solani, F. culmorum*) are the best examples of fungi that combine the advantages of both passive and active possession. *Fusarium culmorum* thoroughly invades the basal portions of wheat stems. During moist, moderate weather, macroconidia produced aerially on the surface of the wheat stubble are washed into the soil and converted to chlamydospores. The fungus in the parenchyma does not rapidly exhaust or destroy the straw, at least not in central Washington, where pure cultures can be obtained from such tissues after they have been 2 years in the field (11). A few chlamydospores form in the tissues, but active possession of the straw (exclusion of other fungi) depends upon active hyphae within the residue, not upon dormant chlamydospores.

Apparently the system of food management (residue possession) is highly significant in determining many characteristics of soil-borne fungal pathogens. In herbaceous, short-lived plants under cultivation, those fungi that exploit and hoard (passive possession) are widespread and usually difficult to control. The use of finite propagules (chlamydospores, oospores, sclerotia) has an advantage in that the food is contained in many propagules; it is physically finely divided. Being divided more than in active possession, in which the root or stem piece remains intact, achieves a greater spatial distribution with little or no sacrifice in longevity. Increased spatial distribution increases encounters between propagules and host structures. Increased target encounter is critical when the stage of susceptibility is short (e.g., when limited to germination of the seed and juvenile stages of the seedling in seedling blights). In contrast, finely divided food reserves in finite resting structures are less common among pathogens of woody tissues. Here we have little need for quick response because of the length of time suscept tissues are available and, accordingly, there is less advantage in being in a finely divided state. In fact, Garrett (21, p. 196) emphasizes that pathogens attacking woody structures require a large food base or high inoculum potential, and these conditions are best met by active possession of large woody structures (e.g., infested roots and stumps) in the soil. It may be a safe generality that passive possession is most highly developed and most advantageous among pathogens of fragile structures, whereas active possession is most highly developed and most advantageous among pathogens of woody structures.

MECHANISMS.—*Mechanisms of passive possession.*—Why are most oospores, microsclerotia, chlamydospores, and sclerotia such secure repositories of food? Are their membranes so perfect no food escapes to arouse or attract would-be enemies? Do the spore walls or coating substances repel other organisms? What dormancy mechanisms are functioning to enhance their efficacy in the waiting game?

Sclerotia of soil-borne organisms are usually considered resistant to adverse environments, but those of *Phymatotrichum omnivorum* are weakened or killed by desiccation (28), and deep tillage of dry soil is a control measure. Microsclerotia of *Verticillium dahliae* are resistant to drying, but once they have germinated, subsequent drying may kill them (35).

The physicochemical nature of the outer covering alone does not necessarily protect sclerotia from attack. Ferguson (19) cut sclerotia of *Sclerotinia sclerotiorum* and *Sclerotium delphinii* and buried them in soil. Even though these sclerotia were rich in food, they were not colonized by fungi. Sclerotia with rinds intact, if killed by heat or chemicals, were rapidly destroyed. Resistance to rotting did not reside in the rind. Smith (39, 40) reported that sclerotia of *Sclerotinia* and *Sclerotium* spp. taken from agar culture and dried at 30°C at 30% relative humidity (RH) to 10% of their original water content rotted quickly in soil. If sclerotia of *Sclerotium rolfsii* were not dried, were washed to remove leakage substances, and then buried, few rotted. The drying treatment apparently resulted in subsequent leakage of sugars and amino acids, which leads to rotting. Whether drying followed by leakage can be used successfully in control may depend on whether slower drying of the sclerotia has the same consequence. It may be difficult to obtain such rapid desiccation in nature. Smith (*personal communication*) found that desiccation above 60% RH does not lead to leakage and death in soil. Slower desiccation may permit physiologic and morphologic alterations or maintenance of membrane integrity that guard against leakage subsequent to desiccation. Dube (16) found that mycelium of *Rhizoctonia solani* grown under moist conditions died when exposed to sudden desiccation; mycelium grown first under moderate drought and exposed more gradually to the same degree of desiccation survived.

A recent concept runs counter to theoretical maximum efficiency of sclerotia or resting structures as instruments of passive possession. Bacteria are more abundant in the vicinity of sclerotia, and they qualitatively differ from populations in the soil at large. This has led to the hypothesis that the bacteria in the mycosphere produce substances that inhibit germination of the resting propagule or prevent its germination by removing food from its surface. Dormancy is thus imposed upon the propagule by bacteria about the surfaces (31). Germination would presumably occur when secretion of the bacteria-sustaining or -attracting substance ceased, or when external stimuli overcame this imposed fungistasis. A long-lived sclerotium in soil must then secrete something over a long period of time. This would be a wasteful system.

But there is another possibility. When sclerotia of *Typhula* spp. form, food is concentrated in them from surrounding hyphae. Droplets develop upon the developing sclerotia, particularly during maturation. Food is transported into the sclerotium in solution or suspension. Dry matter of the mature sclerotium exceeds that of the mother substance. Excess liquid is expelled during dehydration (maturation), and this fluid contains some foods. Thus, the surfaces of new sclerotia would receive the solids (food for the surface bacteria) as the water escaped.

Sclerotia of *Typhula idahoensis* form on rotting leaves of wheat in late winter or early spring before the snow melts. If they germinated at that time, they would be wasted, as this snow mold fungus ceases to function when temperatures rise and snow is gone. True dormancy is not present in the young, immature sclerotia. Possibly the exudates now sustain bacteria whose growth prevents their germination during cool, wet periods of early spring, enabling the sclerotia to mature. I am not convinced that continued exudation sustains a germination-inhibiting microflora, enabling sclerotia to remain unspent in soil for an undetermined period (at least 2 years). Sclerotia are normally formed on exhausted host structures. Trial and error (evolution) favored those forms that did not anticipate rapid renewal of a food source and favorable conditions to attack it. Just possibly, the exudate provides a temporary enforced dormancy that is subsequently replaced by a deeper dormancy. Until further research is done, I prefer to visualize the dormant structure in nature as being as quiescent as possible and leaking as little as possible.

Mechanisms of, or factors affecting, active possession of residues.—Time of establishment in residue.—Possession of residue prior to death of the tissue is probably the greatest advantage a plant pathogen has. Prior establishment by some fungi is so strong other fungi are virtually totally excluded. Being first confers an advantage to any fungus. This fact is significant, both in experiments and in nature. If clean, bright (virgin) wheat straw is introduced into soil infested with chlamydospores of *F. culmorum*, that pathogen can compete with soil-borne saprophytes for some of the residue, and some saprophytic colonization of the residue occurs. If old gray straw already occupied by weak parasites (*Cladosporium, Alternaria*) or saprophytes is introduced into the same infested soil under the same favorable conditions, *F. culmorum* is unable to colonize that straw (10, 36). Similarly, *Typhula idahoensis* can colonize virgin straw but not weathered straw (12). In humid climates, most straw and stubble is already colonized by molds (and by some pathogens) at plowing time, and it is unlikely that much saprophytic colonization of this residue by potential parasites occurs.

Finding perithecia of *Gibberella zeae* on old corn stalks in the field does not mean that *Fusarium graminearum* is necessarily an aggressive saprophytic colonizer of old corn stalks. It probably means that this parasite became established in the living stalk tissue, expanded its area of colonization during decline of the host, and then retained possession of the invaded area by virtue of prior colonization. *Septoria passerini* is a very weak competitive saprophyte, yet it retains tissues invaded as a parasite and produces spores and even new pycnidia in the invaded area, but only in that area occupied during parasitism (32).

Aphanomyces cochlioides has colonized fresh green

barley leaf fragments (H. S. MacWithey, *unpublished*); Trujillo and Hine (43) reported extensive increase of *Pythium aphanidermatum* and *Phytophthora parasitica* in papaya residues. In both cases, the tissues were virgin and alive at the time of burial. The thick leaves of pineapple plants make tons of residue per acre that can be colonized upon burial. *Thielaviopsis parodoxa* preferred mature green tissues; *Phytophthora cinnamomi* preferred the white tissues (42). These examples of colonization of residues point out the need for care in interpreting both field observations and laboratory experiments. In my opinion, they are not examples of saprophytic colonization but of rapid parasitic colonization of weakened, still-living tissues. The condition of the residue at the time the experiment begins should be stated.

Experiments already in the literature demonstrating competitive saprophytic colonization of residues by plant pathogens should be carefully interpreted in respect to the condition of the residue at inception of the study, especially when the structure is autoclaved prior to use as bait.

The above emphasis upon the significance of the condition of the substrate to invasion by pathogens should be no surprise. Pathologists have long used baiting with various living tissues as an aid to selecting pathogens from among the many soil fungi (27, p. 59 and 66-67). A recent interesting demonstration of the selective powers of living tissues is that of Grimm and Alexander (25) in that bits of citrus leaves favored *Phytophthora parasitica* more than did leaves of other plants.

Degree of establishment in tissues before death of the host.—*Cephalosporium gramineum* in such a true xylem-invader that it occupies little parenchyma in most wheats until death of the host. As tissues decline, the fungus invades parenchyma, but much of the peripheral parenchyma is not occupied. These areas can then be invaded by miscellaneous fungi with little or no opposition from *C. gramineum*. In contrast, *F. culmorum* invades parenchyma and is strongly established throughout the wheat stem, including the periphery. *Fusarium culmorum* is in possession of more of the tissue at death of the host. Even among fungi, possession (being first and well-established) is nine points of the law.

Competitive relationships are always influenced by the degree of establishment of one or several organisms ahead of the challengers (29). This is clearly recognized in the process of aerated steam pasteurization of soil in which only a portion of the microflora is killed. The remaining beneficial organisms retain possession of the treated soil and restrict the increase of most pathogens that might be introduced (37).

Menzies (35, p. 17) states that *Verticillium dahliae* forms many more microsclerotia in potato stems in dry climates than in humid climates. Presumably, internal invasion of the xylem was the same, but in humid climates secondary organisms more successfully invaded the as yet unoccupied peripheral cortical tissues of the potato stem and restricted microsclerotia formation.

Rate of residue destruction.—If active possession is to substitute for the long-lived resting propagules characteristic of passive possession, the residue must necessarily be preserved as long as possible. Soft-rot type organisms obviously do not qualify. The dry-rotters do. Herbaceous residues seldom provide for long-term survival, but Macer (33) and Garrett and his associates (14, 22) have emphasized that the fungi that survive longest in wheat straw produce little cellulase. *Cercosporella herpotrichoides*, *Ophiobolus graminis*, and *C. gramineum* are quite dependent upon the integrity of the residue, and all are relatively poor digestors of cellulose. Macer (33) noted that straw infested with *C. herpotrichoides* rotted somewhat more slowly than other straw, and the same thing occurs in the case of *C. gramineum*. This comparative preservation of the residue probably results from exclusion of more strongly cellulolytic organisms.

Theoretically, the lower the rate of metabolism required to maintain active possession, the longer-lived the mycelium should be in a given substrate. *Cercosporella herpotrichoides* survives longer in wheat straw than does *O. graminis*, and the former utilizes cellulose more slowly than the latter. The addition of nitrogen stimulates hydrolysis of cellulose in straw by *O. graminis* but not by *C. herpotrichoides* (14). Deacon (14) interprets the increased hydrolysis of cellulose by *O. graminis* with added nitrogen as proof of survival by saprophytic activity, and the lack of response to nitrogen by *C. herpotrichoides* as evidence that this fungus survives passively. Full acceptance of this interpretation requires caution, however, because the survival of *C. gramineum* in straw was not strongly influenced by addition of nitrogen to the straw (R. Millar and G. W. Bruehl, *unpublished*), yet it seems well-established that *C. gramineum* survives primarily by remaining active in residue.

Fox (20, p. 182-183), reporting on longevity in woody tree roots, stated that *Fomes lignosis* persists about 4-5 years, *Fomes noxius* about 10 years, and *Ganoderma pseudoferreum* about 30 years, and these differing survival periods were related to the speed of destruction of the substrate. Longevities such as these, along with the enormous food reserves of the unrotted wood, must give these fungi considerable inoculum potential. Active possession by parasites of woody structures may be more effective than passive possession.

Competition.—Actually, everything we have said so far, and everything we will say, is about competition. Farmers understand competition. They attempt to prevent weeds from going to seed. They plow and till a piece of land before seeding. They use enough seed so the crop will most efficiently utilize the fertility and climatic factors. In short, they do everything possible to give possession of the substrate to the crop plant. A well-established crop (like a good lawn) restricts weeds fairly well, depending on the crop and the weed.

Virgin substrate (living host tissue or tissue free from other organisms) is comparable to the plowed land free from weed seed prepared for seeding. The seeding is the pathogen entering and establishing itself ahead of the hoard of miscellaneous fungi. As long as the pathogen is active in the substrate, it presents some degree of competition.

It is unnecessary to visualize complex interactions. Surely all weeds do not exude harmful substances. Potatoes can be a serious weed in a plot of sweet corn, and corn can be a serious weed in a gladiolus planting. Barton (3, 4) demonstrated that *Pythium mamillatum* maintained a degree of possession (restricted competitors) on a piece of sugar-infiltrated wood if it was established in the wood block prior to burial in soil. Because *Pythium* spp. are not known to produce antibiotics, he concluded that possession was possible without specific antibiotics. Macer (33) believed that *C. herpotrichoides* established within wheat straw possessed that straw by utilizing available food as it was released by slow hydrolysis.

Staling.—Everyone who has grown fungi in petri dishes is familiar with staling. The apparently nonspecific substances produced by most fungi reduce the suitability of a substrate for growth (17). Fouling, staling, or whatever undoubtedly contributes to maintenance of active possession on a wider scale than do specific, demonstrable antibiotics (44).

How potent does a substance have to be to influence a competitive situation? If humans were microorganisms, an observer might ponder interactions resulting from halitosis or body odor. Although we do not attribute such sensitivity to microorganisms, seemingly neglible deterrents may be significant in a minimal medium like soil.

Wilhelm (45) stressed the effect of growth within a substrate upon that substrate. The residue is not the same when occupied by A as it is when occupied by B. Not only do differing organisms digest the substrate at differing rates and in differing ways, but they differentially alter it by their metabolites. In animal behavior, part of efficient territorial possession is to advertise your presence (13, p. 66). Staling is probably the most universal form of territorial advertising among microorganisms. Among animals, advertising avoids many conflicts and works as an energy-conserving mechanism. Maybe negative advertising (staling) and lack of positive advertising (balancing hydrolysis of substrate with food consumption so little soluble food escapes) works to minimize actual conflicts among organisms within the soil.

Antibiotic production.—There is widespread reluctance to accept antibiotic production as a factor in the maintenance of active possession of a substrate by an organism (24, p. 67). This skepticism may be an overreaction to the failure of efforts to extract antibiotics from the soil mass rather than from bits of residue, coupled with the knowledge that antibiotics can be adsorbed by soil particles or destroyed by microorganisms. Wright (46, 47) extracted antibiotic from infested debris in soil, and Brian (5), Park (38), and Wright stressed the significance of antibiotic production within debris rather than in the soil mass. Our experiments demonstrated that small amounts of antibiotic produced within the substrate enhance the saprophytic possession of residue by *C. gramineum* (6, 8, 26).

Cephalosporium gramineum invades the stems of wheat systemically via conidia in the xylem. It is restricted to xylem while host vitality is high, so it is seldom strongly based within the parenchyma tissues. It lacks true resting structures and depends for survival as a saprophyte upon slow but sustained metabolism that does not rapidly exhaust the substrate. It has a low growth rate and produces little cellulase (41); straws occupied by it decompose less slowly than control straws (26). Its general characteristics (lack of effective resting structures, low vigor, lack of thorough establishment within the living host) make it particularly dependent upon its antibiotic for saprophytic survival in the substrate.

Of several hundred isolates obtained from nature, all produced antibiotic. Low producers or nonproducers were obtained only from cultures maintained for 2-5 years on potato-dextrose agar in a refrigerator. As no mutagen was needed, we assumed that such isolates were produced but failed to survive in nature. Nonproducers tolerated the antibiotic; hence they were not eliminated through coexistence with their mothers. Nonproducers were pathogenic to wheat and produced typical symptoms; hence they were not eliminated directly by the host or by our selection (we isolated only from plants with symptoms, not from healthy plants). Nonproducers grew as fast as producers, so lack of vigor did not account for their demise in nature. We concluded that elimination must occur in the saprophytic phase (8). To test this, antibiotic-producers and nonproducers were established in straws that were then incubated in and upon soil.

Antibiotic-producing isolates survived longer in straw in moist, acid soil than did nonproducers (26). In neutral or alkaline soils, both types of isolates diminished at equal rates. The antibiotic was inactive in alkaline media in vitro (7), and if antibiotic was produced in alkaline soil, it did not enhance survival. In acid soils, producer isolates enjoyed a strong advantage at 15° C at water contents from about −10 to about −80 bars. When straws were placed on water-logged soil, all isolates survived well. This was interpreted as evidence that bacteria and fungi of wet situations were not active in killing *C. gramineum*. At water potentials highly restrictive to *C. gramineum* but at which xerophytic penicillia could grow, the advantage of antibiotic-producing capability diminished greatly (6). This indicated that *C. gramineum* must maintain a certain minimum metabolic level to survive under competitive conditions (critical to active survival).

When the soil was too dry (near −258 bars or less) for penicillia to grow, all isolates survived well (6). Antibiotic capability is no advantage to a dormant organism, and, in the absence of competition, it is not needed anyway. With this organism, then, nonproducers survived essentially the same at the wet and dry extremes, but producers survived better in intermediate environments than did nonproducers. The survival advantage of producers in many soil conditions was attributed directly to the presence of low levels of the wide-spectrum antifungal antibiotic within the occupied substrate.

Dependence to this degree upon antibiotic production for survival is surely rare among soil fungi. Such organisms are "rogues" within nature's scheme of nutrient recycling. Soil organisms utilize a multitude of differing enzymatic capabilities and tolerance to environmental factors (pH, temperature, O_2, etc.) to work in sequences to destroy debris. A vigorous antibiotic-producer could maintain possession of the substrate and block the

normal sequence. Nature has not favored "dogs in the manger" microorganisms.

Marx (34) demonstrated that mycorrhizae protected pine roots from *Phytophthora cinnamomi*. The fungal mantle may exclude the pathogen, or, in the case of *Leucopaxillus cerealis* var. *piceina*, an antibiotic produced by the symbiont may help protect the pine roots. The evidence seems to indicate that mycorrhizal fungi protect the host largely by being there first, by being thoroughly established at would-be infection courts, and by a combination of effects including modifications of the root cortical cells. It has even been postulated that *Rhizobium* spp. may increase their beneficial effect by producing antifungal antibiotics (15).

Cylindrocarpon radicicola is common on the root surfaces of many plants, especially on unthrifty roots of strawberry plants (45). It produces an antibiotic (nectrolide) identical to brefeldin A (produced by *Penicillium brefeldianum*) that is toxic to plant roots (18). Evans et al. (18) did not explore its effect on fungi that might otherwise grow on the root surface. Possibly the toxic substance strengthens the possession of the rhizoplane by *Cylindrocarpon*.

Goldberg and Luckey (23, p. 11) comment that antibiotics may be useful to "establish territorial rights rather than as offensive weapons". There seems little reason to doubt that they play a role in active possession of substrates. Antibiotic is used in this section to refer to powerful chemicals that restrict competition even when present in very low concentrations.

Some attributes of the ideal active possesser.—A) It establishes itself thoroughly within the substrate the easiest way; i.e., by entering living, virgin tissue as a parasite. It invades as much tissue as possible; thus, in herbaceous plants, it invades mostly parenchyma.

B) It sustains minimal metabolism with minimal enzyme production, thus preserving the substrate as long as possible and avoiding the liberation of surplus sugars and amino acids into the environment. (Escaping foods would advertise the presence of the substrate.)

C) It produces minimum staling or antibiotic, only enough to counteract the message conveyed by foods escaping from the substrate. If we visualize gradients of food in the soil as messages (intelligence), we can visualize staling products and antibiotics as message-cancellers or counterintelligence. The ideal passive possessor produces only enough staling, as an excess is a waste of substance.

D) It is both flexible and stable in its responses to environmental changes. Because active possession depends upon continued activity, it must be competitive with all challengers at high and low temperatures at high and low moisture levels. Thus, active possessors are stressed most severely in surface soil layers, where changes are greatest, and are stressed least in roots deep within the soil, where environmental changes are less severe (as in tree roots).

Response to favorable conditions should be adequate, but only adequate. Added nitrogen, improved water or temperature, etc., should not stimulate excessive and wasteful activity, as this would waste substance.

The active possessor with such a sophisticated array of physiological characteristics probably never will exist, at least not among fungal pathogens of herbaceous plants.

Passive possessors that exploit and store are more easily achieved by evolution, and they are spared the need to compete with all challengers at all conditions, as in attribute D above.

The general importance of possession, especially being in the substrate first, as it relates to control is discussed in greater detail by Baker and Cook (1). But it is obvious that most soil-borne fungal plant pathogens owe their lives to this advantage. For a detailed review of control mechanisms of soil-borne pathogens, see Baker (2), and for a different idea about competition among soil organisms, see Clark (9).

An article on ants provides a striking comparison of some aspects of active possession of substrate by ants and fungi. Below are several comments from that article (30, p. 471, paragraph numbering from original).

1. A species that finds a food supply first can mobilize and surround the food in such a way that few foragers of another species ever reach the food. Therefore the second species does not lay an odor trail or signal the presence of food and fails to mobilize in large numbers.

2. It is our impression that when a few foragers encounter large numbers of another species they are timid, more easily disoriented by unfriendly contacts, and more easily defeated even in individual combat. It is as if a pheromone aura sets the local conditions for the encounter.

5. Established colonies have a strong advantage over incipient colonies.

LITERATURE CITED

1. BAKER, K. F., and R. J. COOK. 1974. Biological control of plant pathogens. W. H. Freeman and Co., San Francisco. (In press.)

2. BAKER, R. 1968. Mechanisms of biological control of soil-borne pathogens. Annu. Rev. Phytopathol. 6:263-294.

3. BARTON, R. 1960. Antagonism amongst some sugar fungi, p. 160-167. *In* D. Parkinson and J. S. Waid (eds.), The ecology of soil fungi. Liverpool Univ. Press, Liverpool.

4. BARTON, R. 1961. Saprophytic activity of Pythium mamillatum in soils. II. Factors restricting P. mamillatum to pioneer colonization of substrates. Trans. Br. Mycol. Soc. 44:105-118.

5. BRIAN, P. W. 1960. Antagonism and competitive mechanisms limiting survival and activity of fungi in soil, p. 115-129. *In* D. Parkinson and J. S. Waid (eds.), The ecology of soil fungi. Liverpool Univ. Press, Liverpool.

6. BRUEHL, G. W., B. CUNFER, and M. TOIVIAINEN. 1972. Influence of water potential on growth, antibiotic production, and survival of Cephalosporium gramineum. Can. J. Plant Sci. 52:417-423.

7. BRUEHL, G. W., P. LAI, and O. HUISMAN. 1964. Isolation of Cephalosporium gramineum from buried, naturally infested host debris. Phytopathology 54:1035-1036.

8. BRUEHL, G. W., R. L. MILLAR, and B. CUNFER. 1969. Significance of antibiotic production by Cephalosporium gramineum to its saprophytic survival. Can. J. Plant Sci. 49:235-246.

9. CLARK, F. E. 1965. The concept of competition in microbial ecology, p. 339-345. *In* K. F. Baker and W. C. Snyder (eds.), Ecology of soil-borne plant pathogens. Univ. Calif. Press, Berkeley and Los Angeles.

10. COOK, R. J. 1970. Factors affecting saprophytic colonization of wheat straw by Fusarium roseum f. sp. cerealis 'Culmorum'. Phytopathology 60:1672-1676.

11. COOK, R. J., and G. W. BRUEHL. 1968. Relative significance of parasitism versus saprophytism in colonization of wheat straw by Fusarium roseum 'Culmorum' in the field. Phytopathology 58:306-308.

12. CUNFER, B. M., and G. W. BRUEHL. 1973. Role of basidiospores as propagules and observations on sporophores of Typhula idahoensis. Phytopathology 63:115-120.

13. DAVIS, D. E. 1966. Integral animal behavior. Macmillan Co., New York. 118 p.

14. DEACON, J. W. 1973. Behaviour of Cercosporella herpotrichoides and Ophiobolus graminis on buried wheat plant tissues. Soil Biol. Biochem. 5:339-353.

15. DRAPEAU, R., J. A. FORTIN, and C. GAGNON. 1973. Antifungal activity of Rhizobium. Can. J. Bot. 51:681-682.

16. DUBE, A. J. 1971. Studies on the growth and survival of Rhizoctonia solani. Ph.D. thesis, Univ. Adelaide, Adelaide, Australia. 144 p.

17. DWIVEDI, R. S., and S. D. GARRETT. 1968. Fungal competition in agar plate colonization from soil inocula. Trans. Br. Mycol. Soc. 51:95-101.

18. EVANS, G., J. B. CARTWRIGHT, and N. H. WHITE. 1967. The production of a phytotoxin, nectrolide, by some root-surface isolates of Cylindrocarpon radicicola Wr. Plant Soil 26:253-260.

19. FERGUSON, J. 1953. Factors in colonization of sclerotia by soil organisms. Phytopathology 43:471 (Abstr.).

20. FOX, R. A. 1970. A comparison of methods of dispersal, survival, and parasitism in some fungi causing root diseases of tropical plantation crops, p. 179-187. *In* T. A. Toussoun et al. (ed), Root diseases and soil-borne pathogens. Univ. Calif. Press, Berkeley and Los Angeles.

21. GARRETT, S. D. 1956. Biology of root-infecting fungi. Cambridge Univ. Press, Cambridge. 293 p.

22. GARRETT, S. D. 1972. Factors affecting saprophytic survival of six species of cereal foot-rot fungi. Trans. Br. Mycol. Soc. 59:445-452.

23. GOLDBERG, H. S., and T. D. LUCKEY. 1959. Introduction, p. 1-57. *In* H. S. Goldberg (ed.), Antibiotics, their chemistry and non-medical uses. D. Van Nostrand Co., Inc., Princeton, N.J. 608 p.

24. GRIFFIN, D. M. 1972. Ecology of soil fungi. Syracuse Univ. Press, Syracuse, N.Y. 193 p.

25. GRIMM, G. R., and A. F. ALEXANDER. 1973. Citrus leaf pieces as traps for Phytophthora parasitica from soil slurries. Phytopathology 63:540-541.

26. HOPP, A. D. 1972. The influence of antibiotic production and soil pH on survival of Cephalosporium gramineum in infested wheat straw. Ph.D. thesis, Wash. State Univ., Pullman. 28 p.

27. JOHNSON, L. F., and E. A. CURL. 1972. Methods for research on the ecology of soil-borne plant pathogens. Burgess Publ. Co., Minneapolis, Minn. 247 p.

28. KING, C. J., H. F. LOOMIS, and C. HOPE. 1931. Studies on sclerotia and mycelial strands of the cotton root-rot fungus. J. Agr. Res. 42:827-840.

29. LAI, P., and G. W. BRUEHL. 1968. Antagonism among Cephalosporium gramineum, Trichoderma spp., and Fusarium culmorum. Phytopathology 58:562-566.

30. LEVINS, R., MARY L. PRESSICK, and H. HEATWOLE. 1973. Coexistence patterns in insular ants. Amer. Sci. 61:463-472.

31. LINDERMAN, R. G., and R. G. GILBERT. 1973. Influence of volatile compounds from alfalfa hay on microbial activity in soil in relation to growth of Sclerotium rolfsii. Phytopathology 63:359-362.

32. LUTEY, R. W., and K. D. FEZER. 1960. The role of infested straw in the epiphytology of Septoria leaf blotch of

barley. Phytopathology 50:910-913.

33. MACER, R. C. F. 1961. The survival of Cercosporella herpotrichoides Fron in wheat straw. Ann. Appl. Biol. 49:165-172.

34. MARX, D. H. 1973. Growth of ectomycorrhizal and nonmycorrhizal shortleaf pine seedlings in soil with Phytophthora cinnamomi. Phytopathology 63:18-23.

35. MENZIES, J. D. 1970. Factors affecting plant pathogen population in soil, p. 16-21. *In* T. A. Toussoun et al. (ed.), Root diseases and soil-borne pathogens. Univ. Calif. Press, Berkeley and Los Angeles.

36. NYVALL, R. F., and T. KOMMEDAHL. 1973. Competitive saprophytic ability of Fusarium roseum f. sp. cerealis 'Culmorum' in soil. Phytopathology 63:590-597.

37. OLSEN, C. M., and K. F. BAKER. 1968. Selective heat treatment of soil and its effect on the inhibition of Rhizoctonia solani by Bacillus subtilis. Phytopathology 58:79-87.

38. PARK, D. 1960. Antagonism—the background of soil fungi, p. 148-159. *In* D. Parkinson and J. S. Waid (eds.), The ecology of soil fungi. Liverpool Univ. Press, Liverpool.

39. SMITH, A. M. 1972. Nutrient leakage promotes biological control of dried sclerotia of Sclerotium rolfsii Sacc. Soil Biol. Biochem. 4:125-129.

40. SMITH, A. M. 1972. Biological control of fungal sclerotia in soil. Soil Biol. Biochem. 4:131-134.

41. SPALDING, D. H., G. W. BRUEHL, and R. J. FOSTER. 1961. Possible role of pectinolytic enzymes and polysaccharide in pathogenesis by Celphalosporium gramineum in wheat. Phytopathology 51:227-235.

42. TRUJILLO, E. E. 1969. Relationship of crop residues to increased persistence and inoculum density of soil-borne pathogens, p. 23-25. *In* R. J. Cook and R. D. Watson (eds.), Nature of the influence of crop residues on fungus-induced root diseases. Wash. Agr. Exp. Stn. Bull. 716.

43. TRUJILLO, E. E., and R. B. HINE. 1965. The role of papaya residues in papaya root rot caused by Pythium aphanidermatum and Phytophthora parasitica. Phytopathology 55:1293-1298.

44. WAKSMAN, S. A. 1932. Principles of soil microbiology. 2nd ed. Williams and Wilkins, Baltimore. 894 p.

45. WILHELM, S. 1959. Parasitism and pathogenesis of root disease fungi, p. 356-366. *In* C. S. Holton, et al. (ed.), Plant pathology: problems and progress, 1908-1958. Univ. Wisc. Press, Madison.

46. WRIGHT, J. M. 1956. The production of antibiotics in soil. III. Production of gliotoxin in wheat straw buried in soil. Ann. Appl. Biol. 44:461-466.

47. WRIGHT, J. M. 1956. The production of antibiotics in soil. IV. Production of antibiotics in coats of seeds sown in soil. Ann. Appl. Biol. 44:561-566.

Survival and Multiplication of Soil-Borne Plant Pathogens as Affected by Plant Tissue Amendments

J. A. LEWIS and G. C. PAPAVIZAS—*Soil Scientist and Microbiologist, Plant Protection Institute, Agricultural Research Service, U.S. Department of Agriculture, Beltsville, Maryland.*

INTRODUCTION.—Although amendment decomposition is an important factor in the complex ecological environment affecting survival of a soil-borne plant pathogen, this subject has received little attention during recent years compared to the substantial contributions made during the 1950's and early 1960's. Some reasons for this decline in interest have been mentioned by Papavizas in his introductory remarks to this section. The feelings of frustration and disappointment in attempts to make practical use of biological control are understandable. But discouragement does not serve the aims of science. The study of plant tissue residues in relation to disease is important because of the possibility of applying these materials in place of fungicides for the reduction of soil-borne diseases and because of the efficient use in soil management practices of accumulated organic matter.

LITERATURE REVIEW.—The following review was drawn from reports issued since the First International Congress of Plant Pathology was held in London in 1968. Although most of the work concerns in vitro experimentation, principles of biological control that might be utilized in a practical application continue to be established. Several recent monographs contain specific sections on the effects of decomposing plant residues on soil-borne plant pathogens (6, 22).

Papavizas (20) reported one of several instances in which biological control of a pathogen by plant tissues was related to reduction of the inoculum density of the pathogen. Addition of oat straw, corn stover, or alfalfa hay to *Thielaviopsis basicola*-infested soil caused a population reduction of about 90% and almost complete elimination of the root rot of bean caused by this pathogen. The population reduction was attributed to the stimulation of germination of endoconidia and chlamydospores of *T. basicola*, with subsequent germling lysis occurring before formation of secondary endoconidia or chlamydospores (2, 29). Disease was reduced because decomposing tissues increased soil fungistasis to such an extent that root exudates were unable to stimulate propagule germination for infection (29). These observations were confirmed with subsequent studies of *T. basicola* on tobacco (27) and sesame (1). Because nongerminated spores from soils previously amended with alfalfa hay or corn stover were nonviable, it was suggested (23) that, in addition to fungistasis, the increased microbial activity in the amended soils may have produced fungitoxic substances that decreased the inoculum level.

In contrast to that of *T. basicola*, survival or saprophytic activity of *Rhizoctonia solani* was not reduced with alfalfa hay or corn stover (21). These results, however, did not agree with earlier ones (7, 25) which had suggested that a reduction in survival was caused by antagonists stimulated by decomposing oat straw. In considering the discrepancy, the differences of the carbon:nitrogen (C:N) ratio of the amendment as well as the nutritional and textural properties of the various soils were recognized (21, 28). Manning and Crossan (16) demonstrated the importance of the nature of the plant tissue as well as its maturity in disease suppression. For example, decomposing mature corn tissue in soil adversely affected *R. solani* and its hypocotyl rot of snap beans in the field, whereas decomposing rye-, barley-, oat-, and wheat-straws had no effect.

Papavizas et al. (24) demonstrated a consistent three- to fourfold increase in *Fusarium solani* f. sp. *phaseoli* populations in soils amended with dried alfalfa, bean, corn, potato, sugarbeet, and barley tissues. Similar tissues, previously extracted with water before addition to soil, increased the population to a smaller extent. Chlamydospores constituted more than 70% of the population. Decomposing oat straw also resulted in a slight increase in propagule numbers. Although macroconidia of the pathogen germinated well, regardless of the stage of oat-straw decomposition, chlamydospores germinated poorly, presumably because of fungistasis (3, 28). Wheat straw turned under in wheat fields increased survival and inoculum density of *F. roseum* and *F. solani*. This was attributed to the growth of the pathogens on the residue (31). Cook (5) indicated that populations of *F. roseum* f. sp. *cerealis* 'Culmorum' were greater in wheat fields cropped with oats in a rotation than in fields continually cropped with wheat because the pathogen sporulated better on oat than on wheat tissue. The conidia, entering the soil, were converted to chlamydospores. In Asia, the decomposition of various plant oil cakes like linseed-, groundnut-, and sesame-cake reduced the wilt of coriander caused by *F. oxysporum* f. sp. *corianderi* (30). The maximum reduction in population observed was 50% in amended soils compared with the control.

Green and Papavizas (11) studied the effect of dry,

mature alfalfa, oat, and turnip tissues on survival of microsclerotia of *Verticillium albo-atrum* and observed that alfalfa and oat straw, but not turnip leaf tissue, reduced the number of viable propagules by 50-60%. Jordan et al. (13) showed that decomposing wheat straw and clover in soil decreased germination of both conidia and microsclerotia of *Verticillium dahliae*. Inhibition of germination, however, decreased with increasing periods of residue decomposition. Microsclerotia survived better than hyphae in the amended soils. The initial inhibitory effect on growth of the pathogen in soil as a result of wheat-straw decomposition was also gradually lost. Although *Verticillium* populations decreased in the amended soils, severity of the strawberry wilt caused by the pathogen was not reduced by the decomposing plant tissues.

Decomposing alfalfa meal and barley straw reduced a root rot of cotton caused by *Macrophomina phaseoli*, presumably because of stimulation of antagonistic bacteria and actinomycetes (8). The use of legumes like clover, pea, and alfalfa for the reduction of snowmold of winter cereals caused by *Typhula idahoensis* has been attributed to the reduction in survival of sclerotia in the rhizosphere of these legumes (12). The effect of extracts of soil containing decomposing immature and mature peanut, soybean, corn, sorghum, and cotton plants on the mycelial growth of *Sclerotium rolfsii* also has been studied (4). Extracts from decomposing soybean and peanut tissues in soil significantly reduced mycelial growth of the pathogen, whereas the extracts from the other tissues did not.

Although a discussion of the effect of volatiles on soil-

TABLE 1. Effect of immature and mature plant tissues incorporated into soil on bean root rot caused by *Fusarium solani* f. sp. *phaseoli* and N_i[a] content of soils at time of planting

Plant tissue[b]	Carbon:nitrogen	Disease severity index[c]	N_i (*mg/kg soil*)
Rye	94:1	0.3 a[d]	37 a
Oat	83:1	0.6 a	46 ab
Sorghum	62:1	0.6 a	63 cd
Buckwheat	70:1	1.5 b	49 bc
Corn	9:1	1.8 bc	96 e
Rye	9:1	1.9 bc	64 cd
Oat	9:1	2.0 bc	68 d
Sorghum	13:1	2.0 bc	70 d
Buckwheat	11:1	2.1 c	62 cd
Corn	81:1	2.6 d	52 bc
Control		3.0 d	73 d

[a]Steam-distillable inorganic nitrogen from soil.
[b]Added to soil at rate of 1% (w/w).
[c]Based on a scale in which 0 indicates all roots apparently healthy and 4 indicates all roots and hypocotyl covered by lesions.
[d]Numbers followed by same letter are significantly different at 5% level.

TABLE 2. Effect of plant tissue decomposition in soil on germination of chlamydospores and macroconidia of *Fusarium solani* f. sp. *phaseoli* on agar disks placed on soil-agar mixture[a]

Plant tissue[b]	Spore germination (%) at indicated weeks of amendment decomposition			
	Agar disks with potato-dextrose powder[c]		Water agar disks	
	3	14	3	14
Chlamydospores				
Control	91	95	35	7
Corn (immature)	88	70	3	3
Corn (mature)	83	11	2	7
Rye (immature)	90	69	4	10
Rye (mature)	58	8	5	9
Macroconidia				
Control	96	93	70	75
Corn (immature)	87	85	40	60
Corn (mature)	93	82	37	55
Rye (immature)	92	90	29	68
Rye (mature)	90	79	31	73

[a]Twenty g of soil mixed with 10 ml agar.
[b]Added to soil at rate of 1% (w/w).
[c]Disk allowed to remain on soil-agar mixture of same composition 2 days at 5°C before spore addition.

borne plant pathogens is included in this volume (Linderman and Gilbert), we will present some of our recent work in this area. In greenhouse studies, the reduction in pea root rot, caused by *Aphanomyces euteiches*, with decomposing crucifer tissues (cabbage, kale, collards, Brussels sprouts, kohlrabi, mustard, turnip, and cress) appeared to be caused by the adverse effect on the fungus of toxic volatiles arising from the decomposing tissue (17, 26). Kale was also effective when used in the field. We obtained direct evidence that volatiles like methanethiol, dimethyl sulfide, and dimethyl disulfide, formed in soil during crucifer decomposition, were toxic to *A. euteiches* (14). Many of the crucifers also suppressed both pre- and postemergence damping-off of sugarbeets caused by *A. cochlioides* (15).

Ammonia produced from decomposing plant tissues in soil was suggested by Gilpatrick (9, 10) to be the active volatile principle responsible for the control of some root diseases. He found that *Phytophthora cinnamomi* was eliminated from avocado roots placed in alfalfa-amended soil during maximum NH_3 production. Moreover, NH_3 prevented zoospore germination and killed the pathogen mycelium in vitro. The NH_3 released during decomposition from plant tissues (cotton waste, soybean meal, wheat-, bean-, and soybean-straw) may reduce the amount of inoculum by preventing growth of the pathogen, reducing zoospore production, or killing the pathogen in the root area.

A major objective of our recent research with plant tissues and soil-borne plant pathogens has been to study (a) the effect of decomposing tissues of various maturities on survival of *F. solani* f. sp. *phaseoli* and (b) the effect of volatiles from decomposing tissues on growth and survival of *R. solani*. The following report is a condensation of the work done at the Soilborne Diseases Laboratory, Beltsville Agricultural Research Center.

FUSARIUM.—During initial greenhouse experiments to determine the effect of decomposing plant tissues of various maturities on bean root rot caused by *F. solani* f. sp. *phaseoli*, we observed that all the mature plant tissues used, with the exception of corn (C:N, 81:1), significantly reduced root rot (Table 1). Immature tissues reduced disease slightly. Steam-distillable inorganic nitrogen (N_i) determinations of the amended soils at the time of planting indicated that, in general, there was more N_i in soils amended with decomposing immature plant tissues than in soils amended with mature tissues. In two cases (rye and oat), significant disease suppression was correlated with N_i reduction, whereas there was no such correlation in two other cases (sorghum and buckwheat). Since mature rye (C:N, 94:1) was the most effective tissue in disease suppression and mature corn was the least suppressive, these two plant tissues at both maturities were investigated further to determine the effects of their decomposition in soil on the behavior and subsequent survival of *F. solani* f. sp. *phaseoli*.

Fusarium populations were determined by the dilution plate method (18). Germination of macroconidia and chlamydospores of *F. solani* f. sp. *phaseoli* was determined (a) by adding spores to agar disks placed on soils mixed with dilute agar, (b) by the propagule assay

method (19), and (c) by a modification of the buried slide method.

Population studies in the laboratory performed over a 15-week period showed that decomposing mature rye, which was most effective in suppressing bean root rot, resulted in the lowest *Fusarium* populations in soil. Immature rye (C:N, 9:1) and corn of both maturities (C:N, 9:1 and 81:1) reduced the population slightly compared with that of the nonamended soil. *Fusarium* populations per g of soil were: control, 19.8×10^4; immature corn, 14×10^4; mature corn, 12.3×10^4; immature rye, 16×10^4; and mature rye, 3.9×10^4.

An experiment was recently completed to determine the long-term effect of decomposing immature and mature rye, as well as of mature sorghum and soybean tissues, on *Fusarium* populations. The tissues were added to soils infested with either of two isolates (FS2R and FS16), and the soils were placed in nylon mesh bags and buried in the field at Beltsville. After 58 weeks of amendment decomposition, the population of FS2R in immature rye-amended soil was 102% of the control, whereas it was 42% of the control in mature rye-amended soil. The population in sorghum- and soybean-amended soils was 78% and 23% of the control, respectively. Nonamended soil contained 2.4×10^4 propagules/g of soil, which was 50% less than at the initiation of the experiment. Similar results were obtained with isolate FS16.

The propagule assay method was one of several methods used to determine spore germination in *F. solani* f. sp. *phaseoli*. In this system, no nutrients besides the plant tissue studied were added to soil heavily infested with chlamydospores. Germination 1 week after amendment addition was: control, no germination; immature corn, 43%; mature corn, 3%; immature rye, 39%; and mature rye, no germination. With continued incubation on a nutrient agar, the inhibited chlamydospores germinated, indicating they were affected by soil fungistasis.

Leaching the plant tissues with water reduced their ability to support germination of chlamydospores. Corn and rye tissues of both maturities were extracted with water or ether before their addition to soil. Germination for the various treatments, determined by the propagule assay method, was: control, no germination; ether-extracted immature and mature rye, 80% and 76%; ether-extracted immature and mature corn, 68% and 66%; water-extracted immature and mature rye, 14% and 7%; and water-extracted immature and mature corn, 2% and 3%, each respectively. The data reaffirm the importance of soil fungistasis, since water extraction removed the nutrients necessary for germination.

A modification of the agar disk method with potato-dextrose powder incorporated into the agar was also used to determine spore germination. Chlamydospore, but not macroconidial, germination gradually decreased as the length of time for plant tissue decomposition increased (Table 2). The greatest inhibition occurred with mature rye and mature corn. Water extracts of agar disks containing potato-dextrose powder from the surface of amended soils were analyzed for anthrone- and ninhydrin-positive materials. Although there was little chlamydospore germination on these agar disks placed on mature corn- or mature rye-amended soils after 14 weeks of decomposition, the chemical tests indicated there were sufficient nutrients present to support germination. Inhibited spores germinated when disks were placed on fresh nutrient-containing agar. This suggested the presence of an inhibitor from decomposing mature corn and rye, the effect of which was overcome or reversed by added nutrients or dilution. When germinability of macroconidia, and especially chlamydospores, was assayed with water agar disks, fungistasis prevented germination in soils amended with corn or rye, irrespective of C:N ratio (Table 2). Fungistasis appeared to decrease with increases in time of amendment decomposition.

Inhibition of chlamydospore germination due to inhibitory material was also observed when spores and agar containing potato-dextrose powder were placed on slides and the slides were buried in soil amended with mature corn or mature rye (Table 3). There was no inhibition of germination at the time of amendment, but on slides placed in soil after 7 weeks of tissue decomposition, chlamydospore germination was reduced to 2% with mature corn and to 1% with mature rye. This method also permitted observation of germling lysis. Lysis was increased when plant tissues were incorporated

into the soil. No replacement chlamydospores were observed on the slides. Extent of germination of macroconidia was similar to that of chlamydospores, except that after 7 weeks of tissue decomposition, germination on slides in soil with mature corn and mature rye was 46% and 40%, respectively (Table 3). At the 7-week assay, we observed chlamydospores developing from hyphae originating from macroconidia in the control and in soils containing immature corn and rye. Slides containing either chlamydospores or macroconidia in water agar were also buried in soil after several weeks of tissue decomposition. Less than 5% germination occurred in all cases because of soil fungistasis resulting from nutrient deficiency. The extraction of mature corn- and rye-amended soils with chloroform:methanol (1:1, v/v) indicated the presence of a material that inhibited chlamydospore germination or stimulated germling lysis. Both the propagule assay and buried slide methods were used to determine chlamydospore germination in soil to which the extracts had been added, along with adequate N and glucose to ensure germination in the control soil. With the propagule assay method, using extracts of amended soils incubated for 3 and 6 weeks, germination in the control was 89% and 88%; in mature corn, 63% and 70%; and in mature rye, 23% and 33%, each respectively. Since no further germination occurred with increased incubation, the presence of a toxic material(s), especially from decomposing mature rye, was indicated. With the buried slide method, there was appreciable germination in all treatments; however, with mature rye, many germ tubes from germinating chlamydospores lyzed so completely that the germling contents did not stain well and the walls were barely visible.

We have attempted to stress the importance of the maturity of plant tissues on disease and the pathogen population using *F. solani* f. sp. *phaseoli* and the root rot it causes on bean. We demonstrated the presence of soil fungistasis and an inhibitor in soil as a result of mature rye decomposition and correlated the levels of fungistasis and inhibition with a reduction in bean root rot and in inoculum density. However, we still cannot determine why decomposing mature corn, the effects of which on germination appear to be quite similar to those of

TABLE 3. Effect of plant tissue decomposition in soil on germination of chlamydospores and macroconidia of *Fusarium solani* f. sp. *phaseoli* with the buried slide method

Plant tissue[a]	Spore germination (%) at indicated weeks of amendment decomposition[b]			
	Chlamydospores		Macroconidia	
	0	7	0	7
Control	64	73	70	74*[c]
Corn (immature)	70	54	80	48*
Corn (mature)	60	2	69	46
Rye (immature)	76	70	86	54*
Rye (mature)	63	1	72	40

[a]Added to soil at rate of 1% (w/w).
[b]Germination determined after 24 hours of incubation.
[c]* indicates formation of chlamydospores by germling.

TABLE 4. Effect of decomposing plant tissue vapors on mycelial weight and pigment production in *Rhizoctonia solani* growing on potato-dextrose agar

Plant tissue[a]	Wet weight of mycelia and optical density (O.D./100 mg wet wt)[b] of extracts caused by vapors of plant tissues of indicated C:N			
	C:N < 15		C:N > 50	
	Wet wt	O.D.	Wet wt	O.D.
Oat[c]	98 a[d]	0.16 ab	170 de	0.11 a
Barley	212 f	0.21 bc	118 abc	0.20 bc
Sorghum	155 d	0.25 c	142 cd	0.11 a
Rye	192 ef	0.33 de	158 d	0.12 a
Buckwheat	216 f	0.35 e	140 bcd	0.12 a
Corn	190 e	0.38 e	108 ab	0.13 a

[a]Added to soil at rate of 10%.
[b]Optical density at 460 mμ.
[c]Wet weight of control was 142 cd; O.D./100 mg mycelium was 0.11 a.
[d]Numbers followed by same letter are not significantly different at 5% level.

decomposing rye, neither suppresses disease nor reduces fungus populations.

RHIZOCTONIA.—Volatiles arising from the decomposition of organic matter may involve not only stimulation of germination of propagules, but also fungitoxicity. We have some evidence that a toxic, volatile decomposition product from plant tissues adversely affects the survival of *R. solani*. We observed that passing vapors from several decomposing plant tissues into flasks containing a pure culture of *R. solani* growing on a nutrient agar caused the fungus mycelium to darken with increased exposure to the vapors. The tissues that, upon decomposition, caused the most darkening had low C:N ratios. Immature rye, corn, barley, oat, and sorghum tissues as well as bean, pea, and crucifer residues produced this effect. The addition of ammonium or nitrate to residues of high C:N ratio to reduce the C:N ratio did not cause them to produce effective volatiles.

The darkening of the mycelium significantly affected survival of *R. solani* in soil. Vapors from nonamended soil, bean residue-amended soil, and oat straw-amended soil were passed into flasks containing Perlite moistened with potato-dextrose broth and inoculated with a mycelial suspension of *R. solani*. After 3 weeks of incubation, only the vapors from bean residue-amended soil darkened the fungus mycelium growing over the Perlite. We used the three Perlite preparations as inocula for soil and autoclaved beet seed as bait for *R. solani*. Mycelium of *R. solani* grew from 27% and 41% of the beet seed retrieved from soil that contained Perlite through which vapors from nonamended or oat straw-amended soils, respectively, were passed. Because of decomposing bean residue volatiles, darkened mycelium on Perlite was not viable, as indicated by the absence of beet seed colonization.

We also added *Rhizoctonia*-infested buckwheat stems to natural soils and passed into these preparations vapors from nonamended soil or from soils amended with cabbage (C:N, 11:1) or corn (C:N, 81:1) tissues. After suitable incubation, the buckwheat was retrieved and plated on agar. *Rhizoctonia solani* grew out on the water agar from 91% and 88% of buckwheat pieces through which nonamended soil vapors and decomposing corn-stover vapors were passed, respectively. However, *R. solani* grew out from only 42% of the buckwheat pieces through which decomposing cabbage vapors had been passed. These data indicated the adverse effect on survival from volatiles that darken the fungus mycelium.

The dark pigment was not associated with the formation of sclerotia in culture. After mycelium had been separated from the agar, the pigment was extracted from the mycelium with NaOH. Its solubility in alkali and the production of a linear absorption spectrum between 400 and 600 mμ are characteristics of melanin. The material, however, could not be precipitated from NaOH by neutralization, which prevented its quantitative assay and brought into question the assumption that the pigment was melanin. Consequently, results in Table 4 are given as optical density (O.D.)/100 mg wet weight of mycelium. The higher the O.D., the greater is the pigment concentration. The data indicate that, in most instances, the vapors from decomposing immature tissues darken

the mycelia and cause a significant increase in their weight. By using a series of trapping solutions, we showed that the volatile is a low molecular weight amine or NH_3. The effect of the vapors was eliminated when they were passed through solutions of boric acid or dilute H_2SO_4. It is unknown whether the unexpected increase in the weight of *R. solani* mycelium demonstrated with several of the plant tissues (Table 4) was caused by the added growth stimulus provided by the N of the NH_3 or was due to the increased alkalinity of the medium in which the mycelium was growing. To further implicate NH_3 as the effective volatile and stress the importance of plant tissue maturity in causing pigmentation, we determined the NH_3 and CO_2 evolved from decomposing corn tissue of C:N ratios 9, 17, 20, 33, and 81:1. The O.D./100 mg mycelium decreased from a value of 0.47 with corn (C:N,9:1) to 0.13 with corn (C:N, 81:1). This was directly correlated with a decrease in NH_3 production from 45 mg/5 g tissue with corn (C:N, 9:1) to no NH_3 with corn (C:N, 81:1). There was no correlation between pigment production and CO_2 evolution. Moisture content of the soil containing decomposing plant tissues also affected pigmentation, presumably because of the amount of NH_3 released. Corn (C:N, 9:1) decomposing at 50% water-holding capacity (WHC), for example, gave an O.D./100 mg of 0.52, whereas the same tissue at 100% WHC gave a value of 0.81. The three pathogenic isolates of *R. solani* used in the above studies essentially responded similarly to the volatiles from decomposing plant tissues.

CONCLUSIONS.—An explanation of those mechanisms for pathogen survival and destruction that are intimately associated with the decomposition of a particular residue appears to involve volatile and toxicant production, fungistasis, lysis, antagonism, and antibiosis. We expect it may be possible to find plant-tissue soil amendment that will be effective against a given plant pathogen, taking into consideration the type and maturity of the plant tissue, the nutrient status of the soil, and the physiology of the pathogens.

LITERATURE CITED

1. ADAMS, P. B. 1971. Effect of soil temperature and soil amendments on Thielaviopsis root rot of sesame. Phytopathology 61:93-97.
2. ADAMS, P. B., and G. C. PAPAVIZAS. 1969. Survival of root-infecting fungi in soil. X. Sensitivity of propagules of Thielaviopsis basicola to soil fungistasis in natural and alfalfa-amended soil. Phytopathology 59:135-138.
3. ADAMS, P. B., G. C. PAPAVIZAS, and J. A. LEWIS. 1968. Survival of root-infecting fungi in soil. III. The effect of cellulose amendment on chlamydospore germination of Fusarium solani f. sp. phaseoli in soil. Phytopathology 58:373-377.
4. BOYD, H. W., and D. V. PHILLIPS. 1973. Toxicity of crop residue to peanut seed and Sclerotium rolfsii. Phytopathology 63:70-71.
5. COOK, R. J. 1968. Influence of oats on soil-borne populations of Fusarium roseum f. sp. cerealis 'Culmorum.' Phytopathology 58:957-960.
6. COOK, R. J., and R. D. WATSON (ed.). 1969. Nature of the influence of crop residues on fungus-induced root diseases. Wash. Agr. Exp. Stn. Bull. 716. 32 p.
7. DAVEY, C. B., and G. C. PAPAVIZAS. 1963. Saprophytic activity of Rhizoctonia as affected by the carbon-nitrogen

balance of certain organic soil amendments. Proc. Soil. Sci. Soc. Amer. 27:164-167.

8. GHAFFAR, A., G. A. ZENTMYER, and D. C. ERWIN. 1969. Effect of organic amendments on severity of Macrophomina root rot of cotton. Phytopathology 59:1267-1269.

9. GILPATRICK, J. D. 1969. Role of ammonia in the control of avocado root rot with alfalfa meal soil amendment. Phytopathology 59:973-978.

10. GILPATRICK, J. D. 1969. Effect of soil amendments upon inoculum survival and function in Phytophthora root rot of avocado. Phytopathology 59:979-985.

11. GREEN, R. J., and G. C. PAPAVIZAS. 1968. The effect of carbon source, carbon to nitrogen ratios, and organic amendments on survival of propagules of Verticillium albo-atrum in soil. Phytopathology 58:567-570.

12. HUBER, D. M., and H. C. MC KAY. 1968. Effect of temperature, crop, and depth of burial on the survival of Typhula idahoensis sclerotia. Phytopathology 58:961-962.

13. JORDAN, V. W. L., B. SNEH, and B. P. EDDY. 1972. Influence of organic soil amendments on Verticillium dahliae and on the microbial composition of the strawberry rhizosphere. Ann. Appl. Biol. 70:139-148.

14. LEWIS, J. A., and G. C. PAPAVIZAS. 1971. Effect of sulfur-containing volatile compounds and vapors from cabbage decomposition on Aphanomyces euteiches. Phytopathology 61:208-214.

15. LEWIS, J. A., and G. C. PAPAVIZAS. 1971. Damping-off of sugarbeets caused by Aphanomyces cochlioides as affected by soil amendments and chemicals in the greenhouse. Plant Dis. Reptr. 55:440-444.

16. MANNING, W. J., and D. F. CROSSAN. 1969. Field and greenhouse studies on the effects of plant amendments on Rhizoctonia hypocotyl rot of snapbean. Plant Dis. Reptr. 53:227-231.

17. PAPAVIZAS, G. C. 1966. Suppression of Aphanomyces root rot of peas by cruciferous soil amendments. Phytopathology 56:1071-1075.

18. PAPAVIZAS, G. C. 1967. Evaluation of various media and antimicrobial agents for isolation of Fusarium from soil. Phytopathology 57:848-852.

19. PAPAVIZAS, G. C. 1967. Survival of root-infecting fungi in soil. I. A quantitative propagule assay method of observation. Phytopathology 57:1242-1246.

20. PAPAVIZAS, G. C. 1968. Survival of root-infecting fungi in soil. VI. Effect of amendments on bean root rot caused by Thielaviopsis basicola and on inoculum density of the causal organism. Phytopathology 58:421-428.

21. PAPAVIZAS, G. C. 1969. Survival of root-infecting fungi in soil. XI. Survival of Rhizoctonia solani as affected by inoculum concentration and various soil amendments. Phytopathol. Z. 64:101-111.

22. PAPAVIZAS, G. C. (ed.). 1974. The relation of soil microorganisms to plant diseases. So. Coop. Series Bull. 183. 98 p.

23. PAPAVIZAS, G. C., and P. B. ADAMS. 1969. Survival of root-infecting fungi in soil. XII. Germination and survival of endoconidia and chlamydospores of Thielaviopsis basicola in fallow soil and in soil adjacent to germinating bean seed. Phytopathology 58:371-378.

24. PAPAVIZAS, G. C., P. B. ADAMS, and J. A. LEWIS. 1968. Survival of root-infecting fungi in soil. V. Saprophytic multiplication of Fusarium solani f. sp. phaseoli in soil. Phytopathology 58:414-420.

25. PAPAVIZAS, G. C., C. B. DAVEY, and R. S. WOODARD. 1962. Comparative effectiveness of some organic amendments and fungicides in reducing activity and survival of Rhizoctonia solani in soil. Can. J. Microbiol. 8:915-922.

26. PAPAVIZAS, G. C., and J. A. LEWIS. 1971. Effect of amendments and fungicides on Aphanomyces root rot of peas. Phytopathology 61:215-220.

27. PAPAVIZAS, G. C., and J. A. LEWIS. 1971. Black root rot of bean and tobacco caused by Thielaviopsis basicola as affected by soil amendments and fungicides in the greenhouse. Plant Dis. Reptr. 55:352-356.

28. PAPAVIZAS, G. C., J. A. LEWIS, and P. B. ADAMS. 1968. Survival of root-infecting fungi in soil. II. Influence of amendment and soil carbon-to-nitrogen balance on Fusarium root rot of beans. Phytopathology 58:365-372.

29. PAPAVIZAS, G. C., J. A. LEWIS, and P. B. ADAMS. 1970. Survival of root-infecting fungi in soil. XIV. Effect of amendments and fungicides on bean root rot caused by Thielaviopsis basicola. Plant Dis. Reptr. 54:114-118.

30. SRIVASTAVA, U. S., and S. SINHA. 1971. Effect of various soil amendments on the wilt of coriander (Coriandrum sativum L.). Indian J. Agr. Sci. 41:779-782.

31. WARREN, H. L., and T. KOMMEDAHL. 1973. Fertilization and wheat refuse effects on Fusarium species associated with wheat roots in Minnesota. Phytopathology 63:103-108.

Influence of Volatiles of Plant Origin on Soil-Borne Plant Pathogens

R. G. LINDERMAN and R. G. GILBERT—*Research Plant Pathologist and Research Microbiologist, U.S. Department of Agriculture, Agricultural Research Service, Corvallis, Oregon, and Phoenix, Arizona.*

Plant tissue, in various states from living to dead and decomposing, is a major component of the soil environment. Chemicals emanating or diffusing from these tissues into the soil influence the soil microflora and microfauna in many ways. The effects of plant residues may be either beneficial or detrimental to successive crops, depending upon environmental conditions and upon the length of time the residues have been decomposing (41, 48, 49). Considerable effort has been expended to gain an understanding of the chemical and biological processes associated with decomposition and the effects of the decomposition products released into the soil. Hopefully, such an understanding will enable us to decrease the incidence or severity of plant diseases.

Root exudates have been studied extensively by many workers, and the subject has been thoroughly reviewed (54, 55, 57). Unfortunately, as Rovira (55) has pointed out, most of the information on root exudates has been obtained from plants grown aseptically in solutions or sterile sand; very little is known about root exudates from plants grown in soil under more natural conditions. Also, many questions concerning quantities of root exudates released into soil and distances exudates diffuse out from roots remain unanswered.

The problem of isolating, purifying, and identifying substances that are released into soil has made it difficult to elucidate the biological and chemical processes associated with the decomposition of plant material or with root exudates. Rovira (55) defined such problems as variable production under different environmental conditions; loss of activity in soil; loss of activity during storage, extraction, and purification; and standardization of bioassay procedures. Recent research has resulted in the identification of at least the dominant phytotoxic organic substances resulting from decomposing plant residues in soil (22, 62). Seedling bioassays of concentrated or fractionated water solutions from the soil-residue mixtures were used to associate the toxic activity with specific compounds. Chemical analyses of substances exuded from roots into aseptic water solutions have also resulted in identification of many different classes of compounds.

The procedures used to collect, concentrate, partially purify, and assay for specific biological activities of root exudates or plant residue decomposition products may well have eliminated many biologically active volatile compounds. Most of the research on the effects of volatiles on microbial activity has focused primarily on either wood or soil (15). In the following discussion, we will consider only soil, specifically the direct and indirect effects of volatiles in the soil on soil-borne fungus pathogens.

Organic volatile compounds in the soil probably originate from plant roots, plant residues, soil microbes, and the aboveground atmosphere. Since almost nothing is known about the effects of atmospheric volatiles on soil fungi, they will not be included.

UNIQUE CHARACTERISTICS OF VOLATILES.—Fries (15) has thoroughly reviewed the effects of volatile substances on fungi. He emphasized that volatile compounds have a relatively high vapor pressure at physiologic temperatures, that they are capable of rapid movement as gases through soil, and that they may be biologically active in extremely small quantities. Volatiles may contact an organism through either the liquid or gas phase. They can induce biological activity at a greater distance from their source than can nonvolatile, strictly water-soluble compounds. Hutchinson (27) pointed out that substances in solution are more likely to be chemically inactivated than are gaseous materials. Because of the microenvironments of microbial action, large quantities of material over large areas are unnecessary in producing a significant effect on microbial ecology (6, 27). Further, the concentrations of volatiles necessary to induce biological responses need not persist for very long periods of time. In addition, volatiles are readily metabolized by soil microbes. These characteristics have contributed to the elusiveness of volatiles faced by investigators seeking the cause of some biological phenomenon.

VOLATILE ROOT EXUDATES.—Little is known about organic volatile root exudates, because most work has dealt only with water-soluble, nonvolatile components (54, 55, 57).

Volatiles from Allium spp.—Coley-Smith and his colleagues (7-11, 13, 28) studied the volatiles from *Allium* spp., which stimulate sclerotial germination of *Sclerotium cepivorum*. Germination of *S. cepivorum* sclerotia in nonsterile soil is a highly specific response, induced only by root exudates from members of the genus *Allium*. The active compounds apparently diffuse primarily from the root tips. The active volatile compounds in extracts of *Allium* spp. are alkyl sulfides (10) characteristic of *Allium* spp., although small quantities also occur in the Cruciferae. An apparent contradiction existed in these studies, however, in that distillands of onion and garlic possessed more stimulatory capacity than did the distillates, and few or no stimulatory volatile allyl sulfides, which were highly active components of extracts, were evolved by intact *Allium* plants. Recently, however, Coley-Smith and Cooke (7) reported that intact *Allium* plants and distillands of onion and garlic contained small quantities of nonvolatile, water-insoluble alkyl cysteine sulfoxides that diffuse into soil. When these compounds were metabolized by soil bacteria, a stimulatory mixture of volatile alkyl sulfides was produced (28). This work suggests that stimulation of *S. cepivorum* sclerotial germination in soil by *Allium* spp. is mediated by soil bacteria. These workers (8, 13) also have shown that sclerotia exude into the soil nutrients that stimulate the microbial activity presumed to be responsible for inhibition of sclerotial germination. Yet their results indicate that the volatiles in some way reverse the inhibitory effect of the soil microflora on the sclerotia, removing the restraint on an endogenous metabolic pathway responsible for the initiation of germination. Thus, components of the soil microflora are apparently responsible for the production of volatile germination stimulants as well as inhibitors. How and where these functions are performed in the sequence of events leading to germination has not been clearly established.

Volatiles from mycorrhizal roots.—The involvement of volatile compounds in the ectomycorrhizal association between *Pinus sylvestris* and *Boletus variegatus* was recently reported by Krupa and Fries (29). They identified the major volatile compounds produced by *B. variegatus* mycelium in culture as ethanol, isobutanol, isoamyl alcohol, acetoin, and isobutyric acid. Further, they analyzed the volatiles produced by *P. sylvestris* roots with and without the fungal symbiont. Roots infected with *B. variegatus* produced two to eight times more volatiles than did noninoculated roots. Those volatiles, however, were not the same as those produced by *B. variegatus* in culture, but were primarily monoterpenes and sesquiterpenes. Since many of the latter compounds have fungistatic properties, Krupa and Fries (29) proposed the hypothesis, also suggested by Marx (40), that the volatile (or nonvolatile) substances produced by roots infected with mycorrhizal fungi may serve two functions. They may restrict the growth and development of the mycorrhizal fungus in the host and thus maintain the state of symbiosis, or they may be released into the rhizosphere, where they may inhibit the activity of soil-borne root pathogens and other rhizosphere organisms. Melin and Krupa (43) supported the first part of this hypothesis when they reported inhibition of mycelial growth in vitro of two mycorrhizal fungi, *B. variegatus* and *Rhizopogon roseolus*, by the same volatiles identified from mycorrhizal roots of *P. sylvestris*. Further, Krupa and Nylund (30) demonstrated that the same volatiles inhibited vegetative growth in vitro of two soil-borne fungus pathogens of pine, *Phytophthora cinnamomi* and *Fomes annosus*, though to a lesser degree than the mycorrhizal fungi. Of the volatiles produced by *B. variegatus* in culture, all except ethanol were inhibitory to *P. cinnamomi* and *F. annosus*. Ethanol vapors stimulated both pathogens. Marx (40) and Park (47) demonstrated

the production in vitro of potent antibiotics by ectomycorrhizal fungi against a wide range of soil-borne fungus pathogens. However, they tested inhibition of one organism against the other in vitro and did not demonstrate whether inhibition was attributable to some volatile factor. Nor did they demonstrate that the inhibitory factor was released from mycorrhizal roots.

Although work on volatile substances evolved from mycorrhizal roots is just beginning, there is strong suggestion that they play a role in the widespread protective action of mycorrhizae against root pathogens, as suggested by Zak (66). Further work is needed to determine if and how such mechanisms can be exploited in efforts to control soil-borne fungus pathogens.

Volatiles from wheat roots.—Several workers (42, 45, 51) have described the volatile root exudates from cereals, including ethylene (51) and acetaldehyde (45), but nothing has been reported on the biological activity of those volatiles on soil-borne fungus pathogens in the soil. We know from other studies, however, that ethylene and acetaldehyde can have significant effects on higher plants and on soil microbes. Research is needed to characterize the activity of these and other gaseous exudates on soil-borne fungus pathogens in soil.

VOLATILES FROM PLANT RESIDUES.—Relatively few studies have been reported on specific effects of volatile plant decomposition products that influence soil-borne fungus pathogens. Even fewer have demonstrated both the release of volatiles from residues decomposing in soil and the response to them of a soil fungus pathogen in soil. Generally, organic compounds released into soil are immediately metabolized by soil microbes. Thus, it is difficult to know what compound, volatile or nonvolatile, may actually be responsible for the observed biological response. Nevertheless, several instances will be considered in which volatiles from plant residues either have a direct effect in vitro on a soil-borne fungus pathogen or initiate an indirect effect in soil mediated by other soil microbes.

S-containing volatiles from crucifers.—Lewis and Papavizas (32, 33) identified volatile sulfur-containing compounds from decomposing crucifer residues and showed they were strongly inhibitory to all stages of *Aphanomyces euteiches* in vitro. The S-containing compounds emanating from cruciferous amendments decomposing in soil suppressed or prevented mycelial growth, zoospore formation, and zoospore motility and germination. Their study offers a plausible explanation for the control of *Aphanomyces* root rot of pea in the field by using cruciferous amendments.

Enhancement of respiration and growth of soil microbes by alfalfa volatiles.—Menzies and Gilbert (44) observed that fungus mycelium grew from soil in sealed glass tubes across an air gap toward moist alfalfa hay in response to some volatile factor emanating from the alfalfa. The increased activity of the total microflora was demonstrated further by increased O_2 uptake and increased bacterial numbers in soils exposed to vapors from a distillate of the alfalfa hay. Gilbert and Griebel (18) showed that respiration could be enhanced in 12 different soils by the alfalfa volatiles, and that the volatiles were neutral organic compounds that rapidly penetrated the soil. The volatiles induced an immediate respiration increase by the soil microflora up to an optimum concentration, beyond which toxicity occurred. Qualitative changes (20) in the soil microflora occurred in response to high concentrations of the volatiles; e.g., some genera of fungi were quite sensitive and were killed, whereas others were tolerant and persisted after the treatment.

Using soil respiration enhancement as the bioassay, coupled with gas chromatographic separation, Owens et al. (46) identified the major active components of the alfalfa distillate as acetaldehyde, isobutyraldehyde, isovaleraldehyde, 2-methylbutanal and/or valeraldehyde, methanol, and ethanol. Experiments using a reconstituted mixture of these compounds enhanced microbial respiration 75% as much as did the crude alfalfa distillate. They further demonstrated that acetaldehyde and methanol were the major active volatile components. Acetaldehyde preferentially increased numbers of soil fungi; methanol had little effect on fungi, but preferentially increased numbers of bacteria. Mixtures of acetaldehyde and methanol synergistically increased bacteria, including actinomycetes.

Effects of alfalfa volatiles on Verticillium dahliae.—On the basis of the previously described studies on respiration enhancement by alfalfa volatiles, it was easily theorized that propagules of specific soil-borne pathogens in soil exposed to the volatiles might also be directly or indirectly affected. Gilbert and Griebel (17) found that exposing soil containing microsclerotia of *Verticillium dahliae* for 24 hours to vapors from a concentrated alfalfa hay distillate eliminated that pathogen from the soil. However, when the concentration of the distillate was reduced or when the exposure time at the higher concentration was shortened to 6 hours, a two- to fivefold population increase occurred, apparently because of vegetative growth from the microsclerotia. Farley et al. (14) reported that such population increases were caused primarily by conidial production after microsclerotial germination. Although no direct observations were made, it appears likely that conidia, or some other nonresistant propagule, contributed to the population increase in soil exposed to the alfalfa volatiles. Further incubation of the soil showed that the expanded *V. dahliae* population ultimately decreased to less than the initial population level. The quantitative and

Fig. 1-3. Fig. 1. System used to study the sclerotial germination response of *Sclerotium rolfsii* exposed to vapors emanating from hanging droplets of alfalfa distillate (AD) across an air gap to the sclerotia pressed into soil. Fig. 2. Sclerotial growth response ratings indicating the degree of stimulation of sclerotial germination and growth of *S. rolfsii* on soil exposed to vapors of alfalfa distillate. One drop of alfalfa distillate was approximately 0.037 ml. Fig. 3. Comparative population changes of bacteria, fungi, and actinomycetes in soil exposed for varying time periods to vapors from 0.15 ml of alfalfa distillate (lines labeled 100) or water (lines labeled 0).

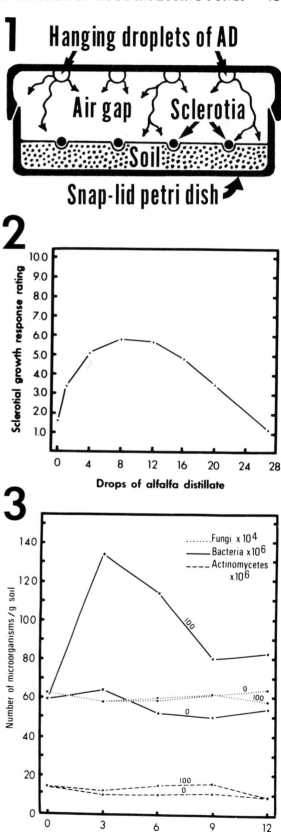

qualitative microbial changes that also occurred in the soil in response to the volatiles suggested that increased competition or antagonism contributed to the eventual decline of *V. dahliae* in the treated soil.

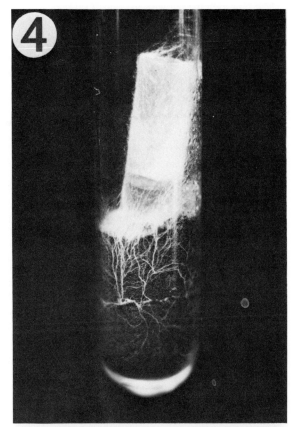

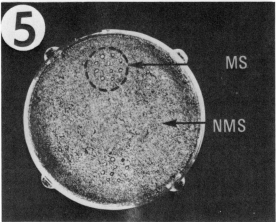

Fig. 4-5. Fig. 4. Mycelial growth from sclerotia of *Sclerotium rolfsii* buried in a sealed soil column in response to vapors emanating from alfalfa distillate in the internal vial. Fig. 5. Illustration of the soil dish mycosphere system used to examine population changes of microbes in soil adjacent to sclerotia of *S. rolfsii*, mycosphere (MS), or away from sclerotia, nonmycosphere (NMS).

Effects of alfalfa volatiles on Sclerotium rolfsii: sclerotial germination.—The desire to directly observe propagules of a soil-borne pathogen respond to the alfalfa volatiles led to our selection of *Sclerotium rolfsii* as a test organism. Not only were the sclerotia large enough to observe directly in soil, but the pathogen is characterized as necrotrophic and as such shows a high dependency on the presence of ample plant debris in the soil in order for significant disease to occur (16). Boyle (4) suggested that the sclerotia of *S. rolfsii* do not contain enough energy to parasitize the host directly, but must first colonize plant residues to obtain enough energy to initiate pathogenesis. Our studies offer another explanation for this dependency on plant residues: It may also be a dependency on the volatiles that emanate from the residues through the soil to trigger germination of the resting sclerotia. Subsequently, vegetative growth from the sclerotia reaches the residue, either by chance or by chemotactic orientation of the hyphae toward the residue. The volatiles may bridge the gap between the resting sclerotia and the residue food base, and thus ultimately between the sclerotia and the host. The following discussion describes our experiments and data in support of this explanation.

In our studies on *S. rolfsii* and alfalfa volatiles, we have looked beyond the macroecology of the pathogen-plant residue-soil interactions, viewed on a field disease-incidence basis, to the microecology of the soil-sclerotium interface when it is exposed to volatiles diffusing through the soil from a more distant residue source. The volatile source droplets were separated from the soil supporting the test sclerotia by an air gap, guaranteeing that observed effects were induced by gases passing across the air gap (Fig. 1). The plastic snap-lid petri dishes used provided a closed yet not absolutely airtight chamber.

When sclerotia were pressed into moist soil and hanging droplets of a distillate of alfalfa hay (AD) were placed on the lid and closed, the vapors emanating from the droplets stimulated vigorous sclerotial germination within 48 hours, whereas sclerotia exposed to water vapor germinated poorly or not at all. The vigor of mycelial growth from each of the sclerotia was subjectively rated after 5 days using a scale of 0 to 10, where 0 indicated no germination and 10 indicated maximum mycelial growth and total exhaustion of the sclerotial contents. Because this was a subjective rating system, responses within each experiment were relative to each other and especially to the water controls. The concentration of AD vapors to which the soil and sclerotia were exposed was varied by increasing the number of hanging AD droplets placed on the dish lid. Sclerotial germination vigor was increased over that of water vapor controls by vapors from only 1 drop of AD (approximately 0.037 ml), reached a maximum at 8-10 drops, and, beyond that, toxicity was indicated by reduced germination vigor (Fig. 2) (36). At 27 drops or 1.0 ml of AD/dish, sclerotial germination and mycelial growth were nearly completely inhibited. At that vapor concentration, some sclerotia were killed, but the survivors germinated vigorously when the AD droplets were removed. A similar dosage-response phenomenon occurred in vitro on 2% purified water agar where vegetative growth was stimulated at lower dosages, whereas at higher dosages growth was initially inhibited, but later resumed with even more vigor (more dense growth). Thus the response of *S. rolfsii* sclerotia to volatiles from alfalfa was similar to that of *V. dahliae*, in that lower dosages were stimulatory and higher dosages were inhibitory or lethal.

The fate of sclerotia triggered to germinate by the volatiles was also similar in that a net reduction in total numbers resulted, even though by a different mechanism. For example, the number of viable sclerotia of *S. rolfsii* buried for 3 weeks in soil contained in sealed tubes also containing a vial with 1.0 ml of AD on the soil surface was reduced by 50%. This net reduction in numbers was the result of germination (Fig. 4) and complete exhaustion of sclerotial contents, followed by hyphal lysis. In addition, germinated sclerotia whose contents were not completely exhausted were more readily colonized by saprophytic and/or antagonistic organisms. Among those fungi most frequently observed colonizing sclerotia after exposure to AD vapors were species of *Chaetomium*, *Trichoderma*, and *Fusarium* (36).

The germination response to the crude alfalfa distillate was nearly duplicated by a synthetic mixture of the major volatiles identified in the distillate by Owens et al. (46), namely methanol, acetaldehyde, isobutyraldehyde, and isovaleraldehyde. Tested individually, these compounds were much less stimulatory, and isobutyraldehyde was completely ineffective (36).

Fries (15) has pointed out that, in general, volatiles function primarily as metabolic regulators, and small quantities are sufficient to induce a response. We found (36) that exposure of sclerotia on soil to AD vapors for only 15 minutes stimulated as much germination response as did exposure for the whole 5-day assay. This evidence indicates that the volatiles are stimulants or metabolic regulators, not nutrient substrates. However, we cannot say that the volatiles in a 15-minute exposure act as stimulants directly and only on the sclerotia themselves. In fact, the volatiles induced some increased germination response with a 2-minute exposure, during which the sclerotia had not yet completely imbibed water. Thus, it seems that the soil supporting the sclerotia may be involved in the germination response by physically and/or biologically absorbing the volatiles during short exposures, possibly even more than the sclerotia absorb.

We tested the hypothesis that stimulation of sclerotial germination by short exposures to the volatiles was mediated by the microorganisms in the soil surrounding the sclerotia. Sclerotia pressed into soil were exposed to AD vapors for 15, 30, 60, or 180 minutes, after which the sclerotia were removed with forceps, transferred to fresh soil in dishes, and incubated the remainder of the 5-day assay in the presence of water vapors only (Table 1, treatment B). New sclerotia that had not been exposed to the AD vapors were placed in the depressions in the AD-exposed soil left by the removed sclerotia. This AD-exposed soil and new sclerotia (treatment A) were also incubated the remainder of the 5 days with water vapors only. Control sclerotia were exposed to AD vapors for the various time periods, the AD droplets were replaced by water droplets, and the dishes were incubated for the remainder of the assay (treatment C). The results of this experiment (Linderman and Gilbert, *unpublished data*) indicated that germination vigor was less in either treatment A or B than in the control treatment C, but that sclerotial germination in A at all treatment times was consistently higher than in B (Table 1). The action of the volatiles in stimulating sclerotial germination of *S. rolfsii*, therefore, may be on the microflora in the soil surrounding the sclerotia, as well as on the sclerotia themselves. In this regard, Griebel and Owens (21) have demonstrated the transient nature of microbial utilization of the alfalfa volatiles, ethanol and acetaldehyde, using radioactively labeled carbon. When soil was exposed to the labeled volatiles, 80-90% of the CO_2 evolved during the respiration increase was radioactive, indicating that the soil microbes metabolized the labeled volatiles and did not use their reserves. During the increased respiratory activity, no changes in numbers of soil microbes occurred, and none occurred during the subsequent 2 weeks.

Interaction of volatiles and soil microbes on Sclerotium rolfsii.—Mycelial growth of *S. rolfsii* through soil may be inhibited by antagonistic microorganisms (2, 5), or antibiotic-producing organisms associated with the

TABLE 1. Influence of alfalfa volatiles that stimulate germination of sclerotia of *Sclerotium rolfsii* on soil relative to their site of action on soil or sclerotia

Treatment time (min)[b]	Sclerotial growth response rating[a]		
	Treatment A[c]	Treatment B[d]	Treatment C[e]
15	2.5	1.8	3.6
30	3.4	1.5	4.6
60	2.7	1.5	4.5
180	3.7	2.0	3.9

[a]Averaged sclerotial growth ratings (0 to 10 scale) from three experiments with two replications/treatment and 10 sclerotia/replication.

[b]Exposure to vapors from 0.15 ml alfalfa distillate (AD) for time period; then AD droplets replaced with water droplets for duration of 5-day assay.

[c]Treatment A: sclerotia and soil exposed to AD vapors, then sclerotia removed to soil not exposed to AD vapors (treatment B), and new sclerotia placed in remaining depressions (germination data from the new sclerotia).

[d]Treatment B: sclerotia exposed to AD vapors in treatment A placed into soil not exposed to AD.

[e]Treatment C: sclerotia exposed to AD vapors for various time periods and left in place after AD was removed.

sclerotia at the soil-sclerotium interface may inhibit sclerotial germination (24). These inhibitory phenomena have been correlated with the addition to the soil of nitrogenous amendments known to decrease the incidence of disease caused by *S. rolfsii* (31). Exposure of soil to alfalfa volatiles also enhances the activity of saprophytic or antagonistic organisms that contribute to the decrease in sclerotial numbers of both *S. rolfsii* and *V. dahliae* (17, 36, 37). These reports suggest that soil treatments that enhance specific microbial activity might inhibit sclerotial germination of *S. rolfsii*. Yet our studies demonstrate that alfalfa volatiles stimulate germination in soil (36). The explanation for this apparent discrepancy seems to lie in the time of exposure of the soil to the volatiles. When we exposed soil and sclerotia simultaneously to the alfalfa volatiles, germination was stimulated. But when we exposed the soil to the volatiles first and then assayed sclerotial germination 7 days later, again using the volatiles as a germination stimulant, the vigor of mycelial growth from sclerotia on AD-treated soil was markedly reduced compared to the response on soil previously exposed only to water vapors (37). This response was also duplicated with a synthetic mixture of some of the alcohols and aldehydes found in the crude alfalfa distillate (46). Thus, prior exposure of soil to the volatiles apparently affected the organisms capable of inhibiting sclerotial germination or reducing germination vigor, presumably by increasing their numbers and/or metabolic activity.

Our population studies indicated that numbers of fungi and actinomycetes in AD-treated soil were relatively unchanged, but bacterial numbers had significantly increased within 7 days at the dosage used (Fig. 3). The expanded bacterial population declined by 9 days, however, although it stabilized at a level higher than the water control soil. One might conclude from these results that the volatiles enhanced the bacterial population in treated soil and that bacterial activity was inhibitory to sclerotial germination. However, when we lowered the dosage of volatiles to only one drop of AD (0.037 ml) for 1 week, sclerotial germination was inhibited relative to the water control, yet bacterial populations in that soil were not significantly increased. Further, the addition of numerous substrates to the soil greatly increased the numbers of bacteria, but did not increase the capacity of the soil to inhibit sclerotial germination.

It is possible that the volatiles induced qualitative, but not necessarily quantitative, changes in the microflora or that the volatiles induced both qualitative and quantitative changes in the soil microflora, but only in the immediate vicinity of the sclerotia, during the germination assay. To test this hypothesis, we exposed sclerotia on soil to AD vapors for 1 week, removed the germinated sclerotia and attached mycelium, and assayed the inhibitory capacity of the soil with new sclerotia and AD vapors. The combination of fungus plus AD vapors decreased the vigor of sclerotial germination more than with AD vapors alone or with the fungus alone (37). These data suggest that *S. rolfsii* sclerotia in some way contributed to their own inhibition, especially in the presence of alfalfa volatiles.

Sclerotium rolfsii: the mycosphere effect.—Several of

our studies mentioned thus far, plus reports of others (24), strongly suggest that soil microorganisms that may influence sclerotia of *S. rolfsii* probably do so by their activity at the soil-sclerotium interface. The net effect of the exchanges at that interface may result in either stimulation or inhibition. Therefore, we compared the qualitative and quantitative microbial changes in soil adjacent to and surrounding sclerotia of *S. rolfsii* with those in soil away from the sclerotia. Antecedent exposure of soils to vapors from AD or water was also used to determine their effects on the comparative microbial populations. Populations of fungi, bacteria, and actinomycetes in soil collected around groups of 10 sclerotia (18-mm diameter) (Fig. 5) were compared with soil collected about 1 cm away from the group using standard dilution plate techniques. We found that the microbial activity in soil near sclerotia, referred to as the "mycosphere" (MS), increased markedly within 24 hours after sclerotia were pressed into the soil, whereas no changes occurred in soil not near the sclerotia, "nonmycosphere" (NMS) (19). We called this increased microbial activity in the MS soil the "mycosphere effect," in the same sense as the "rhizosphere effect" applies to plant roots. The MS effect primarily involved an increase in the numbers of bacteria. Further, specific bacteria seemed to be selectively increased, as indicated by increased numbers of bacteria tolerant of 100 ppm streptomycin sulfate and oxgall bile salts in the plating medium. The energy substrate for the MS effect was apparently of sclerotial origin and not from culture nutrients adhering to the sclerotia, because washing sclerotia for 15 minutes did not decrease the effect, and 18-hour sclerotial leachate (concentrated 10×) also induced the MS effect. Antecedent exposure of the soil to AD vapors increased the MS effect when either sclerotia or sclerotial leachate was used. After 72 hours the total bacterial population was higher in MS soil previously exposed to AD vapors than it was in soils exposed to water vapors. The AD vapors particularly increased the numbers of streptomycin- and oxgall-tolerant bacteria in the MS (19).

Based on the above data and that reported by Lingappa and Lockwood (39) and others (9, 13, 58, 59), one might again conclude that nutrients diffusing from sclerotia into soil at the soil-sclerotium interface increase the activity of inhibitory organisms. Further, the activity of inhibitory organisms apparently is also increased by antecedent exposure to alfalfa volatiles, so that when new sclerotia are introduced into the soil in the germination assay, these organisms may multiply to inhibitory population levels very rapidly and in some way reduce sclerotial germination vigor.

Coley-Smith and Dickinson (8) reported that treatment of *S. cepivorum* sclerotia with calcium hypochlorite increased sclerotial exudation, which presumably would increase the nutrient substrate on which inhibitory organisms might multiply. Smith (58, 59) recently reported that drying sclerotia of *S. rolfsii* increased their leakage of nutrients, on which sclerotia-colonizing organisms grew, and as a result the sclerotia were quickly rotted. Drying also induced increased sclerotial germination, but not necessarily rotting. In our studies, treatment of sclerotia of *S. rolfsii* with sodium hypochlorite enhanced sclerotial germination, even in the absence of any alfalfa-volatile stimulants (38). These data suggest, as did data presented earlier, that sclerotial exudation enhances the activity of specific microbes, which under certain conditions may function in a stimulatory rather than an inhibitory manner.

When we isolated bacteria from both MS and NMS soil and assayed their in vitro activity on several media in dual cultures with *S. rolfsii* (19), we found that MS bacteria affected *S. rolfsii* more than did NMS bacteria; e.g., MS bacteria as a group were more stimulatory or more inhibitory (depending on the medium) than were NMS bacteria. Most of the MS bacteria inhibited *S. rolfsii* growth on Difco plate-count agar, had no effect on its growth on potato-dextrose agar, and stimulated its growth on Czapek-Dox agar. Thus, the same MS bacterium may respond to different nutritional conditions by producing metabolites that either inhibit or stimulate *S. rolfsii*. And the qualitative and quantitative nutritional changes that take place in the MS, either from nutrients in sclerotial exudate or from external sources like plant residues or root exudates, may determine what metabolic message the MS organisms give to the sclerotium. This could mean that MS organisms, in effect, sense the environment around *S. rolfsii* sclerotia by metabolically reacting to changes in the nutritional regime in the MS and subsequently releasing inhibitors or stimulators that affect sclerotial germination.

VOLATILE METABOLITES OF SOIL MICROBES.

—In the discussion above, we have considered evidence suggesting that soil microbes, whether in or out of the rhizosphere or mycosphere, may utilize various volatile or nonvolatile substrates and release volatile metabolites into the soil that may influence the behavior of the propagules of soil-borne pathogens. Coley-Smith and his colleagues have demonstrated that precursor materials exuded from *Allium* spp. into the soil are metabolized by soil bacteria, which in turn release volatiles that stimulate germination of sclerotia of *S. cepivorum*. Our own studies suggest that bacteria in the mycosphere are capable of either stimulating or inhibiting *S. rolfsii* sclerotial germination in soil or vegetative growth in vitro.

Recent reports suggest that volatile metabolites of microbial activity released into the soil atmosphere may be involved in the widespread phenomenon of soil fungistasis (3, 12, 25, 26, 53, 60, 65). Others report significant inhibition of fungal growth in vitro by volatiles from soil microbes (52). Hora and Baker (25, 26) have associated volatile fungistatic factors with the metabolic activities of actinomycetes, especially in alkaline soils. They have also implicated *Trichoderma* spp. as possible sources of volatile inhibitors of spore germination. Dennis and Webster (12) have also reported the production of volatile inhibitors by *Trichoderma* spp. as assayed in vitro against several soil-borne pathogens. One of the main volatiles they identified was acetaldehyde, reported earlier to be a major component of the alfalfa volatiles that affected *S. rolfsii* sclerotia. Acetaldehyde has also been reported by others to have sporostatic properties in its vapor phase on several fungi (52). Wells et al. (64) recently reported *Trichoderma harzianum* as a

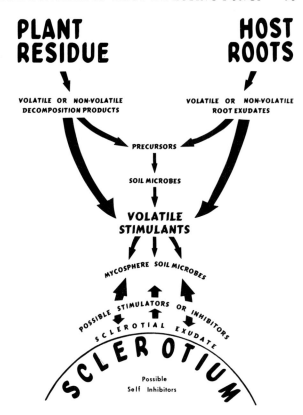

Fig. 6. Hypothetical scheme describing three major sources of volatile organic compounds in soil (plant residues, plant roots, and soil microbes) and their influence and possible modes of action on soil microbes and, specifically, propagules (sclerotia) of soil-borne fungus pathogens in soil.

biological amendment to field soil to reduce the disease caused by *S. rolfsii*. Whether acetaldehyde or some other volatile metabolites evolved by *T. harzianum* might be involved in some part of the control mechanism is unknown.

A classic example in which soil microbes release volatile metabolites is the induction of commercial mushroom (*Agaricus bisporus*) sporophores by metabolites from soil bacteria like *Pseudomonas putida* in the casing soil (23). Although the effects of volatiles in this instance do not involve a soil-borne fungus pathogen, the example emphasizes the metabolic exchanges that can take place between soil microbes and a specific fungus via the gaseous phase. Volatile compounds evolved by the mushroom mycelium (ethanol, ethyl acetate, acetone, and acetaldehyde) (23) induced qualitative changes in the soil microflora until bacterial stimulators like *P. putida* were the dominant organisms. Bacterial metabolites (apparently still unidentified) in turn induced sporophore production. The significant point here is that volatiles (alcohols and aldehydes) are produced by the fungus and released into the MS, and they inhibit some microbes and stimulate others. Thus the mycelium itself selects a highly adapted microflora dominated by sporophore stimulators, which in nature effectively aid its reproduction and dispersion (23).

If one were to substitute a basidiomycete soil-borne pathogen like *Armillaria mellea* for *A. bisporus*, then volatiles would play a significant role in the disease epidemiology. For *A. mellea*, however, the mycelial metabolites presumably released into the soil to establish its MS microflora apparently have not been identified. Ethanol, however, is released as a gaseous metabolite into soil by soil microbes and induces increased rhizomorph production (50, 63). A significant point made by Pentland (50) is that *A. mellea*, in a natural soil environment, would be exposed rather continuously to low concentrations of ethanol evolved by 'soil microbes. In her tests (50), exposure to low concentrations of ethanol at regular intervals was as stimulatory for rhizomorph production as was one higher initial concentration (63).

Ethylene.—Whereas ethylene is probably a significant gas in soil microbiology, its elusiveness, like that of other volatiles, probably accounts for the dearth of information about its biological activity on soil microbes, including pathogenic fungi. Abeles (1) points out that plants are the largest source of natural ethylene, but other sources include soil microbes. Ethylene produced in infected plant material in some instances is thought to alter susceptibility to certain soil-borne fungus pathogens, a reaction that might ultimately influence their survival in soil. Such effects, however, are caused by activity of ethylene within the host tissue.

Ethylene from various sources, such as the aboveground atmosphere, is readily absorbed by the soil microflora (1). Soil microbes are also a source of ethylene in soil, especially under conditions of restricted aeration and high organic matter. Smith and Russell (61) reported that ethylene accumulated in waterlogged or poorly aerated soil in sufficient concentrations to alter the growth of wheat plants. Rovira and Vendrell (56) detected physiologically significant concentrations of ethylene in soil, especially in radiation-sterilized soil. The atmosphere over their unsterilized soil controls, however, contained nearly 0.5 ppm ethylene. Smith (60) reported that ethylene occurred in aerated soils at significant concentrations as a result of microbial activity. His studies also demonstrated that purging the soil of its ethylene with water-saturated air nullified soil fungistasis. Conversely, ethylene introduced back into air-purged soil at concentrations less than 1 ppm restored soil fungistasis. He demonstrated the efficacy of ethylene as a volatile fungistatic factor by inhibiting germination of *S. rolfsii* sclerotia, which otherwise would germinate spontaneously on soil.

In recent studies, Linderman (*unpublished data*) found that significant amounts of ethylene were produced when green azalea leaves were colonized by *Cylindrocladium* spp. in soil (34, 35). Smith (60) also showed a correlation between soils with high organic matter and the ability to produce ethylene. These studies suggest that colonized green residues or other organic matter may be the primary source of ethylene in soil. More research is needed to define the specific effects of ethylene from all sources on fungi in soil.

SUMMARY AND CONCLUSIONS.—In the preceding discussion, we have presented thoughts and experimental evidence concerning the biological activity of organic volatile compounds on soil-borne pathogenic fungi (summarized in Fig. 6). We hope the reader can now identify certain unique characteristics of volatiles, compared to nonvolatile, water-soluble compounds, which make their potential for biological activity great. By moving through the air spaces of soil, volatile compounds can diffuse rapidly from their source to their site of action. Although we know that they induce biological responses at very low concentrations and often in extremely short times, their mode of action as apparent metabolic regulators is not yet understood.

We can point to several major sources of organic volatiles. We have concentrated on those emanating from plant roots, plant residues, and soil microbes, but the aboveground atmosphere is also an important source. Regardless of the source, however, there seems to be a common denominator in many of the effects induced by volatiles. The volatiles (and nonvolatiles) diffusing through soil are all exposed to physical and biological pressures. The soil microbes, especially bacteria, apparently absorb and metabolize organic volatiles rapidly, probably more rapidly than do propagules of soil-borne fungi. Further, because of the mycosphere effect around fungus propagules, the mycosphere organisms probably are exposed to the volatiles before the fungus propagule is. Therefore, many of the net reactions of organic compounds on soil-borne fungus pathogens that researchers observe and report are undoubtedly mediated by other soil microbes. We have mentioned examples where volatiles stimulated soil respiration, stimulated or inhibited germination of fungus propagules, increased fungistatic potential of soil to pathogen activity, accentuated the metabolic exchanges between pathogen propagule and mycosphere organisms, and were involved as a selective pressure in the establishment of the mycosphere.

We visualize soil microorganisms as a biological filter for organic compounds (volatile or nonvolatile) that are diffusing through soil. These compounds must pass through the filter before they reach the fungus propagule and induce an effect. Interestingly enough, the fungus propagule itself is a source of organic substrate that must pass, though in the opposite direction, through the filter to be metabolized by the microbes therein. Products of that metabolism, whether volatile or nonvolatile, then may regulate the activity of that fungus propagule during its immediate future.

LITERATURE CITED

1. ABELES, F. B. 1973. Ethylene in plant biology. Academic Press, New York and London. 302 p.
2. AVIZOHAR-HERSHENZON, Z., and P. SHACKED. 1969. Studies on the mode of action of inorganic nitrogenous amendments on Sclerotium rolfsii in soil. Phytopathology 59:288-292.
3. BALIS, C., and V. KOUYEAS. 1968. Volatile inhibitors involved in soil mycostasis. Ann. Inst. Phytopathol. Benaki, N.S. 8:145-149.
4. BOYLE, L. W. 1961. The ecology of Sclerotium rolfsii with emphasis on the role of saprophytic media. Symp. on Sclerotium rolfsii. Phytopathology 51:117-119.
5. BROADBENT, P., K. F. BAKER, and Y. WATERWORTH. 1971. Bacteria and actinomycetes

antagonistic to fungal root pathogens in Australian soils. Austral. J. Biol. Sci. 24:925-944.

6. BURGES, A. 1963. Some problems in soil microbiology. Trans. Br. Mycol. Soc. 46:1-14.

7. COLEY-SMITH, J. R., and R. C. COOKE. 1971. Survival and germination of fungal sclerotia. Annu. Rev. Phytopathol. 9:65-92.

8. COLEY-SMITH, J. R., and D. J. DICKINSON. 1971. Effects of sclerotia of Sclerotium cepivorum Berk. on soil bacteria. The nature of substances exuded by sclerotia. Soil Biol. Biochem. 3:27-32.

9. COLEY-SMITH, J. R., D. J. DICKINSON, J. E. KING, and R. W. HOLT. 1968. The effect of species of Allium on soil bacteria in relation to germination of sclerotia of Sclerotium cepivorum Berk. Ann. Appl. Biol. 62:103-111.

10. COLEY-SMITH, J. R., and J. E. KING. 1969. The production by species of Allium of alkyl sulphides and their effect on germination of sclerotia of Sclerotium cepivorum Berk. Ann. Appl. Biol. 64:289-301.

11. COLEY-SMITH, J. R., and J. E. KING. 1970. Response of resting structures of root-infecting fungi to host exudates: an example of specificity, p. 130-133. In T. A. Toussoun, et al. (ed.), Root diseases and soil-borne pathogens. Univ. Calif. Press, Berkeley and Los Angeles.

12. DENNIS, C., and J. WEBSTER. 1971. Antagonistic properties of species groups of Trichoderma. II. Production of volatile antibiotics. Trans. Br. Mycol. Soc. 57:41-48.

13. DICKINSON, D. J., and J. R. COLEY-SMITH. 1970. Stimulation of soil bacteria by sclerotia of Sclerotium cepivorum Berk. in relation to soil fungistasis. Soil Biol. Biochem. 2:157-162.

14. FARLEY, J. D., S. WILHELM, and W. C. SNYDER. 1971. Repeated germination and sporulation of microsclerotia of Verticillium albo-atrum in soil. Phytopathology 61:260-264.

15. FRIES, N. 1973. Effects of volatile organic compounds on the growth and development of fungi. Trans. Br. Mycol. Soc. 60:1-21.

16. GARREN, K. H. 1964. Inoculum potential and differences among peanuts in susceptibility to Sclerotium rolfsii. Phytopathology 54:279-281.

17. GILBERT, R. G., and G. E. GRIEBEL. 1969. The influence of volatile substances from alfalfa on Verticillium dahliae in soil. Phytopathology 59:1400-1403.

18. GILBERT, R. G., and G. E. GRIEBEL. 1969. Stimulation of soil respiration by volatiles from alfalfa. Proc. Soil Sci. Soc. Amer. 33:270-273.

19. GILBERT, R. G., and R. G. LINDERMAN. 1971. Increased activity of soil microorganisms near sclerotia of Sclerotium rolfsii in soil. Can. J. Microbiol. 17:557-562.

20. GILBERT, R. G., J. D. MENZIES, and G. E. GRIEBEL. 1969. The influence of volatiles from alfalfa upon growth and survival of soil microorganisms. Phytopathology 59:992-995.

21. GRIEBEL, G. E., and L. D. OWENS. 1972. Nature of the transient activation of soil microorganisms by ethanol or acetaldehyde. Soil Biol. Biochem. 4:1-8.

22. GUENZI, W. D., and T. M. MC CALLA. 1966. Phytotoxic substances extracted from soil. Proc. Soil Sci. Soc. Amer. 30:214-216.

23. HAYES, W. A., P. E. RANDLE, and F. T. LAST. 1969. The nature of the microbial stimulus affecting sporophore formation in Agaricus bisporus (Lange) Sing. Ann. Appl. Biol. 64:177-187.

24. HENIS, Y., and I. CHET. 1968. The effect of nitrogenous amendments on the germinability of sclerotia of Sclerotium rolfsii and on their accompanying microflora. Phytopathology 58:209-211.

25. HORA, T. S., and R. BAKER. 1970. Volatile factor in soil fungistasis. Nature (London) 225:1071-1072.

26. HORA, T. S., and R. BAKER. 1972. Soil fungistasis: microflora producing a volatile inhibitor. Trans. Br. Mycol. Soc. 59:491-500.

27. HUTCHINSON, S. A. 1971. Presidential address: biological activity of volatile fungal metabolites. Trans. Br. Mycol. Soc. 57:185-200.

28. KING, J. E., and J. R. COLEY-SMITH. 1969. Production of volatile alkyl sulphides by microbial degradation of synthetic alliin and alliin-like compounds, in relation to germination of sclerotia of Sclerotium cepivorum Berk. Ann. Appl. Biol. 64:303-314.

29. KRUPA, S., and N. FRIES. 1971. Studies on ectomycorrhizae of pine. I. Production of volatile organic compounds. Can. J. Bot. 49:1425-1431.

30. KRUPA, S., and J. NYLUND. 1972. Studies on ectomycorrhizae of pine. III. Growth inhibition of two root pathogenic fungi by volatile organic constituents of ectomycorrhizal root systems of Pinus silvestris L. Eur. J. For. Pathol. 2:88-94.

31. LEACH, L. D., and A. E. DAVEY. 1942. Reducing southern sclerotium rot of sugar beets with nitrogenous fertilizers. J. Agr. Res. 46:1-18.

32. LEWIS, J. A., and G. C. PAPAVIZAS. 1970. Evolution of sulfur-containing volatiles from decomposition of crucifers in soil. Soil Biol. Biochem. 2:239-246.

33. LEWIS, J. A., and G. C. PAPAVIZAS. 1971. Effect of sulfur-containing volatile compounds and vapors from cabbage decomposition on Aphanomyces euteiches. Phytopathology 61:208-214.

34. LINDERMAN, R. G. 1972. Isolation of Cylindrocladium from soil or infected azalea stems with azalea leaf traps. Phytopathology 62:736-739.

35. LINDERMAN, R. G. 1973. Formation of microsclerotia of Cylindrocladium spp. in infected azalea leaves, flowers, and roots. Phytopathology 63:187-191.

36. LINDERMAN, R. G., and R. G. GILBERT. 1969. Stimulation of Sclerotium rolfsii in soil by volatile components of alfalfa hay. Phytopathology 59:1366-1372.

37. LINDERMAN, R. G., and R. G. GILBERT. 1973. Influence of volatile compounds from alfalfa hay on microbial activity in soil in relation to growth of Sclerotium rolfsii. Phytopathology 63:359-362.

38. LINDERMAN, R. G., and R. G. GILBERT. 1973. Behavior of sclerotia of Sclerotium rolfsii produced in soil or in culture regarding germination stimulation by volatiles, fungistasis, and sodium hypochlorite treatment. Phytopathology 63:500-504.

39. LINGAPPA, B. T., and J. L. LOCKWOOD. 1964. Activation of soil microflora by fungus spores in relation to soil fungistasis. J. Gen. Microbiol. 35:215-227.

40. MARX, D. H. 1969. The influence of ectotrophic mycorrhizal fungi on the resistance of pine roots to pathogenic infections. I. Antagonism of mycorrhizal fungi to root pathogenic fungi and soil bacteria. Phytopathology 59:153-163.

41. MC CALLA, T. M., and F. A. HASKINS. 1964. Phytotoxic substances from soil microorganisms and crop residues. Bacteriol. Rev. 28:181-207.

42. MC DOUGAL, B. M. 1970. Movement of C^{14}-photosynthate into roots of wheat seedlings and exudation of C^{14} from intact roots. New Phytologist 69:37-46.

43. MELIN, E., and S. KRUPA. 1971. Studies on ectomycorrhizae of pine. II. Growth inhibition of mycorrhizal fungi by volatile organic constituents of Pinus silvestris (Scots Pine) roots. Physiol. Plant. 25:337-340.

44. MENZIES, J. D., and R. G. GILBERT. 1967. Responses of

the soil microflora to volatile components in plant residues. Proc. Soil Sci. Soc. Amer. 31:495-496.

45. NANCE, J. F., and L. W. CUNNINGHAM. 1951. Evolution of acetaldehyde by excised wheat roots in solutions of nitrate and nitrite salts. Amer. J. Bot. 38:604-609.

46. OWENS, L. D., R. G. GILBERT, G. E. GRIEBEL, and J. D. MENZIES. 1969. Identification of plant volatiles that stimulate microbial respiration and growth in soil. Phytopathology 59:1468-1472.

47. PARK, J. Y. 1970. Antifungal effect of an ectotrophic mycorrhizal fungus, Lactarius sp., associated with basswood seedlings. Can. J. Microbiol. 16:798-800.

48. PATRICK, Z. A., and T. A. TOUSSOUN. 1965. Plant residues and organic amendments in relation to biological control, p. 440-459. *In* K. F. Baker and W. C. Snyder (ed.), Ecology of soil-borne plant pathogens. Univ. Calif. Press, Berkeley and Los Angeles.

49. PATRICK, Z. A., T. A. TOUSSOUN, and L. W. KOCH. 1964. Effect of crop-residue decomposition products on plant roots. Annu. Rev. Phytopathol. 2:267-292.

50. PENTLAND, G. D. 1967. Ethanol produced by Aureobasidium pullulans and its effect on the growth of Armillaria mellea. Can. J. Microbiol. 13:1631-1640.

51. PLHAK, F., and V. URBANKOVA. 1969. Study of the effect of volatile substances from cereal roots. Biol. Plant. 11:226-235.

52. ROBINSON, P. M., and D. PARK. 1966. Volatile inhibitors of spore germination produced by fungi. Trans. Br. Mycol. Soc. 49:639-649.

53. ROMINE, MAUREEN, and R. BAKER. 1973. Soil fungistasis: evidence for an inhibitory factor. Phytopathology 63:756-759.

54. ROVIRA, A. D. 1965. Plant root exudates and their influence upon soil microorganisms, p. 170-186. *In* K. F. Baker and W. C. Snyder (ed.), Ecology of soil-borne plant pathogens. Univ. Calif. Press, Berkeley and Los Angeles.

55. ROVIRA, A. D. 1969. Plant root exudates. Bot. Rev. 35:35-57.

56. ROVIRA, A. D., and M. VENDRELL. 1972. Ethylene in sterilized soil: its significance in studies of interactions between microorganisms and plants. Soil Biol. Biochem. 4:63-69.

57. SCHROTH, M. N., and D. C. HILDEBRAND. 1964. Influence of plant root exudates on root infecting fungi. Annu. Rev. Phytopathol. 2:101-132.

58. SMITH, A. M. 1972. Drying and wetting sclerotia promotes biological control of Sclerotium rolfsii Sacc. Soil Biol. Biochem. 4:119-123.

59. SMITH, A. M. 1972. Nutrient leakage promotes biological control of dried sclerotia of Sclerotium rolfsii Sacc. Soil Biol. Biochem. 4:125-129.

60. SMITH, A. M. 1973. Ethylene as a cause of soil fungistasis. Nature (London) 246:311-313.

61. SMITH, K. A., and R. S. RUSSELL. 1969. The occurrence of ethylene and its significance in anaerobic soil. Nature (London) 222:769-771.

62. TOUSSOUN, T. A., A. R. WEINHOLD, R. G. LINDERMAN, and Z. A. PATRICK. 1968. Nature of phytotoxic substances produced during plant residue decomposition in soil. Phytopathology 58:41-45.

63. WEINHOLD, A. R. 1963. Rhizomorph production by Armillaria mellea induced by ethanol and related compounds. Science 142:1065-1066.

64. WELLS, H. D., D. K. BELL, and C. A. JAWORSKI. 1972. Efficacy of Trichoderma harzianum as a biocontrol for Sclerotium rolfsii. Phytopathology 62:442-447.

65. WILLIS, G. M., and L. E. WILLIAMS. 1968. Fungistatic properties of soils exposed to different antecedent environments. Can. J. Microbiol. 14:755-761.

66. ZAK, B. 1964. Role of mycorrhizae in root disease. Annu. Rev. Phytopathol. 2:377-392.

Effect of Inorganic Amendments and Soil Reaction on Soil-Borne Plant Diseases

Y. HENIS and J. KATAN—*Department of Plant Pathology and Microbiology, Faculty of Agriculture, The Hebrew University of Jerusalem, Rehovot, Israel.*

Inorganic soil amendments are widely used in agriculture as fertilizers and to change the physicochemical properties of the soil. The long- and short-term effects of these minerals frequently influence the organisms present in the environment, and, consequently, they also affect diseases. It is therefore important to understand the effect of these compounds on disease incidence and severity.

Inorganic amendments (including urea, which may be regarded as such because of its rapid decomposition to NH_3 and CO_2 in soil) differ in several ways from the commonly used organic amendments of plant origin. The composition of the inorganic amendments is simpler and more consistent than that of the organic ones. With the exception of lime and sulfur, inorganic amendments are applied at much lower rates than organic ones. Inorganic amendments are applied before sowing, and, frequently, also during the growing season, whereas organic amendments are usually incorporated into soil before sowing. The effects that inorganic amendments have on the plant are either immediate, without the mediation of soil microorganisms, or indirect during biological transformation. The effect of organic amendments is usually exerted during microbial decomposition. The increase in microbial populations following soil amendment is much more pronounced after application of organic substances, which also serve as an energy source and favor zymogenous soil microorganisms. Inorganic amendments are applied in several ways: by incorporation into soil, as foliar sprays, in irrigation water, and as seed dressing.

Various agricultural practices may change the inorganic balance in soil, yielding results similar to those obtained by direct application of the amendments. Such changes are expected upon soil fumigation or sterilization, the use of water with a high mineral content for irrigation, mineralization, and immobilization of nutrients and mycorrhizal activity. These indirect effects will not be discussed here.

The literature dealing with the effect of inorganic amendments on soil-borne diseases is abundant, and the subject is covered in reviews and books (13, 19, 27, 28, 64); we shall deal mainly with specific topics.

EFFECT OF INORGANIC AMENDMENTS ON DISEASE INCIDENCE AND SEVERITY.—Application of inorganic amendments often results in an increase or decrease in disease incidence or disease severity. The changes in disease resulting from soil amendment may be different under various conditions, even with the same disease.

The effect of nitrogenous fertilizers on the severity of soil-borne diseases has received much attention, and more reports have been published on this subject than on other elements. Although disease incidence and severity caused by various pathogens (e.g., fusarial diseases) are generally known to increase under the influence of nitrogen fertilizers (27), cases of decrease also have been reported (Table 1). The sources of nitrogen used by different workers and the different nitrogen requirements of the pathogen, soil microflora, and host involved, as well as other factors, may explain the contradictory results summarized in Table 1. For example, severity of Verticillium wilt of cotton was increased twofold by NH_4^+-N or NO_3^--N as compared with NH_4NO_3 (63). In one case, calcium nitrate reduced disease progress in tomato by *Sclerotium rolfsii* (57), but not in other cases where severe infection occurred (82, 84). Nitrogenous fertilizers reduced *S. rolfsii* disease in various crops (4), but not in apple (48). Take-all disease of wheat caused by *Ophiobolus graminis* increased with application of ammonium sulfate in the fall, but decreased when the same fertilizer was applied in the spring (39). Secondary factors accompanying nitrogen fertilization that may also affect disease severity include changes in soil pH due to nitrification or denitrification and changes in water penetration in soil due to the dispersion effect of the NH_4^+ ion.

The idea that potassium increases resistance to diseases due to its effect on cell wall thickness should not be accepted as a general rule. The differences in results obtained in some cases may be related to the level of other elements present in the soil. For example, when nitrogen level was low, increased concentrations of potassium resulted in a decrease in Gibberella rot of corn, whereas when nitrogen levels were high, the opposite was true (70). In the case of Fusarium wilt of red clover, minimum disease was obtained at optimal concentrations of NPK (14). Walker and Foster (72) found that Fusarium disease of tomatoes was favored by high nitrogen and low potassium. When applied before inoculation, on the other hand, disease was more rapid with low nitrogen and high potassium (25). Potassium chloride did not affect Fusarium disease of tomato when it was sprayed before inoculation, but it did increase it when it was sprayed after inoculation (9).

The effect of calcium on disease was studied when the element was applied as a nutrient at relatively low levels and when it was applied at much higher levels as lime for

TABLE 1. Examples of effects of nitrogen on soil-borne diseases[a]

Pathogen	Host	Amendment	Reference
Disease increased			
Fusarium oxysporum f. sp. *lycopersici*	Tomato	NO_3^-	72
F. moniliforme	Sorghum	$NaNO_3$, NH_4NO_3	24
F. roseum	Carnation	NO_3^-	21
F. solani f. sp. *phaseoli*	Bean	NH_4^+	76
Ophiobolus graminis	Wheat	$(NH_4)_2SO_4$	39
Phytophthora parasitica var. *nicotianae*	Tobacco	NO_3^-	2
Verticillium albo-atrum	Cotton	$(NH_4)_2SO_4$, $Ca(NO_3)_2$, KNO_3	63
Streptomyces scabies	Potato	$NH_4NO_3 + CaCO_3$	47
Disease decreased			
F. oxysporum f. sp. *cubense*	Banana	Urea (nitrite)	65
F. solani f. sp. *phaseoli*	Bean	KNO_3	54
O. graminis	Wheat	$(NH_4)_2SO_4$	39
P. cinnamomi	Avocado	KNO_2	86
Sclerotium rolfsii	Tomato	$Ca(NO_3)_2$	57
S. rolfsii	Sugarbeet	NH_3, $(NH_4)_2SO_4$, $Ca(NO_3)_2$	50

[a]Additional examples can be found in other references (13, 19, 27, 64).

adjustment of soil reaction. Nutritional and pH factors were studied separately in only a few cases. Fusarium wilt of tomato is controlled by liming as a result of immobilization of microelements at high pH values (42). In contrast, however, liming has been shown to increase the severity of Verticillium wilt (43). The addition of micronutrients had no effect on Verticillium wilt of tomato at any soil pH, but it increased Fusarium wilt at pH 6.0 (43). Thus, Jones and Woltz (43) suggested a discrete use of micronutrient amendments on virgin land where race 2 of *Fusarium* has not been established, in order to prevent its increase.

The relationship between trace elements and disease severity has been established for Fusarium wilt of various hosts, especially tomato (33, 42, 79, 80) and pigeon pea, as well as for other diseases (64). On the basis of the contradictory information available (33, 43), it is difficult to draw a general rule regarding the effect of trace elements on wilt diseases. More work should be done on this subject, since trace elements are very easy to handle and control due to the relatively small quantities needed.

Common scab of potato and clubroot of crucifers illustrate the effect of soil reaction on soil-borne diseases. Soil pH may affect disease incidence through the direct effect of the hydrogen ion concentration or of the amendment used for its adjustment on the pathogen, host, or soil microorganisms; or indirectly by changing the availability of nutrients like manganese (42, 55). Soil pH may also affect the efficiency of pesticides in soil, which may in turn affect diseases.

MECHANISMS INVOLVED IN DISEASE CHANGES.—Changes in disease incidence or disease severity due to the application of inorganic amendments or adjustment of soil reaction result from effects on each of the living components involved in disease: the pathogen, the host, and soil microorganisms. The effect on the organisms may be direct or indirect; i.e., through changes in the chemical or physical properties of the soil. Either the amendment itself or one or more of its degradation or transformation products could be responsible for these changes. For example, in some cases, the toxic effects of nitrogenous fertilizers have been attributed to nitrite production (see below). The sum of all the effects on the organisms finally determines if and to what extent a change in disease will occur. It is possible, therefore, that a certain inorganic amendment can be beneficial to the pathogen or can increase host susceptibility, but its application still results in reduction in disease incidence or severity. Thus, root rot of poinsettia caused by *Thielaviopsis basicola* was highest at pH 6.7-8.0, whereas optimal growth of the pathogen occurred at a pH 5.0 (5).

The various possible interactions between organisms and inorganic amendments are summarized in Fig. 1, based on the scheme suggested by Park (62).

Effect on the pathogen.—Inorganic amendments may affect disease through their effect on germination of pathogen propagules, chemotaxis of zoospores or bacteria, mycelial growth in soil or on the host surface, penetration and colonization of host tissues, toxin and enzyme production, reproduction, formation of resting structures, and autolysis. An increase in pathogen population in a natural soil brought about by inorganic amendments need not be the result of the amendment's

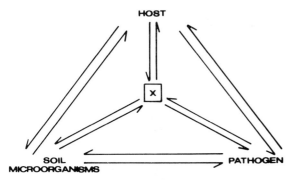

Fig. 1. Effect of inorganic amendments and soil reaction (x) on the organisms involved in disease. Effects may be direct or indirect.

effect on the pathogen per se, but could be the result of suppression of antagonists.

Numerous studies have been conducted in vitro (and to a lesser extent in sterile soil) in which the effects of mineral nutrients per se and pH on soil pathogens have been demonstrated (28, 41, 60). Davey and Danielson (19) raised the possibility that exchangeable cations maintain resting bodies in the soil in a dormant condition. Inorganic nutrients may also affect fungal growth and sporulation independently. For example, the various stages of asexual reproduction in *Aphanomyces euteiches* are affected in a different way by various mineral nutrients (56), and calcium increased mycelial growth but decreased sporulation in *Fusarium roseum* (8). Chlamydospore formation by *Thielaviopsis basicola* in culture depends, among other factors, on pH level (5) and on the amount and form of nitrogen, ammonium sulfate being superior to other forms (69). Growth of *Fusarium* in the conducting vessels of tomato stems was enhanced when calcium in the vessels was low (16). Since Ca^{++} inhibits the activity of polygalacturonase produced by the pathogen, it may interfere with the release of pectic substances that provide the fungus with carbon and energy.

The effect of a certain mineral nutrient when added alone may differ from its effect when added in combination with other organic or inorganic ingredients. Ammonium sulfate alone did not induce germination of *F. solani* f. sp. *phaseoli* in soil (15) or in vitro (30) but had a significant effect when combined with a source of carbon. In contrast, *Fusarium* chlamydospores were formed in appropriate mineral salt solutions only in an environment deficient in energy sources (38). In sterile soil, the respiration rate of *S. rolfsii* was much higher when glucose and $NaNO_3$ were added together than when each nutrient was applied alone (18). Thus, the effect of inorganic amendments might be different in soils that have a different nutrient content. Pathogens may differ in their ability to utilize different forms of minerals. Ammonium-N had a better effect on various fungi than did NO_3^--N (10, 11, 53), whereas with other fungi, the reverse was true (32). *Fusarium* of beans utilized NO_3^--N for growth and formation of conidia better than did NH_4^+-N; the latter was better utilized for pathogenesis (52). Differences regarding mineral utilization have been found between species of the same genus (1, 10, 45). Other factors affecting utilization of inorganic nutrients are the presence of buffers, organic acids, and inoculum concentration (31, 53, 58).

Certain inorganic materials are toxic to pathogens at relatively low levels. For example, the toxic effect of ammonia has been demonstrated in culture or in soil for *S. rolfsii*, *F. roseum* f. sp. *cerealis*, *F. solani* f. sp. *pisi*, *Phytophthora cinnamomi*, and nematodes, and it has been considered a possible means for disease control (29, 34, 46, 49, 66, 67). Other nitrogenous fertilizers have also been shown to be toxic to certain fungi (3, 49, 50). Toxicity of ammonia to *S. rolfsii* has been attributed to an increase in the pH of the soil (34). Warren (74) reviewed ammonia toxicity to organisms belonging to all systematic groups and concluded that it is related to pH level due to changes in the proportion between the nontoxic ionized and toxic nonionized forms of ammonia. Nitrites, which under certain conditions accumulate in soils upon addition of nitrogenous compounds, are highly toxic to some pathogens, resulting in the suppression of *Pythium, Phytophthora,* and *Fusarium* (32,65,66,86). In *S. rolfsii*, toxicity of NH_4NO_3 was not correlated with accumulation of nitrite (3). Soluble manganese suppressed *Streptomyces scabies* (55, 59). Release of this element occurs in soil at low pH levels; this might be the explanation for the disease control found at these pH values (55). Low concentrations of trace elements increased sporulation of three species of *Fusarium* but were toxic at higher levels (83).

Changes in virulence due to inorganic nutrients have been shown for many pathogens, including *F. solani* f. sp. *phaseoli*, *F. oxysporum* f. sp. *lycopersici*, and *Rhizoctonia solani* (52, 71, 75, 76, 79). Toxins and enzymes involved in the disease syndrome are produced by the pathogen during pathogenesis. Alterations in pathogen metabolism through inorganic nutrition may consequently alter its virulence. Sadasivan (64) extensively reviewed the literature regarding the effects of inorganic nutrition on production of toxins and of pectolytic and cellulotytic enzymes by fusarial pathogens. Suppression of the production of these fungal metabolites by trace elements at certain concentrations may at least partially explain the beneficial effect trace elements have in reducing wilt diseases.

Calcium has a pronounced effect on polygalacturonase (PG), which plays a role in the syndrome of diseases caused by many soil fungi. The formation of lesions in beans inoculated with *R. solani* was suppressed by Ca^{++}. This was attributed to the formation of calcium pectate, which is less available to PG than is pectic acid (6). Moreover, the rise of pH level occurring in the tissues also has an adverse effect on PG activity. In contrast, the activity of another pectolytic enzyme produced by plant pathogens, pectin lyase, is enhanced by calcium and by high pH levels. Thus, in a disease where this enzyme is of major importance, the effect calcium has on the disease might be different from that reported for PG. Calcium has been shown to affect southern blight caused by *S. rolfsii* by neutralizing oxalic acid produced by the fungus (7).

In applying the results obtained in culture to soil conditions, we should take into consideration the actual concentration of available minerals in soil (which depends on adsorption on soil colloids, leaching, immobilization of the soil microflora and by plants, as well as on oxidation, reduction, and chelating processes), the propagule type of the pathogen, and the presence or absence of other nutrients.

Effect on the host.—Inorganic nutrients may change host structure, composition, and metabolism. These changes may affect the host as a medium supporting growth and reproduction of the pathogen and the formation of its toxins or enzymes, or they may affect host response to pathogen invasion and to the action of their toxins. Foliar or root application of nutrients may also affect root exudation, which in turn affects the germination of propagules and their penetration into the host.

Growing plants under a certain nutrient regime prior to

inoculation may predispose them to disease, as shown with tomato Fusarium wilt (25). This type of effect may be exclusively on the host. However, the often-claimed increase or decrease in host susceptibility due to inorganic amendments cannot in all cases be exclusively attributed to the amendment's effect on the plant. The pathogen or soil microorganisms are also affected, and in certain cases it is difficult to distinguish between these various effects. Studies on the effect of mineral nutrients on the host usually involve growing the plant in sand or in mineral solution culture and studying the anatomy, composition, and metabolism of the treated plant in connection with changes in disease, as well as treating the plants with nutrients before and after inoculation. Mineral nutrition of plants is usually reflected in the composition of the plant tissues. Therefore, plant analysis for mineral content may be used for predicting possible changes in host resistance (13, 21).

Mineral nutrition affects plant vigor and development, but the correlation between these phenomena and susceptibility varies with different diseases. Hoagland solution at half normal concentration was most favorable for red clover development and least favorable for Fusarium disease (14). Pinto bean plants grown in sand poor in nitrogen were more susceptible to *Fusarium* infection than were vigorous plants grown in a nitrogen-rich medium (52). Increasing calcium concentrations to 250 ppm, the optimum for tobacco growth, resulted in an increase in the black shank disease (78). Increasing nutrient concentration in sand culture resulted in more vigorous tomato growth as well as in an increase in Verticillium wilt disease (73). The opposite was observed with the same plant species inoculated with *Fusarium*, another wilt pathogen (72). Nitrogen fertilization reduced disease caused by *O. graminis* in wheat, although it increased the intrinsic susceptibility of individual roots (26). This was attributed to disease escape, since nitrogen enhances new root production and this outweighs the increased susceptibility. Nitrogen increased common scab of potatoes by delaying tuber production and exposing them to infection under better environmental conditions for the disease (47).

Histological studies showed that the decrease in Fusarium disease found in red clover growing under a balanced nutrition regime was correlated with the faster rate of root and stem maturation. The tissues were firm and had well-developed vascular elements that were strongly lignified and contained abundant fibers (14). Calcium deficiency in sand culture increased severity of infection of soybeans by *R. solani* (12). The cortical tissue of these plants was poorly organized, the cells had thin walls with poorly developed middle lamelleae, and the intercellular spaces were large.

Response of plants to disease is greatly affected by cell wall and membrane structure. These, as well as regulation of membrane permeability, monovalent cation uptake, and membrane integrity, are affected by calcium nutrition. Calcium deficiency has frequently been reported to increase susceptibility to root diseases (23, 64, 81). Tomato plants deficient in calcium were attacked more severely by *Fusarium*. They contained more soluble pectins that were degraded easily by enzymes (23). In oats, calcium suppressed symptoms induced by the toxin

victorin. This may be attributed to toxin deactivation or to the decrease of its uptake and transport through cell membranes (22). A complicated relationship among disease, hydrogen ion concentration, and calcium was found in common scab of potato (37). Disease was related to the calcium content of the tuber, and calcium content of the tuber depends on the replaceable calcium content of the soil, which in turn depends on hydrogen ion concentration and calcium dosage. All this may finally affect tuber response to the pathogen. Other mineral elements like boron are also associated with cell wall synthesis, and changes in their concentration may affect resistance (81).

Disease resistance of different cultivars may change differently in response to mineral nutrition. Thus, an increase in the concentration of mineral solution resulted in a decrease in Fusarium disease of tomato in one susceptible and two moderately resistant varieties, but not in a monogenic, highly resistant one (72). Mineral solutions induced formation of pisatin, a phytoalexin involved in the resistance mechanism of plants to various diseases (17). Mineral nutrition affected the sensitivity of tomato to fusaric acid; plants under normal nutrition were more sensitive to this wilt toxin than were those that were under- or overnourished (85).

Effect on soil microorganisms.—Inorganic amendments may affect the interaction between soil-borne pathogens and microorganisms in the soil, in plant debris, and in the rhizosphere. Consequently, a change in disease or in the survival ability of the pathogen (28) may occur. Nitrogen affected the saprophytic activity of *Rhizoctonia* (20) and *S. rolfsii* (3) in soil. Both saprophytic activity of and disease caused by *Rhizoctonia* were increased by ammonium sulfate (36). Changes in pathogen population, however, are not necessarily correlated with respective changes in disease if at the same time the aggressiveness of the pathogen or the susceptibility of the host is changed. Following applications of nitrogen, changes in population density of *Fusarium* and in disease incidence in beans (76) or sorghum (24) were not correlated.

Activity and population density of antagonists are also affected. Ammonium increased production of antibiotics by streptomycetes in culture (77). Nitrogenous compounds decreased the number of antibiotic-producing organisms in corn- but not in wheat-soil (44), and increased it in the vicinity of *S. rolfsii* sclerotia incorporated into soil, resulting in a decrease in germinability (35). Nitrogen that enhances saprophytic activity of *Rhizoctonia* (20) decreased the populations of antagonists in the bean rhizosphere as well as the lytic properties of the amended soil (61). Competition between the pathogen and its antagonists in favor of the latter may be promoted if the pathogen is a weak competitor, as when the pathogen is unable to utilize the supplied nutrient.

Combining organic and inorganic amendments so as to increase the carbon nitrogen (C:N) ratio has frequently been suggested as a means for disease suppression. Fusarium root rot of beans was suppressed by using barley, cellulose, or other amendments with a high C:N ratio. Adding nitrogen nullified this effect (52, 54, 68).

The mechanism involved in this case of disease control was attributed to immobilization by the soil microorganisms of nitrogen required for pathogen activity. Huber et al. (40) concluded that the C:N hypothesis is not consistent with all results known in the literature. They suggested that amendments modify disease severity through their effect on nitrification, which, in turn, determines the form of nitrogen available. Papavizas (61) suggested that not only is the C:N ratio per se involved, but the absolute amount and form of C and N may determine the effectiveness of the amendments.

An interesting indirect form of biological control was reported by Sequeira (65). Urea suppressed *F. oxysporum* f. sp. *cubense* due to accumulation of nitrite, which resulted from the inhibition of nitrite-oxidizing bacteria.

In the former sections, the effects of individual inorganic amendments on each component of the pathogen-host-soil microorganisms complex were discussed separately. In nature, however, these amendments act simultaneously on all the components, sometimes in a contradictory manner (52).

INORGANIC AMENDMENTS AND CONTROL OF SOIL-BORNE DISEASES.

—In order to avoid undesirable effects and to take advantage of beneficial ones, fertilizing practice should be considered with respect to the side effects fertilizers have on many diseases. For example, shortening pathogen contact with the fertilizer (e.g., by application after plant emergence) may be considered (75). The information available regarding effect of fertilizers on disease is voluminous but scattered, and it is often difficult to compare results. Whenever new fertilizers or methods of application are introduced, especially in important crops, experiments regarding effect on diseases should be conducted.

Inorganic amendments may also be used to control certain soil-borne diseases. This can be done whenever their use fulfills the usual criteria for disease control: efficiency of control, economic value, effect on the environment, etc. Unlike control of foliar diseases, soil-borne disease control is usually carried out before planting. Those inorganic amendments that can also be used after planting (foliar and soil applications) offer special opportunities. Fertilizers like calcium, which may affect pathogen or host metabolism, can be considered therapeutants.

Most experiments on the control of soil-borne diseases by inorganic amendments have been conducted empirically. Better understanding of the mechanisms involved may enable us to select the fertilizers for studies in disease control. For example, by revealing the inorganic nutritional requirements of the pathogen, a fertilizer available only to its antagonists may be selected (51). Advantageous effects on host-parasite relationships can be utilized to reduce disease whenever these are clarified. Hopefully, a suitable combination of fungicides and fertilizers will greatly improve disease control, crop yield, and quality with a minimum of undesirable effects on the environment.

LITERATURE CITED

1. AGNIHORTI, U. D., and O. VAARTAJA. 1967. The influence of nitrogenous compounds on growth of Pythium species. Can. J. Microbiol. 13:1509-1613.
2. APPLE, J. L. 1961. The development of black shank in tobacco as influenced by host nutrition. Phytopathology 51:386-389.
3. AVIZOHAR-HERSHENZON, Z., and P. SHACKED. 1969. Studies on the mode of action of inorganic nitrogenous amendments on Sclerotium rolfsii in soil. Phytopathology 59:288-292.
4. AYCOCK, R. 1966. Stem rot and other diseases caused by Sclerotium rolfsii. N. C. Agr. Exp. Stn. Tech. Bull. 174. 202 p.
5. BATEMAN, D. F. 1962. Relation of soil pH to development of poinsettia root rots. Phytopathology 52:559-566.
6. BATEMAN, D. F. 1964. An induced mechanism of tissue resistance to polygalacturonase in Rhizoctonia-infected hypocotyls of bean. Phytopathology 54:438-445.
7. BATEMAN, D. F., and S. V. BEER. 1965. Simultaneous production and synergistic action of oxalic acid and polygalacturonase during pathogenesis by Sclerotium rolfsii. Phytopathology 55:204-211.
8. BEAN, G. A. 1967. Influence of calcium and sulfur on mycelial growth and sporulation of Fusarium roseum. Phytopathology 57:644.
9. BLOOM, J. B., and J. C. WALKER. 1955. Effect of nutrient sprays on Fusarium wilt of tomato. Phytopathology 45:443-444.
10. BYTHER, R. 1965. Ecology of plant pathogens in soil. V. Inorganic nitrogen utilization as a factor of competitive saprophytic ability of Fusarium roseum and F. solani. Phytopathology 55:852-858.
11. CARLEY, H. E., and T. H. KING. 1968. Influence of nitrogen form on the incidence and severity of Aphanomyces root rot of peas. Phytopathology 58:1046 (Abstr.).
12. CASTANO, J. J., and M. F. KERNKAMP. 1956. The influence of certain plant nutrients on infection of soybeans by Rhizoctonia solani. Phytopathology 46:326-328.
13. CHAPMAN, H. D. 1965. Chemical factors of the soil as they affect microorganisms, p. 120-141. *In* K. F. Baker and W. C. Snyder (ed.), Ecology of soil-borne plant pathogens. Univ. Calif. Press, Berkeley and Los Angeles.
14. CHI, C. C., and E. W. HANSON. 1961. Nutrition in relation to the development of wilts and root rots incited by Fusarium in red clover. Phytopathology 51:704-711.
15. COOK, R. J., and M. N. SCHROTH. 1965. Carbon and nitrogen compounds and germination of chlamydospores of Fusarium solani f. phaseoli. Phytopathology 55:254-256.
16. CORDEN, M. E. 1965. Influence of calcium nutrition on Fusarium wilt of tomato and polygalacturonase activity. Phytopathology 55:222-224.
17. CRUICKSHANK, I. A. M., and D. R. PERRIN. 1963. Studies in phytoalexins. VI. Pisatin: the effect of some factors on its formation in Pisum sativum L., and the significance of pisatin in disease resistance. Austral. J. Biol. Sci. 16:111-128.
18. CURL, A. E., R. RODRIGUEZ-KABANA, and H. H. FUNDERBURK, JR. 1968. Influence of atrazine and various carbon and nitrogen amendments on growth of Sclerotium rolfsii and Trichoderma viride in soil. Phytopathology 58:323-328.
19. DAVEY, C. B., and R. M. DANIELSON. 1968. Soil chemical factors and biological activity. Phytopathology 58:900-908.
20. DAVEY, C. B., and G. C. PAPAVIZAS. 1963. Saprophytic activity of Rhizoctonia as affected by the carbon-nitrogen balance of certain organic soil amendments. Proc. Soil Sci. Soc. Amer. 27:164-167.

21. DORWORTH, C., and J. TAMMEN. 1969. Influence of nutrition, soil moisture, and soil temperature on proneness of Dianthus caryophyllus to attack by Fusarium roseum. Phytopathology 59:1703-1705.

22. DOUPNIK, B., JR. 1968. The suppression of victorin-induced disease by calcium. Phytopathology 58:215-218.

23. EDGINGTON, L. V., M. E. CORDON, and A. E. DIMOND. 1961. The role of pectic substances in chemically induced resistance to Fusarium wilt of tomato. Phytopathology 51:179-182.

24. ERINLE, I. D., and L. K. EDMUNDS. 1970. Effect of nitrogen on development of seedling disease caused by Fusarium moniliforme in sorghum. Phytopathology 60:1291 (Abstr.).

25. FOSTER, R. E., and J. C. WALKER. 1947. Predisposition of tomato to Fusarium wilt. J. Agr. Res. 74:165-185.

26. GARRETT, S. D. 1948. Soil conditions and the take-all disease of wheat. VI. The effect of plant nutrition upon disease resistance. Ann. Appl. Biol. 28:14-18.

27. GARRETT, S. D. 1956. Biology of root-infecting fungi. Cambridge Univ. Press, London and New York. 293 p.

28. GARRETT, S. D. 1970. Pathogenic root-infecting fungi. Cambridge Univ. Press, London and New York. 294 p.

29. GILPATRICK, J. D. 1969. Role of ammonia in the control of avocado root rot with alfalfa meal soil amendment. Phytopathology 59:973-978.

30. GRIFFIN, G. J. 1964. Long-term influence of soil amendments on germination of conidia. Can. J. Microbiol. 10:605-612.

31. GRIFFIN, G. J. 1970. Exogenous carbon and nitrogen requirements for chlamydospore germination by Fusarium solani: dependence on spore density. Can. J. Microbiol. 16:1366-1371.

32. GROVER, R. K., and J. S. SIDHU. 1966. Effect of nitrogen on the growth of Pythium aphanidermatum (Edson) Fitz. Sydowia 19:231-237.

33. HART, J. H. 1967. Manganese nutrition and Fusarium wilt development. Phytopathology 57:814 (Abstr.).

34. HENIS, Y., and I. CHET. 1967. Mode of action of ammonia on Sclerotium rolfsii. Phytopathology 57:425-427.

35. HENIS, Y., and I. CHET. 1968. The effect of nitrogenous amendments on the germinability of sclerotia of Sclerotium rolfsii and on their accompanying microflora. Phytopathology 58:209-211.

36. HENIS, Y., B. SNEH, and J. KATAN. 1967. Effect of organic amendments on Rhizoctonia and accompanying microflora in soil. Can. J. Microbiol. 13:643-650.

37. HORSFALL, J. G., J. P. HOLLIS, and H. G. M. JACOBSON. 1954. Calcium and potato scab. Phytopathology 44:19-24.

38. HSU, S. C., and J. L. LOCKWOOD. 1972. Chlamydospore formation by Fusarium in model systems. Phytopathology 62:766-769.

39. HUBER, D. M. 1972. Spring versus fall nitrogen fertilization and take-all of spring wheat. Phytopathology 62:434-436.

40. HUBER, D. M., R. D. WATSON, and G. W. STEINER. 1969. Crop residues, nitrogen, and plant disease. Soil Sci. 100:302-308.

41. HUTNER, S. H. 1972. Inorganic nutrition. Annu. Rev. Microbiol. 26:313-346.

42. JONES, J. P., and S. S. WOLTZ. 1970. Fusarium wilt of tomato: interaction of soil liming and micronutrient amendments on disease development. Phytopathology 60:812-813.

43. JONES, J. P., and S. S. WOLTZ. 1972. Effect of soil pH and micronutrient amendments on Verticillium and Fusarium wilt of tomato. Plant Dis. Reptr. 56:151-153.

44. KAUFMAN, D. D., and L. E. WILLIAMS. 1965. Influence of soil reaction and mineral fertilization on numbers and types of fungi antagonistic to four soil-borne plant pathogens. Phytopathology 55:570-574.

45. KRAFT, J. M., and D. C. ERWIN. 1967. Effects of nitrogen sources on growth of Pythium aphanidermatum and Pythium ultimum. Phytopathology 57:374-376.

46. LAHOVER, D., and Z. AVIZOHAR-HERSHENZON. 1961. The use of aqueous ammonia for the control of Sclerotium rolfsii and its residual value in Israel soils. Israel J. Agr. Res. 11:141-150.

47. LAPWOOD, D. H., and P. W. DYSON. 1966. An effect of nitrogen on the formation of potato tubers and the incidence of common scab (Streptomyces scabies). Plant Path. 15:9-14.

48. LAVEE, S. 1962. The effect of ammonium sulfate and farmyard manure on young M-II apple rootstocks infected with Sclerotium rolfsii Sacc. Israel J. Agr. Res. 12:89-90.

49. LEACH, L. D., and A. E. DAVEY. 1935. Toxicity of low concentrations of ammonia to mycelium and sclerotia of Sclerotium rolfsii. Phytopathology 25:957-959.

50. LEACH, L. D., and A. E. DAVEY. 1942. Reducing southern Sclerotium rot of sugar beets with nitrogenous fertilizers. J. Agr. Res. 64:1-18.

51. LU, K. C., J. M. TRAPPE, and W. B. BOLLEN. 1967. Selective nitrogen assimilation by Poria weirii. Nature (London) 213:814.

52. MAIER, C. R. 1968. Influence of nitrogen nutrition on Fusarium root rot of pinto bean and on its suppression by barley straw. Phytopathology 58:620-625.

53. MALCA, I., and D. C. ERWIN. 1966. Effect of pH and carbon and nitrogen sources on the growth of Verticillium albo-atrum. Phytopathology 56:401-406.

54. MAURER, C. L., and R. BAKER. 1965. Ecology of plant pathogens in soil. II. Influence of glucose, cellulose, and inorganic nitrogen amendments on development of bean root rot. Phytopathology 55:69-72.

55. MC GREGOR, A. J., and G. C. S. WILSON. 1966. The influence of manganese on the development of potato scab. Plant Soil 25:3-16.

56. MITCHELL, J. E., and C. Y. YANG. 1966. Factors affecting growth and development of Aphanomyces euteiches. Phytopathology 56:917-922.

57. MOHR, H. C., and G. M. WATKINS. 1959. The nature of resistance of southern blight on tomatoes and the influence of nutrition on its expression. Proc. Amer. Soc. Hort. Sci. 74:484-493.

58. MORTON, A. G., and A. MAC MILLAN. 1954. The assimilation of nitrogen from ammonium salts and nitrate by fungi. J. Exp. Bot. 5:232-252.

59. MORTVEDT, J. J., M. H. FLEISCHFRESSER, K. C. BERGER, and H. M. DARLING. 1961. The relation of soluble manganese to the incidence of common scab in potatoes. Amer. Potato J. 38:95-100.

60. NICHOLAS, D. J. D. 1965. Utilization of inorganic nitrogen compounds and amino acids by fungi, p. 343-376. In G. C. Ainsworth and A. S. Sussman (ed.), The Fungi, Vol. 1. Academic Press, New York and London.

61. PAPAVIZAS, G. C. 1963. Microbial antagonism in bean rhizosphere as affected by oat straw and supplemental nitrogen. Phytopathology 53:1430-1435.

62. PARK, D. 1963. The ecology of soil-borne fungal disease. Annu. Rev. Phytopathol. 1:241-258.

63. RANNEY, C. D. 1962. Effects of nitrogen source and rate on the development of Verticillium wilt of cotton. Phytopathology 52:38-41.

64. SADASIVAN, T. S. 1965. Effect of mineral nutrients on soil microorganisms and plant disease, p. 460-469. In K. F. Baker and W. C. Snyder (ed.), Ecology of soil-borne plant pathogens. Univ. Calif. Press, Berkeley and Los Angeles.

65. SEQUEIRA, L. 1963. Effect of urea applications on survival of Fusarium oxysporum f. cubense in soil.

Phytopathology 53:332-336.

66. SMILEY, R. W., R. J. COOK, and R. I. PAPENDICK. 1970. Anhydrous ammonia as a soil fungicide against Fusarium and fungicidal activity in the ammonia retention zone. Phytopathology 60:1227-1232.

67. SMILEY, R. W., R. J. COOK, and R. I. PAPENDICK. 1972. Fusarium root rot of wheat and peas as influenced by soil applications of anhydrous ammonia and ammonia-potassium azide solutions. Phytopathology 62:86-91.

68. SNYDER, W. C., M. N. SCHROTH, and T. CHRISTOU. 1959. Effect of plant residues on root rot of bean. Phytopathology 49:755-756.

69. STOVER, R. H. 1956. Effect of nutrition on growth and chlamydospore formation in brown and gray cultures of Thielaviopsis basicola. Can. J. Bot. 34:459-472.

70. THAYER, P., and L. E. WILLIAMS. 1960. Effect of nitrogen, phosphorus and potassium concentrations on the development of Gibberella stalk- and root-rot of corn. Phytopathology 50:212-214.

71. TOUSSOUN, T. A., S. M. NASH, and W. C. SNYDER. 1960. The effect of nitrogen sources and glucose on the pathogenesis of Fusarium solani f. phaseoli. Phytopathology 50:137-140.

72. WALKER, J. C., and R. E. FOSTER. 1946. Plant nutrition in relation to disease development. III. Fusarium wilt of tomato. Amer. J. Bot. 33:259-264.

73. WALKER, J. C., M. E. GALLEGLY, JR., J. R. BLOOM, and R. P. SCHEFFER. 1954. Relation of plant nutrition to disease development. VIII. Verticillium wilt of tomato. Amer. J. Bot. 41:760-762.

74. WARREN, K. S. 1962. Ammonia toxicity and pH. Nature (London) 195:47-49.

75. WEINHOLD, A. R., R. L. DODMAN, and T. BOWMAN. 1972. Influence of exogenous nutrition on virulence of Rhizoctonia solani. Phytopathology 62:278-281.

76. WEINKE, K. E. 1962. The influence of nitrogen on the root disease of bean caused by Fusarium solani f. phaseoli. Phytopathology 52:757 (Abstr.).

77. WHALEY, J. W., and A. M. BOYLE. 1967. Antibiotic production by streptomyces species from the rhizosphere of desert plants. Phytopathology 57:347-351.

78. WILLS, W. H., and L. D. MOORE. 1969. Calcium nutrition and black shank of tobacco. Phytopathology 59:346-351.

79. WOLTZ, S. S., and J. P. JONES. 1968. Micronutrient effect on the in vitro growth and pathogenicity of Fusarium oxysporum f. sp. lycopersici. Phytopathology 58:336-338.

80. WOLTZ, S. S., and J. P. JONES. 1972. Control of Fusarium wilt of tomato by varying the nutrient regimes in soils. Phytopathology 62:799 (Abstr.).

81. WOOD, R. K. S. 1967. Physiological plant pathology. Blackwell Sci. Pub., Oxford. 570 p.

82. WORLEY, R. E., and D. J. MORTON. 1964. Ineffectiveness of calcium nitrate and other calcium sources in reducing southern blight incidence on Rutgers tomato under epiphytotic conditions. Plant Dis. Reptr. 48:63-65.

83. YOGESWARI, L. 1950. Trace element nutrition of fungi with special reference to Fusarium spp. Ph.D. thesis, Univ. Madras, cited in Sadasivan, 1965.

84. YOUNG, P. A. 1960. Controlling southern blight of tomato with chemicals and crop rotation. Phytopathology 50:578 (Abstr.).

85. ZÄHNER, H. 1955. Über den Einfluss der Ernährung auf die Toxinempfindlichkeit von Tomatenpflanzen. Phytopathol. Z. 23:49-88, cited in Sadasivan, 1965.

86. ZENTMYER, G. A., and F. T. BINGHAM. 1956. Influence of nitrite on the development of Phytophthora root rot of avocado. Phytopathology 46:121-124.

Fomes annosus *in Eastern North America*

ELDON W. ROSS—*Assistant Director, Forest Insect and Disease Research, Northeastern Forest Experiment Station, Upper Darby, Pennsylvania.*

INTRODUCTION.—*Fomes annosus* (Fr.) Karst. is widely distributed in eastern North America. Collections have been reported from as far south as Jamaica, Guatemala, Cuba, and Mexico (29). It may be found throughout the eastern United States, but valid collections in eastern Canada have been made only from southern Ontario (35). Collections of *F. annosus* from eastern North America date back to the late 1800's, when specimens were identified from New York and Alabama (32, 47). Despite the apparent widespread natural occurrence of *F. annosus*, little concern was shown over it as a serious pathogen until the past three decades. It was during this period that many intensively managed coniferous plantations were thinned for the first time. Some of the first reports of severe damage in thinned pine plantations came from North Carolina in 1944 (16), New York in 1948 (45), Georgia and South Carolina in the early 1950's (4), and Ontario, Canada, in 1955 (19).

Past history of *F. annosus* is a matter of record, and further discussion here would serve no useful purpose. The importance placed upon this pathogen by scientists and forest managers is attested to by the more than 1,000 research papers and reports that have been generated on the subject (18, 21). Justice could not be done in this paper to all the research on this pathogen in the eastern United States and Canada. Instead, some of the more important contributions in the past decade will be discussed, and the status of an integrated management system will be described.

RESEARCH IN EASTERN NORTH AMERICA.—*Epidemiology.*—During early experiences with *F. annosus*, forest managers relied heavily on information developed by European scientists (6, 36, 37, 39). But they soon discovered that applying European findings to U.S. conditions produced inconsistent results or even failures, particularly when applied to southern pines. Sinclair's work (45) in New York provided us with some of our first sound data on the epidemiology and control of the disease in the United States. He found that damage was severe in thinned plantations of red pine (*Pinus resinosa*) in some localities, but that statewide damage was generally low. Sinclair predicted from his research that *F. annosus* would not become a major limiting factor in producing coniferous tree crops in New York. That prediction has been accurate.

The studies by Sinclair (45) and others (20, 46) in northeastern North America gave some early insight into what could be expected from *F. annosus* root rot, but it soon became apparent that these findings did not apply to the behavior of the pathogen in the southern United States. A survey throughout the South in 1960 revealed that 50% of loblolly (*Pinus taeda*) and 44% of slash (*Pinus elliottii* var. *elliottii*) pine plantations examined had *F. annosus* (33). The rate of spread of the pathogen in the South appeared to be much faster than in the northern part of the country (45), and under some conditions, losses amounted to over 40% of the residual stand. In the early 1960's, the U.S. Forest Service, several universities, and private industry in the South launched an extensive research effort to elucidate the epidemiology and develop controls for the *F. annosus* root rot pathogen in that region.

Kuhlman and Hendrix (26) studied colonization and growth rates of *F. annosus* in the southern pines. More importantly, they discovered that, as in Europe, *Peniophora gigantea* (13, 40) is a powerful competitor of *F. annosus* for the stump and root substrate and can actually replace it.

Original infection of coniferous stands by *F. annosus* occurs mainly through freshly cut stump surfaces with subsequent spread to residual trees through root contacts. Hendrix and Kuhlman (14) found that in slash pine plantations, infection occurs most commonly through stumps but that direct infection of roots also occurs. This finding suggests that regardless of their effectiveness, stump protectants cannot give complete protection against the development of disease centers caused by this pathogen. It has also been generally accepted that mycelium of *F. annosus* cannot compete or survive in soil away from the root substrate; however, Kuhlman (24, 25) found that conidia of *F. annosus* could survive in some soils up to 10 months and that relatively few spores were necessary to cause infection in nonwounded stump roots.

Southern pines are frequently managed on a short rotation, with final harvesting in 25 to 30 years. Most areas are site-prepared and replanted soon after harvest. European experience indicated that there may be limited *F. annosus* in the first planting but that in the second planting infection may increase to more than 50% (28) because of persistent infected stumps from the first planting. Early investigations in the southern United States suggested that we may also have similar problems (9, 15): seedlings killed or infected by *F. annosus* were found in plantations established on sites where the pathogen had been present in the first planting. Early mortality in some plantations ranged up to 6%; however, after 6 years, losses had not increased, and it appears that *F. annosus* may not be a serious problem in regenerating pine plantations in the South (27). Southern pine stumps are generally small, insect activity is vigorous, and year-round warm temperatures contribute to rapid deterioration. Only fragments of stumps may remain after 6 to 8 years, and they can no longer serve as a food base for *F. annosus*.

Much has been learned about the growth and activity of *F. annosus* in infected trees. For example, lateral growth of *F. annosus* is more rapid in roots of suppressed loblolly pine than in roots of dominant trees (30). This may explain the eventual stabilization of *F. annosus* infection centers in pine plantations. A toxin, fomannosin, has been detected in culture filtrates of *F. annosus* (1). Although the toxin has not been found in naturally infected trees, when it is applied to small wounds on seedlings it causes a systemic response similar to that in naturally infected trees. Shain (44) found a phenol-enriched, resin-soaked reaction zone between sound sapwood of loblolly pine and sapwood infected with *F. annosus*. Compounds in the reaction zone, primarily pinosylvins, were accumulated as a dynamic response to injury, but they inhibited the growth of *F. annosus*, thus protecting the wood of living trees from infection and subsequent decay.

Several studies in Europe and the United States have suggested seasonal optima for spore production and stump infection by *F. annosus* (38, 45, 46, 48). In the southeastern United States, Driver and Ginns (8) found pronounced fluctuations in seasonal colonization of stumps and postulated a relation to temperature. Gooding et al. (12) found in laboratory studies that actively growing mycelium of *F. annosus* in wood chips was killed in less than 2 hours at 40°C. Neither spore stage of *F. annosus* is likely to survive more than 1 hour at 45°C on stump surfaces (42).

In 1964, extensive studies were undertaken to determine the relation of physical and biotic factors to stump colonization and eventual losses in the residual stands of pine plantations in the southeastern United States (43). The study examined three major topics: (a) the seasonal availability of spores of *F. annosus, P. gigantea*, and *Trichoderma* spp., (b) the relation of seasonal thinning to colonization of stumps by *F. annosus* in pine plantations, and (c) losses caused by *F. annosus* in the residual stand.

In the first part of the study, deposition of viable spores of *F. annosus, P. gigantea*, and *Trichoderma* spp. was recorded at weekly intervals for 1 year in 18 pine stands from Florida to New Hampshire. At most locations, deposition of *F. annosus* and *P. gigantea* spores was lowest in the summer, but *Trichoderma* spp. followed no apparent seasonal pattern. Few spores were trapped in the most southern locations after daily stand temperatures reached a maximum of 32°C or a mean of 21°C. The number of spores trapped was inversely correlated with temperature at the more southern locations.

From earlier laboratory studies (12, 42), it was discovered that actively growing mycelium, conidia, and basidiospores of *F. annosus* could not withstand 40° to 45°C temperatures for more than 1 to 2 hours. Field studies of stump and air temperatures indicated that these thermal inactivation temperatures were frequently reached from April to September. Accordingly, colonization of pine stumps was lowest during late spring and summer and highest during fall and winter (43).

Control.—In multiple regression analyses, natural inoculum availability, stump and air temperatures, and competing organisms were significantly related to recovery of *F. annosus* from stumps, but none was found to be the sole governing factor (43). However, as much as 92% of the total variation in occurrence of *F. annosus* could be explained by considering these variables. Analyses of these studies (43) led to the conclusion that plantations south of approximately 34°N latitude in the southeastern United States may be thinned in late spring and summer with no danger of stump colonization by *F. annosus*. North of this latitude, however, chemical or biologic protectants should be applied to stumps after thinning at all times of the year.

In one area favorable for spread of *F. annosus*, residual stand losses 5 years after thinning were highly correlated with the degree of stump colonization and with the percentage of stumps infected 2 months after stump surfaces were exposed to infection. Smaller suppressed and intermediate trees were killed first by *F. annosus*. Percentage volume losses were always less than the percentage of stems killed because smaller trees were the first to be killed. After 5 years in the peak infection plots, over 5% of the residual stand volume was lost, with an additional 12% infected by *F. annosus*. Estimated losses after 10 years would approximate 5 standard cords per acre (43).

Seasonal timing of thinning operations cannot be solely relied upon to control *F. annosus*. Early efforts at chemical control of *F. annosus* in this country centered around the use of creosote as a stump protectant because of its reported success in Europe. Sinclair (45) was probably the first to demonstrate experimentally that creosote was not an effective stump protectant in the United States, but several practical tests had already given erratic results. Several chemical stump protectants, including urea, sodium nitrite, creosote, ammonium fluoride, borax (sodium tetraborate decahydrate), and others (2, 7, 17, 22, 34, 45), have been tested, but only borax has proven consistently superior. Borax will persist uniformly at toxic concentrations in the upper 5 cm of the stump for at least 8 weeks (23) and should protect the stump against infection by *F. annosus* for a sufficient time, since pine stumps in the eastern United States are not highly susceptible to infection for more than 2 weeks (5, 41). Currently, borax sprinkled lightly on stump surfaces is the only chemical registered by the U.S. Environmental Protection Agency for the control of *F. annosus*. Sometimes it may be inadvisable to use borax in plantations where *F. annosus* is already present. In any area where use of a chemical may be undesirable, stump surfaces may be artificially inoculated with the biologic agent *P. gigantea*. It will compete successfully for the stump and root substrate and afford adequate control of the pathogen (3, 10, 17, 43).

Because it must be applied in a liquid spore suspension, *P. gigantea* may be especially desirable in mechanical harvesting operations where hydraulic sprayers can be attached to the mechanical shear (Ross and Hodges, *unpublished data*). Steps are being taken to secure registration from the U.S. Environmental Protection Agency for the use of *P. gigantea* for control of *F. annosus*. Knowledge of certain soil and site characteristics may be useful in controlling losses from *F. annosus* root rot. In the southern United States, for example, soil type, moisture, and other soil and site factors may affect the damage it causes (11, 31, 33, 43, Ross and Hodges, *unpublished data*). The effects are variable, and it has not been established whether they are upon the pathogen, the host, or the infected tree. The general tendencies are for *F. annosus* root rot to be more severe on deep, light-textured, and well-drained soils, with little damage on heavy, imperfectly drained, or shallow soils.

RECOMMENDATIONS FOR INTEGRATED CONTROL.—Productive coniferous plantations can be managed in eastern North America to avoid losses from *F. annosus* root rot by employing a number of control techniques. They include:

A) Plant at as wide a spacing as optimum productivity permits without increasing hazard of fusiform rust (*Cronartium fusiforme*) infection.

B) Delay thinning as long as possible, and reduce the number of thinnings.

C) South of 34° N latitude, when possible, thin during late spring and summer. If thinning is done when it is cool (during fall, winter, and early spring), apply stump protectants.

D) North of 34° N latitude, apply stump protectants at all times when thinning stands on high-hazard sites.

E) Use borax as a stump protectant on high-hazard sites, especially in the first thinning where *F. annosus* is not already present.

F) As soon as registration is obtained, use *P. gigantea* as an alternate to borax, especially where a limited amount of *F. annosus* is already present in the stand, where use of chemicals is undesirable, or with mechanical harvesters.

LITERATURE CITED

1. BASSETT, C., R. T. SHERWOOD, J. A. KEPLER, and P. B. HAMILTON. 1967. Production and biological activity of fomannosin, a toxic sesquiterpene metabolite of Fomes annosus. Phytopathology 57:1046-1052.
2. BERRY, F. H. 1965. Treat stumps to prevent Fomes annosus in shortleaf pine plantations. USDA Forest Serv., Central States Forest Exp. Stn. Note CS-34. 4 p.
3. BOYCE, J. S., JR. 1966. Sporulation by Peniophora gigantea with reference to control of F. annosus root rot. Forest Sci. 12:2-7.
4. CAMPBELL, W. A., and G. H. HEPTING. 1954. Fomes annosus on slash pine. Plant Dis. Reptr. 38:217.
5. COBB, F. W., JR., and R. A. SCHMIDT. 1964. Duration of susceptibility of eastern white pine stumps to Fomes annosus. Phytopathology 54:1216-1218.
6. DAY, W. R. 1948. Penetration of conifer roots by Fomes annosus. Quart. J. For. 42:99-101.
7. DRIVER, C. H. 1963. Further data on borax as a control of surface infection of slash pine stumps by Fomes annosus. Plant Dis. Reptr. 47:1006-1009.

8. DRIVER, C. H., and J. H. GINNS, JR. 1964. The effects of climate on occurrence of annosus root rot in thinned slash pine plantations. Plant Dis. Reptr. 48:509-511.
9. DRIVER, C. H., and J. H. GINNS, JR. 1964. Annosus root-rot in young southern pines planted on an infested site. Plant Dis. Reptr. 48:803-807.
10. DRIVER, C. H., and J. H. GINNS, JR. 1969. Ecology of slash pine stumps: fungal colonization and infection by Fomes annosus. Forest Sci. 15:2-10.
11. FROELICH, R. C., T. R. DELL, and C. H. WALKINSHAW. 1966. Soil factors associated with Fomes annosus in the Gulf States. Forest Sci. 12:356-361.
12. GOODING, G. V., JR., C. S. HODGES, JR., and E. W. ROSS. 1966. Effect of temperature on growth and survival of Fomes annosus. Forest Sci. 12:325-333.
13. GREMMEN, J. 1963. Biological control of the root rot fungus Fomes annosus by Peniophora gigantea. Ned. Bosbouw Tijdschr. 35:356-367.
14. HENDRIX, F. F., JR., and E. G. KUHLMAN. 1964. Root infection of Pinus elliottii. Nature (London) 201:55-56.
15. HENDRIX, F. F., JR., E. G. KUHLMAN, C. S. HODGES, JR., and E. W. ROSS. 1964. Fomes annosus—a serious threat to regeneration of pine. USDA Forest Serv., Southeastern Forest Exp. Stn. Res. Note SE-24. 4 p.
16. HEPTING, G. H., and A. A. DOWNS. 1944. Root and butt rot in planted white pine at Biltmore, N.C. J. For. 42:119-123.
17. HODGES, C. S. 1970. Evaluation of stump treatment chemicals for control of Fomes annosus, p. 43-53. Proc. Third Int. Conf. on Fomes annosus (1968), IUFRO. USDA Forest Serv., Washington, D.C.
18. HODGES, C. S., JR., J. W. KOENIGS, E. G. KUHLMAN, and E. W. ROSS. 1971. Fomes annosus, a bibliography with subject index 1960-1970. USDA Forest Serv., Southeastern Forest Exp. Stn. Res. Pap. SE-84. 75 p.
19. JORGENSEN, E. 1956. Fomes annosus on red pine in Ontario. For. Chron. 32:86-88.
20. JORGENSEN, E. 1961. On the spread of Fomes annosus. Can. J. Bot. 39:1437-1445.
21. KOENIGS, J. W. 1960. Fomes annosus. A bibliography with subject index. USDA Forest Serv., Southern Forest Exp. Stn. Occas. Pap. No. 181. 35 p.
22. KOENIGS, J. W. 1969. Growth and survival of Fomes annosus at high concentrations of borax. Phytopathology 59:1717-1721.
23. KOENIGS, J. W. 1971. Borax: its toxicity to Fomes annosus in wood and its diffusion, persistence, and concentration in treated stumps of southern pines. Phytopathology 61:269-274.
24. KUHLMAN, E. G. 1969. Survival of Fomes annosus spores in soil. Phytopathology 59:198-201.
25. KUHLMAN, E. G. 1969. Number of conidia necessary for stump root infection by Fomes annosus. Phytopathology 59:1168-1169.
26. KUHLMAN, E. G., and F. F. HENDRIX, JR. 1964. Infection, growth rate, and competitive ability of Fomes annosus in inoculated Pinus echinata stumps. Phytopathology 54:556-561.
27. KUHLMAN, E. G., and E. W. ROSS. 1970. Regeneration of pine on Fomes annosus infested sites in the Southeastern United States, p. 71-76. Proc. Third Int. Conf. on Fomes annosus (1968), IUFRO. USDA Forest Serv., Washington, D.C.
28. LOW, J. D., and R. J. GLADMAN. 1970. Fomes annosus in Great Britain. An assessment of the situation in 1959. Forest Rec. 41. 22 p.
29. LOWE, J. L. 1957. Polyporaceae of North America. The genus Fomes. State Univ. N.Y. Coll. For. Tech. Pub. 80. 97 p.
30. MILLER, T., and A. KELMAN. 1966. Growth of Fomes

annosus in roots of suppressed and dominant loblolly pines. Forest Sci. 12:225-233.

31. MORRIS, C. L., and D. H. FRAZIER. 1966. Development of a hazard rating for Fomes annosus in Virginia. Plant Dis. Reptr. 50:510-511.

32. PECK, C. H. 1891. Species of plants not before reported, p. 15-30. Annu. Rept. N.Y. State Bot., 1890.

33. POWERS, H. R., JR., and A. F. VERRALL. 1962. A closer look at Fomes annosus. Forest Farmer 21(13):8-9, 16-17.

34. PUNTER, D. 1968. Fomes annosus root rot in Ontario. Can. Dept. Forestry and Rural Dev., Ontario Region Sault Ste. Marie. Info. Rept. 8 0-X-82. 10 p.

35. PUNTER, D. 1970. Fomes annosus in eastern Canada, p. 156-160. *In* T. A. Toussoun et al. (ed.), Root diseases and soil-borne pathogens. Univ. Calif. Press, Berkeley and Los Angeles.

36. RENNERFELT, E. 1947. On butt rot caused by Polyporus (Fomes) annosus in Sweden. Its distribution and mode of occurrence. Statens Skogforskminst Medd. 35:1-88.

37. RISHBETH, J. 1950. Observations on the biology of Fomes annosus, with particular reference to East Anglian pine plantations. I. The outbreaks of disease and ecological status of the fungus. Ann. Bot. 14:365-383.

38. RISHBETH, J. 1951. Observations on the biology of Fomes annosus, with particular reference to East Anglian pine plantations. II. Spore production, stump infection, and saprophytic activity in stumps. Ann. Bot. 15:1-21.

39. RISHBETH, J. 1959. Stump protection against Fomes annosus. I. Treatment with creosote. Ann. Appl. Biol. 47:519-528.

40. RISHBETH, J. 1963. Stump protection against Fomes annosus. III. Inoculation with Peniophora gigantea. Ann. Appl. Biol. 52:63-77.

41. ROSS, E. W. 1968. Duration of stump susceptibility of loblolly pine to infection by Fomes annosus. Forest Sci. 14:206-211.

42. ROSS, E. W. 1969. Thermal inactivation of conidia and basidiospores of Fomes annosus. Phytopathology 59:1798-1801.

43. ROSS, E. W. 1973. Fomes annosus in the southeastern United States. Relation of environmental and biotic factors to stump colonization and losses in the residual stand. USDA Forest Serv. Tech. Bull. 1459. 26 p.

44. SHAIN, L. 1967. Resistance of sapwood in stems of loblolly pine to infection by Fomes annosus. Phytopathology 57:1034-1045.

45. SINCLAIR, W. A. 1964. Root- and butt-rot of conifers caused by F. annosus, with special reference to inoculum dispersal and control of the disease in New York. Cornell Univ. Agr. Exp. Stn. Mem. 391. 54 p.

46. STAMBAUGH, W. J., F. W. COBB, R. A. SCHMIDT, and F. C. KRIEGER. 1962. Seasonal inoculum dispersal and white pine stump invasion by Fomes annosus. Plant Dis. Reptr. 46:194-198.

47. UNDERWOOD, L. M. 1897. Some new fungi, chiefly from Alabama, p. 81-86. Torrey Bot. Club Bull. 24.

48. YDE-ANDERSEN, A. 1962. Seasonal incidence of stump infection in Norway spruce by airborne Fomes annosus. Forest Sci. 8:98-103.

PART V

MICROBIAL ANTAGONISM AS A MECHANISM OF DEFENSE OF THE HOST AGAINST SEED- AND SOIL-BORNE PATHOGENS

The Role of Ectomycorrhizae in the Protection of Pine from Root Infection by Phytophthora cinnamomi

DONALD H. MARX—*Principal Plant Pathologist, USDA Forest Service, Southeastern Forest Experiment Station, Forestry Sciences Laboratory, Athens, Georgia.*

Ectomycorrhizae are essential for the survival and normal growth of pines and other trees (3, 9). The fungal symbionts of ectomycorrhizae are primarily Basidiomycetes, which produce mushrooms or puffballs. These fungi penetrate intercellularly and partially replace the middle lamellae between primary cortical cells of the feeder roots. This hyphal arrangement around the cortical cells is called the Hartig net. Ectomycorrhizal fungi also form a dense, usually continuous, hyphal network over the feeder root surface called the fungal mantle. Thickness of this mantle varies from one or two hyphal diameters to as many as 30 to 40. There are over 2,200 species of ectomycorrhizal fungi of forest trees in North America.

The benefits of ectomycorrhizae to plants are well documented from research conducted in various parts of the world (3, 9). Physiologically, ectomycorrhizae (a) increase the area of active nutrient absorption on the root system, (b) increase the rate and accumulation of essential nutrients, (c) make available essential nutrients from soil minerals and organic complexes normally unavailable to the plant, (d) increase tolerance of the host to adverse soil factors such as extremes of soil pH, temperature, drought, and toxins, and (e) increase longevity of feeder root function.

Another benefit of ectomycorrhizae is their role as biological deterrents to feeder root infections caused by pathogenic fungi like *Phytophthora cinnamomi*.

Zak (25) and, more recently, Marx (13) reviewed the mechanisms of the resistance of ectomycorrhizae to pathogenic root infections. Ectomycorrhizal fungi can (a) utilize various chemicals in the root and at the root surface, thereby reducing the amount of nutrients available to pathogens, (b) provide a mechanical barrier, the fungal mantle, to penetration of primary cortical cells by pathogens, (c) support, along with the root, an antagonistic rhizosphere population of microorganisms, and (d) induce production of inhibitors in the cortical cells that inhibit infection and spread of pathogens.

This report concerns only ectomycorrhizae of pine and their relation to feeder root infection caused by *P. cinnamomi*. See the above reviews for accounts of work relative to other plant hosts and pathogens.

Ectomycorrhizae and root infections of pine by *P. cinnamomi* are both intimately involved with the primary tissues of succulent feeder roots. Propagules of ectomycorrhizal fungi in soil are stimulated by pine feeder roots; they then germinate, infect, and eventually transform feeder roots into dual organs in which the primary cortical cells are enclosed in the Hartig net and isolated from direct contact with the soil by the fungal mantle. Similarly, propagules of *P. cinnamomi* in soil are stimulated by pine feeder roots and then germinate, infect, ramify through meristematic and primary cortical tissues, and eventually cause limited or extensive feeder root necrosis.

If *P. cinnamomi* infects and destroys cortical cells of the pine feeder root prior to infection of this root by an ectomycorrhizal fungus, an ectomycorrhiza will not form. In this instance, the pathogenic infection by *P. cinnamomi* negates symbiotic infection. If the order of infection is reversed; i.e., if the symbiont infects the feeder root and synthesizes an ectomycorrhiza prior to infection of the root by *P. cinnamomi*, can the symbiotic infection negate the pathogenic infection? This question has been answered affirmatively by recent research done on ectomycorrhizae of southern pines and *P. cinnamomi*. Most of the purported mechanisms of resistance are demonstrably effective.

ANTIBIOTIC PRODUCTION.—Over 100 species of ectomycorrhizal fungi produce either antibacterial, antifungal, or antiviral antibiotics in pure culture (13). However, only a few ectomycorrhizal fungi have been tested against pathogenic fungi of feeder roots such as *P. cinnamomi*. *Leucopaxillus cerealis* var. *piceina*, an ectomycorrhizal fungus of shortleaf pine, strongly inhibited *P. cinnamomi* and other related fungi (10). The antibiotic produced by this fungal symbiont was identified as diatretyne nitrile, a polyacetylene (11). It inhibited germination of zoospores of *P. cinnamomi* at 50-70 parts per billion and killed zoospores at 2 ppm. The antibiotic also inhibited bacteria from forest soil at 0.5 ppm. Anchel et al. (1) identified diatretyne antibiotics from culture filtrates of *Clitocybe diatreta*, *C. odora*, and *Lepista nuda*, which are also ectomycorrhizal fungi (24). Differences in biological activities of antibiotics produced by other ectomycorrhizal fungi were found (10). *Laccaria laccata* inhibited 16 of 21 species of *Pythium*, but only one of nine species of *Phytophthora* (not *P. cinnamomi*). *Lactarius deliciosus*, however, did not inhibit any species of *Pythium*, but did inhibit five of six *Phytophthora* spp. tested, including *P. cinnamomi*. Extremes in activity were also found. *Pisolithus tinctorius* did not inhibit any of the 48 root pathogens tested, and *L. cerealis* var. *piceina* inhibited 92%. Intermediate in the biological spectrum of antibiotic activity was *Suillus luteus*, which inhibited over 70% of the pathogens, including *P. cinnamomi*. Pratt (20) also reported that *S. luteus* and *L. deliciosus* inhibited *P. cinnamomi*. Marx and Bryan (15) reported that *Scleroderma bovista*, a symbiont of pine and pecan, inhibited several species of *Pythium*, as well as *P. cinnamomi* and four other species of *Phytophthora*. *Thelephora terrestris*, which is probably the most widespread ectomycorrhizal fungus on pine seedlings in nurseries, inhibited several species of *Pythium* but not *P. cinnamomi* (16). The symbiont *Rhizopogon vinicolor* also inhibited *P. cinnamomi*, as well as related root pathogenic fungi (26).

There is only one report of antibiotic production by a fungal symbiont in ectomycorrhizal association with pine that is effective in inhibiting root infection by *P. cinnamomi*. Marx and Davey (17) extracted diatretyne nitrile and diatretyne 3 from ectomycorrhizae formed by *L. cerealis* var. *piceina* and from the rhizosphere substrate of the ectomycorrhizae on axenic shortleaf pine seedlings. The diatretynes were effective in the resistance of feeder roots to infection by *P. cinnamomi*. Not only were the ectomycorrhizae resistant, but nonmycorrhizal short roots adjacent to the ectomycorrhizae containing diatretyne antibiotics were only 25% susceptible to infection by zoospores of *P. cinnamomi*. It was not determined whether the diatretynes were translocated to short roots from adjacent ectomycorrhizae or were absorbed from the rhizosphere. Short roots on control seedlings and on seedlings with ectomycorrhizae formed by either *Laccaria laccata* or *Pisolithus tinctorius* were completely susceptible to infection. *Phytophthora cinnamomi* infected only 77% of short roots on shortleaf and 85% of short roots on loblolly pine seedlings adjacent to ectomycorrhizae formed by *Suillus luteus*. Antibiotic protection was also inferred in these observations, since *S. luteus* inhibited *P. cinnamomi* in antagonism studies.

MECHANICAL BARRIERS.—The fungal mantles of ectomycorrhizae create a unique obstruction to *P. cinnamomi*. In mature ectomycorrhizae, these mantles are composed of tightly interwoven hyphae, often in well-defined layers, which usually completely cover the root meristem and cortical tissues, isolating them from direct contact with the soil in the rhizosphere. Marx and Davey (17, 18) and Marx (12) presented histological evidence showing that the fungal mantles of ectomycorrhizae are effective physical barriers to penetration by *P. cinnamomi*. Ectomycorrhizae formed by several fungal symbionts of shortleaf and loblolly pines were resistant to infection by either zoospores or vegetative mycelium of *P. cinnamomi*. Nonmycorrhizal short roots of these pine species were highly susceptible. Fungal mantles formed by non-antibiotic-producing fungal symbionts, which passively covered adjacent nonmycorrhizal root initials, protected these initials from infection by *P. cinnamomi*. Short root initials not covered by fungal mantles were heavily infected by the pathogen. There was further evidence that the fungal mantle covering of root meristems is a barrier against infection by *P. cinnamomi*. Meristematic tissues of ectomycorrhizae were readily infected by *P. cinnamomi* when the mantle covering was either incompletely formed over the root tip or was

artificially removed. Infection, however, did not take place in the meristem tissues when the root tips were covered by a complete fungal mantle. The Hartig net surrounding the cortical cells may function as an additional mechanical barrier, since spread of *P. cinnamomi* originating from either infection of unprotected meristem tissue without fungal mantle coverings or from infection through artificially excised root tips was blocked in this region. Separating the possible indirect chemical effect of the Hartig net, i.e., inducing the production of chemical inhibitors in cortical cells, from the suggested mechanical effect of the Hartig net was impossible.

CHEMICAL INHIBITORS PRODUCED BY PINE HOST.—As discussed by Marx (13), there is a biochemical and cytological reaction of cortex cells to infection by ectomycorrhizal fungi. Volatile (6, 7) and nonvolatile (5) organic compounds are produced by host cortical cells in response to infection by ectomycorrhizal fungi. Symbiotic infection of pine roots by *Boletus variegatus* resulted in the nonspecific production and accumulation of volatile terpenes and sesquiterpenes in concentrations up to eight times greater than those found in nonmycorrhizal roots (7). These substances were inhibitory to the vegetative growth of ectomycorrhizal (19) and to root pathogenic fungi, including *P. cinnamomi* (8). Recently, Krupa et al. (6) reported that ectomycorrhizae of shortleaf pine formed by *P. tinctorius*, *Cenococcum graniforme*, or *T. terrestris* contained up to 40 times more terpenes than did nonmycorrhizal roots. The volatile terpenes, myrcene, β-phellandrene, and 3-carene, not only were inhibitory to the fungal symbionts, but were also inhibitory to *P. cinnamomi* and to other root pathogens. Synthesis of these compounds is apparently stimulated by infection from fungal symbionts, and they probably function as inhibitors to the fungal symbionts, resulting in the balanced parasitic state. These compounds also are ideally located for inhibiting infection and for the development of root pathogens.

DIFFERENCES IN CHEMICAL EXUDATION.—There are no published data on differences in exudates between ectomycorrhizal and nonmycorrhizal roots of pine, particularly on how they relate to pathogenic root infection (13). However, circumstantial evidence is available. Marx and Davey (17, 18) observed that encysted zoospores of *P. cinnamomi* germinated faster and more vigorously at the growing tips and at the region of cell elongation on nonmycorrhizal roots. Encysted zoospores on ectomycorrhizae germinated slowly, with weak germtube elongation comparable to responses of zoospores on heavily suberized root parts. This indicated, indirectly, that the ectomycorrhizae were not as chemically stimulating to zoospore germination and germtube growth as were nonmycorrhizal, nonsuberized roots. Zoospores of *P. cinnamomi* responded chemotactically, however, to cut tips of both ectomycorrhizal and nonmycorrhizal roots.

PROTECTIVE RHIZOSPHERE POPULATION.—There are no published data on microbial rhizosphere populations of pine antagonistic to *P. cinnamomi.* Data are available, however, on differences in microbial rhizosphere populations between ectomycorrhizal and nonmycorrhizal roots of pine (9).

INTERACTION OF NEMATODES, PHYTOPHTHORA CINNAMOMI, AND ECTOMYCORRHIZAE.—Many plant-parasitic nematodes found in forest soils and tree nurseries are important in feeder root disease complexes. Nematodes that parasitize roots are normally limited to the feeder roots, the same ones infected by ectomycorrhizal fungi. Nematodes may have several effects on ectomycorrhizae. Mycophagous nematodes may reduce the inoculum potential of fungal symbionts in soil (21); plant-parasitic nematodes may inhibit ectomycorrhizal development by damaging primary cortex cells prior to symbiotic infection, and both may also feed directly on pine ectomycorrhizae (23). Plant-parasitic nematodes can also destroy the resistance of ectomycorrhizae to pathogenic root infections. Barham (2) found that ectomycorrhizae of shortleaf pine parasitized by spiral nematodes (*Helicotylenchus dihystera*) were susceptible to infection by *P. cinnamomi.* Disruption of the structural integrity of the fungal mantle by spiral nematodes created infection courts for *P. cinnamomi.* Intracellular hyphae and vesicles of *P. cinnamomi* were found in cortex cells surrounded by the Hartig net. The interactions of nematodes, *P. cinnamomi,* and ectomycorrhizae of pine are currently under further investigation.

The foregoing discussion of mechanisms concerned only the resistance or susceptibility of individual feeder roots to infection by *P. cinnamomi* and not to the complete syndrome of feeder root disease. Only limited research has been accomplished on the role of ectomycorrhizae of pine in the complete development of feeder root disease. In a greenhouse pot study, Ross and Marx (22) found that seedlings of the Ocala race of sand pine were protected against *P. cinnamomi* by ectomycorrhizae formed by *P. tinctorius.* Nonmycorrhizal pine seedlings infected by *P. cinnamomi* exhibited massive necrosis of feeder roots, and 60% died after 2 months. Only 30% of the ectomycorrhizal seedlings died due to attack by *P. cinnamomi,* even though only 25% of the roots were ectomycorrhizal. Nonmycorrhizal roots on these seedlings, however, were infected by *P. cinnamomi,* whereas the cortical tissues in ectomycorrhizae were free from the pathogen, verifying their resistance. In a similar study, Marx (14) found that shortleaf pine seedlings with ectomycorrhizae did not show symptoms of feeder root disease caused by *P. cinnamomi.* However, nonmycorrhizal seedlings exposed to the pathogen were lighter in dry weight and had fewer new lateral roots than did nonmycorrhizal seedlings grown in the absence of the pathogen. Inoculum densities of *P. cinnamomi* in soil with the nonmycorrhizal seedlings did not significantly change during the experiment. Shortleaf pine seedlings with 70 to 89% ectomycorrhizal feeder roots were not stunted, and development of new lateral roots in the presence of *P. cinnamomi* was not reduced. The inoculum densities of *P. cinnamomi* were reduced in soil with ectomycorrhizal seedlings. Apparently, the abundance of ectomycorrhizal

roots on those seedlings reduced the amount of susceptible tissue available for attack by *P. cinnamomi,* which in turn reduced the development of feeder root disease. Neither of the two ectomycorrhizal fungi used in this experiment produces antibiotics effective against *P. cinnamomi.* These two reports strongly suggest that pine seedlings with significant quantities of ectomycorrhizae do not exhibit reduced top growth, chlorosis, restricted root development, necrosis, and eventual death and are, therefore, more resistant to feeder root disease than are nonmycorrhizal plants.

CONCLUSIONS.—The role of ectomycorrhizae in the protection of pine roots from pathogenic feeder root infections by *P. cinnamomi* is an important aspect of our understanding of the overall significance of ectomycorrhizae. When one considers the nearly worldwide distribution of *P. cinnamomi* in forest and nursery soils, as well as that of numerous other closely related feeder root pathogens like *Pythium* spp. (4), then the ecological value of ectomycorrhizae as deterrents to infections by these pathogens is self-evident. Simply stated, mature trees and seedlings with significant quantities of ectomycorrhizae growing in soils containing feeder root pathogens would have very little susceptible (nonmycorrhizal) root tissue exposed to attack. Additionally, these plants would gain from the nutritional benefits of ectomycorrhizae.

The protective role of ectomycorrhizae is functionally nonsystematic, since the presence of a few mycorrhizae on a root system does not furnish control of pathogenic infections on nonmycorrhizal feeder roots on the same root system. This means, therefore, that the degree of ectomycorrhizal development would correlate directly with the degree of control of feeder root infections. A possible exception to this generalization is the involvement of the purported antibiotic mechanism. An ectomycorrhizal fungus producing in its mycorrhizae a potent antibiotic effective against a feeder root pathogen may afford protection to adjacent nonmycorrhizal feeder roots against this pathogen simply by translocation or diffusion of the antibiotic into the nonmycorrhizal roots. In this circumstance, the presence of a few ectomycorrhizae formed by an antibiotic-producing fungal symbiont may be as valuable in controlling pathogenic root infections as the presence of considerably more ectomycorrhizae formed by a non-antibiotic-producing symbiont.

A point that should be stressed concerning this subject is that the protective role of ectomycorrhizae, based on current information, is only applicable to feeder root diseases. Root diseases initiated by infections of root parts (such as woody parts) other than primary root tissues associated with ectomycorrhizal development may not be directly affected by ectomycorrhizae.

LITERATURE CITED

1. ANCHEL, M., W. B. SILVERMAN, N. VALANJU, and C. T. ROGERSON. 1962. Patterns of polyacetylene production. I. The diatretynes. Mycologia 54:249-257.
2. BARHAM, R. O. 1972. Infectivity of plant-parasitic nematodes and Phytophthora cinnamomi on

ectomycorrhizae of Pinus echinata seedlings. M.S. thesis, Univ. Georgia, Athens. 18 p.

3. HACSKAYLO, E. (ed.). 1971. Mycorrhizae. USDA Misc. Pub. 1189. 255 p.

4. HENDRIX, F. F., JR., and W. A. CAMPBELL. 1973. Pythiums as plant pathogens. Annu. Rev. Phytopathol. 11:77-98.

5. HILLS, W. E., and N. ISHIKURA. 1969. The extractives of mycorrhizas and roots of Pinus radiata and Pseudotsuga menziesii. Austral. J. Biol. Sci. 22:1425-1436.

6. KRUPA, S., J. ANDERSSON, and D. H. MARX. 1973. Studies on ectomycorrhizae of pine. IV. Volatile organic compounds in mycorrhizal and nonmycorrhizal root systems of Pinus echinata. Eur. J. For. Pathol. 4:194-200.

7. KRUPA, S., and N. FRIES. 1971. Studies on ectomycorrhizae of pine. I. Production of volatile organic compounds. Can. J. Bot. 49:1425-1431.

8. KRUPA, S., and J. E. NYLUND. 1972. Studies on ectomycorrhizae of pine. III. Growth inhibition of two root pathogenic fungi by volatile organic constituents of ectomycorrhizal root systems of Pinus sylvestris L. Eur. J. For. Pathol. 2:88-94.

9. MARKS, G. C., and T. T. KOZLOWSKI. 1973. Ectomycorrhizae: their ecology and physiology. Academic Press, New York and London. 444 p.

10. MARX, D. H. 1969. The influence of ectotrophic mycorrhizal fungi on the resistance of pine roots to pathogenic infections. I. Antagonism of mycorrhizal fungi to root pathogenic fungi and soil bacteria. Phytopathology 59:153-163.

11. MARX, D. H. 1969. The influence of ectotrophic mycorrhizal fungi on the resistance of pine roots to pathogenic infections. II. Production, identification, and biological activity of antibiotics produced by Leucopaxillus cerealis var. piceina. Phytopathology 59:411-417.

12. MARX, D. H. 1970. The influence of ectotrophic mycorrhizal fungi on the resistance of pine roots to pathogenic infections. V. Resistance of mycorrhizae to infection by vegetative mycelium of Phytophthora cinnamomi. Phytopathology 60:1472-1473.

13. MARX, D. H. 1972. Ectomycorrhizae as biological deterrents to pathogenic root infections. Annu. Rev. Phytopathol. 10:429-454.

14. MARX, D. H. 1973. Growth of ectomycorrhizal and nonmycorrhizal shortleaf pine seedlings in soil with Phytophthora cinnamomi. Phytopathology 63:18-23.

15. MARX, D. H., and W. C. BRYAN. 1969. Scleroderma bovista, an ectotrophic mycorrhizal fungus of pecan. Phytopathology 59:1128-1132.

16. MARX, D. H., W. C. BRYAN, and L. F. GRAND. 1970. Colonization, isolation, and cultural descriptions of Thelephora terrestris and other ectomycorrhizal fungi of shortleaf pine seedlings grown in fumigated soil. Can. J. Bot. 48:207-211.

17. MARX, D. H., and C. B. DAVEY. 1969. The influence of ectotrophic mycorrhizal fungi on the resistance of pine roots to pathogenic infections. III. Resistance of aseptically formed mycorrhizae to infection by Phytophthora cinnamomi. Phytopathology 59:549-558.

18. MARX, D. H., and C. B. DAVEY. 1969. The influence of ectotrophic mycorrhizal fungi on the resistance of pine roots to pathogenic infections. IV. Resistance of naturally occurring mycorrhizae to infection by Phytophthora cinnamomi. Phytopathology 59:559-565.

19. MELIN, E., and S. KRUPA. 1971. Studies on ectomycorrhizae of pine. II. Growth inhibition of mycorrhizal fungi by volatile organic constituents of Pinus sylvestris L. (Scots pine) roots. Physiol. Plant. 25:337-340.

20. PRATT, B. H. 1971. Isolation of Basidiomycetes from Australian Eucalypt forest and assessment of their antagonism to Phytophthora cinnamomi. Trans. Br. Mycol. Soc. 56:243-250.

21. RIFFLE, J. W. 1971. Effect of nematodes on root-inhabiting fungi, p. 97-113. In E. Hacskaylo (ed.), Mycorrhizae. USDA Misc. Pub. 1189.

22. ROSS, E. W., and D. H. MARX. 1972. Susceptibility of sand pine to Phytophthora cinnamomi. Phytopathology 62:1197-1200.

23. RUEHLE, J. L. 1972. Nematodes and forest trees—types of damage to tree roots. Annu. Rev. Phytopathol. 11:99-118.

24. TRAPPE, J. M. 1962. Fungus associates of ectotrophic mycorrhizae. Bot. Rev. 28:538-606.

25. ZAK, B. 1964. Role of mycorrhizae in root disease. Annu. Rev. Phytopathol. 2:377-392.

26. ZAK, B. 1971. Characteristics and classification of mycorrhizae of Douglas fir. II. Pseudotsuga menziesii and Rhizopogon vinicolor. Can. J. Bot. 49:1079-1084.

Genetic Control of the Rhizosphere Microflora of Wheat

T. G. ATKINSON, J. L. NEAL, Jr., and RUBY I. LARSON—*Research Scientist, Cereal Pathology, Research Scientist, Soil Microbiology, and Senior Research Scientist, Cytogenetics, Research Station, Canada Department of Agriculture, Lethbridge, Alberta.*

The title "Microbial Antagonism as a Mechanism of Defense of the Host Against Seed- and Soil-Borne Pathogens" is somewhat speculative. It suggests that exudates from seeds, from roots and from below-ground stems or leaves of some hosts might selectively stimulate, directly or indirectly, a spermatosphere (35), rhizosphere (15), or laimosphere (25) microflora antagonistic to a potential pathogen. That such a mechanism could be the basis of some forms of host resistance has, indeed, been previously recognized (12, 16, 30-33). With the exception of the special case of ectomycorrhizal associations (26 and Marx, this volume), however, the involvement of antagonists in the resistance of a host to soil-borne pathogens under natural conditions is still debatable (2, 33, 42).

The direct effect of plant exudates in stimulating resting structures of soil-borne pathogens to activity and in providing energy for growth and penetration of the host has been well documented (33; 41, p. 95-137). In contrast, the indirect effect that these same exudates might have by stimulating the growth and metabolism of nonpathogenic microorganisms that could be antagonistic or stimulatory to pathogens has been less well analyzed. There is little doubt that the role of the spermatosphere, laimosphere, and rhizosphere microfloras in influencing the host's reaction to soil-borne diseases is largely unresolved because of the following related problems: the general inapplicability of in vitro results to in vivo relationships; the difficulty of in situ experimentation under natural conditions; and, notwithstanding the use of foliar treatments, the absence of any useful experimental means of manipulating the quantity and quality of plant exudates.

This paper, in part, reviews our attempts to determine whether rhizosphere antagonists are involved in the resistance of wheat to common root rot. More importantly, it focuses attention on our finding that the genotype of the host governs the magnitude and composition of bacterial populations in the rhizosphere with surprising specificity. Presumably, this control is exercised through root exudates acting directly as nutrients and, notwithstanding the importance of ion-exchange at the root surface (Smiley, this volume), by their indirect effect on the physico-chemical environment of the rhizosphere through the influence of subsequent microbial activity (33). In this sense, therefore, we believe that the potential of the host to influence the course of pathogenesis by determining the characteristics of the rhizosphere microflora should not be overlooked as a possible pathway in the biological control of soil-borne pathogens. The formal exclusion of the host by Garrett's (2, 14) definition of biological control rejects this pathway as a possible mechanism (2, Fig. 1).

THE PROBLEM AND OUR APPROACH.—Common root rot is a serious and widespread soil-borne disease of wheat and barley on the Canadian prairies and in many other cereal-growing areas of the world (6, 23). Although no sources of immunity to the principal pathogen, *Cochliobolus sativus* (Ito and Kurib.) Drechs. ex Dastur (stat. conid. *Helminthosporium sativum* Pamm., King and Bakke), have been found, many currently grown varieties have appreciable resistance to common root rot. In contrast, varieties of solid-stemmed wheat developed in Canada for their resistance to the wheat stem sawfly, *Cephus cinctus* Nort., are strikingly susceptible to the disease.

Because of these contrasting disease reactions, chromosome substitution lines originally developed to identify which chromosomes carried genes for stem solidness (21, 22) have proved a source of lines differing in reaction to common root rot. Utilizing these genetically related lines, we are attempting to identify the mechanisms determining their differences in disease reaction in greenhouse and growth cabinet experiments using both field soil naturally infested with the pathogen and "disease-free" field soil.

Two series of disomic whole-chromosome substitution lines were available: one in which chromosomes from the relatively resistant *donor* variety Apex had been individually substituted for their homologues in the highly susceptible *recipient* line S-615; the other a conversely related series in which the root rot-susceptible variety Rescue was the donor parent and the relatively resistant variety Cadet was the recipient parent. Root rot tests based on the extent of lesioning of subcrown internodes (5) have repeatedly shown that chromosome 5B has a major effect in differentiating the root rot resistance of Apex and Cadet from the susceptibility of S-615 and Rescue (20).

Chromosome 5B from Apex substituted for its homologue in S-615 makes the chromosome substitution line S-A5B as resistant as Apex (Fig. 1A), whereas, inversely, chromosome 5B from Rescue substituted for its homologue in Cadet makes the line C-R5B as susceptible as Rescue (Fig. 1B). Moreover, both the resistance and susceptibility mechanisms, whatever they may be, are genetically active: the monosomics of resistant lines show dosage effects, and the F_1 hybrid between the susceptible recipient parent S-615 and its resistance chromosome substitution line S-A5B shows dominance of susceptibility (20).

The identification of these chromosome substitution lines, critical for host reaction to common root rot, provides genetic model systems useful in detecting and analyzing the mechanisms involved. For example, the resistant line S-A5B differs from its susceptible recipient parent S-615 only by the 5B chromosome pair received from its resistant donor parent Apex. Therefore, differences between S-A5B and S-615 in other characteristics are more likely to be associated with resistance, especially if the substitution line resembles the resistant donor parent, than would be differences between S-615 and Apex. As Flor (10) has recently stated,

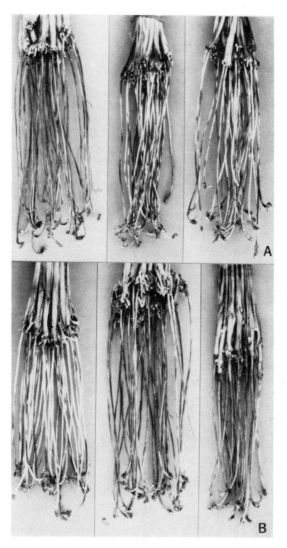

Fig. 1. Root rot lesions on subcrown internodes. *A)* S-615 (susceptible), Apex (resistant), and S-A5B (resistant). *B)* Cadet (resistant), Rescue (susceptible), and C-R5B (susceptible). [From Larson and Atkinson (20).]

"Chemical, morphological, or physiological variations observed between phenotypes that differ only in one gene for resistance are more apt to be related to resistance than variations observed between phenotypes that, in addition to the gene for reaction, differ in many other genes." Of course, chromosome 5B carries many genes besides the one controlling reaction to common root rot, so donor characteristics associated with the 5B substitution line could be due to genes at other loci on the same chromosome.

RELATIONSHIP BETWEEN ROOT ROT REACTION AND RHIZOSPHERE CHARACTERISTICS.—*Incidence of bacteria antibiotic in vitro to Cochliobolus sativus in the rhizosphere of S-A5B.*—An initial study revealed two rhizosphere microbial characteristics in which the substitution line S-A5B was like Apex and different from S-615 (28). The total bacterial rhizosphere populations of S-A5B and Apex were half the magnitude of that of the susceptible line S-615, and, whereas none of the S-615 rhizosphere bacteria exhibited antibiosis to *C. sativus* in vitro, about 20% of the rhizosphere isolates from the resistant lines were antibiotic (Fig. 2). The presence of greater numbers of bacteria in the rhizosphere of the root rot-susceptible line S-615 compared with that of S-A5B, from which it differs by only one chromosome, confirms what others have found using genetically unrelated susceptible and resistant varieties; i.e., root rot-susceptible varieties have a more pronounced rhizosphere effect than have resistant ones. But, at the time, the intriguing correlation between the presence of the critical 5B chromosome from Apex and the occurrence of rhizosphere bacteria antagonistic to *C. sativus* appeared more significant. Nevertheless, we recognized that it was not possible from such data "to associate conclusively resistance to root rot with the occurrence of these antagonistic microbes in the rhizosphere" (28). As Baker (2) has stated, "While such correlations are important, quantitation of the impact of antibiosis on biological control will require sturdier stuff."

Bacterization with rhizosphere antagonists.—A bacterization experiment was carried out in an attempt to verify directly that antibiotic bacteria isolated from the rhizospheres of the variety Apex and the chromosome substitution line S-A5B were responsible for resistance to common root rot (1). Individual lots of greenhouse-grown seed of Apex, S-A5B, and of the susceptible line S-615 were treated separately with a heavy suspension of antagonists isolated from the rhizospheres of Apex and S-A5B. Check lots of seed of each line were similarly treated with sterile, distilled water. When sown in field soil naturally infested with *C. sativus* and grown in a controlled environment, none of the lines showed any change in their characteristic root rot reaction as a result of the bacterization treatments (Table 1). Although bacterization failed to alter disease reactions, population estimates by plate-dilution frequency techniques showed that some significant changes in populations of bacteria antagonistic to *C. sativus* had been effected. Estimates of total bacterial populations and numbers of antagonists were made both on the seminal root rhizosphere soil and on the laimosphere soil surrounding the subcrown internodes. To avoid improper use of the term rhizosphere, laimosphere has been proposed (25) to designate the zone of influence of below-ground stems or leaves on soil microbes. The subcrown internode in wheat is, anatomically, more like a stem than a root.

In the untreated lines, we found, as we had in the initial study using "disease-free" soil (28), that the total number of bacteria in the rhizosphere of S-615 was twice that of Apex, but that the numbers and thus the percentage of bacterial isolates that exhibited antibiosis in vitro to *C. sativus* were significantly fewer (Table 1). As before, the root rot-resistant chromosome substitution line S-A5B was like Apex in both these characteristics. Bacterization with antagonists isolated from the rhizosphere of Apex did not alter any of these relationships, but treatment with the S-A5B culture did increase the numbers of antagonists in the rhizosphere of S-615 to a level comparable with the resistant lines. However, because of the greater rhizosphere effect of S-615, the percentage of these rhizosphere antagonists was not increased to a comparable level.

The subcrown internodes on which the disease ratings were made provided laimosphere data that showed interesting similarities to, and contrasts with, those of the corresponding rhizospheres. Unlike the rhizosphere, the laimosphere of the untreated, susceptible line harbored no greater number of bacteria than did the laimospheres of the resistant lines (Table 1). As in the rhizosphere, however, bacterial populations in the laimosphere of the resistant lines contained much higher percentages of in vitro antagonists of *C. sativus* than did that of S-615. Again, the S-A5B substitution line behaved like its donor parent, Apex. Bacterization with antagonists either from

Apex or S-A5B increased the percentage of antagonists in the laimosphere of S-615 to the same level as those in the resistant lines, but did not decrease its susceptibility. These laimosphere results made us seriously question the significance of the causal relationship between antagonistic bacteria and root rot resistance, especially because the disease ratings were made on the subcrown internodes. The failure of these augmented laimosphere populations of antagonists to reduce root rot could be attributed to the inability of the bacteria originally isolated from the rhizospheres of Apex or S-A5B to function antibiotically in the S-615 environment. But the probability that the antibiotic potential of these microbes is not realized, or at least is not effective, even in the rhizospheres and laimospheres in which they normally flourish, is indicated by the failure of the resistant reaction of S-A5B to be affected by the unexpected but verified decrease in laimosphere antagonists that occurred when the line was bacterized with the Apex culture (Table 1).

Incidence of rhizosphere antagonists in additional lines differential for root rot reaction.—Further evidence that the association between root rot resistance governed by chromosome 5B of Apex and the occurrence of rhizosphere and laimosphere antagonists of *C. sativus* is entirely coincidental was obtained when an attempt was made to verify this relationship by analyzing bacterial rhizosphere populations of additional lines grown in "disease-free" field soil. The behavior of the root rot-susceptible chromosome substitution line S-A5D, carrying the Apex homoeologue of chromosome 5B, was compared with S-615, Apex, and S-A5B. In agreement with previous trends, the total bacterial rhizosphere

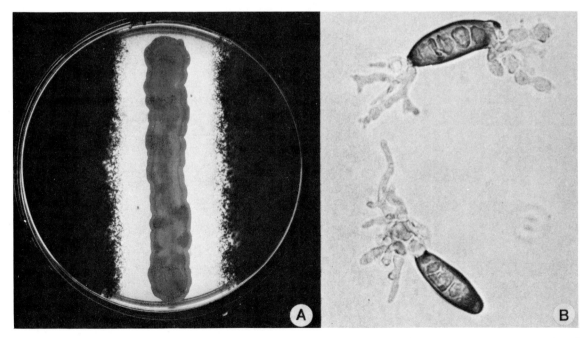

Fig. 2. Antagonism to *Cochliobolus sativus* in vitro by rhizosphere bacterial isolates. *A)* Clear zone showing inhibition of germination and growth. [From Neal et al. (28).] *B)* Distortion of conidial germ tubes.

populations of the susceptible lines, S-A5D and its recipient parent S-615, were twice the magnitude of those enumerated for Apex and S-A5B (Fig. 3). In contrast to S-615, however, the numbers of bacteria antagonistic in vitro to *C. sativus* in the rhizosphere of S-A5D were so great they were statistically comparable with populations

TABLE 1. Effect of bacterizing seed with rhizosphere bacteria antibiotic in vitro to *Cochliobolus sativus* on the root rot reactions of wheat lines and on population estimates of total bacteria and bacteria showing antibiosis to *C. sativus* in the rhizosphere and subcrown internode laimosphere[a]

Wheat line or sample source and origin of antagonists used for bacterization	Root rot ratings	Rhizosphere[b]			Laimosphere[b]	
		Total bacteria	Antagonists		Total bacteria	Antagonists
	$\%^c$	$\times 10^6$	$\times 10^6$	$\%^c$	$\times 10^6$	$\%^c$
S-615						
Untreated	95.0 a[d]	332 a	14 c	4.2 e	114 a	1.6 c
Apex bacteria	96.7 a	324 a	15 c	4.5 de	122 a	12.7 a
S-A5B bacteria	93.0 a	339 a	21 b	6.3 d	123 a	11.0 a
Apex						
Untreated	19.3 b	161 c	25 ab	15.7 b	109 a	11.2 a
Apex bacteria	20.3 b	204 b	33 a	15.8 b	112 a	9.7 a
S-A5B bacteria	20.4 b	160 c	29 ab	18.4 a	118 a	8.6 ab
S-A5B						
Untreated	21.0 b	165 bc	25 ab	15.3 b	112 a	10.4 a
Apex bacteria	19.8 b	166 bc	28 ab	16.8 ab	115 a	3.7 bc
S-A5B bacteria	20.0 b	191 bc	25 ab	13.2 c	111 a	10.1 a
Nonrhizosphere, nonlaimosphere soil		32 d	1 d	3.7 e	32 b	3.7 bc

[a]Adapted from Atkinson et al. (1).
[b]Per gram of soil, oven-dry basis.
[c]Percentage data are arithmetic means, other values are geometric means, of six replicates.
[d]Means in each column followed by the same letter do not differ statistically ($P = 0.01$).

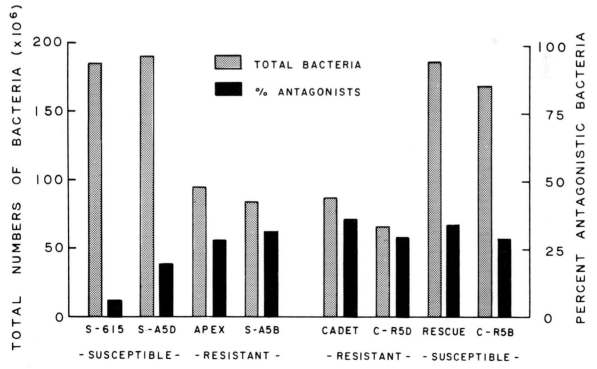

Fig. 3. Populations of total bacteria and of bacteria antagonistic to *Cochliobolus sativus* in the rhizospheres of wheat lines differing in susceptibility to common root rot. [Adapted from Atkinson et al. (1).]

in the rhizospheres of Apex and S-A5B, even when expressed as a percentage of its much greater total population. A similar rhizosphere analysis of appropriate representatives from the Cadet-Rescue chromosome substitution series showed no correlation between disease reactions and incidence of antagonists (Fig. 3). Significantly, the root rot-susceptible lines again showed total bacterial populations in the rhizosphere more than double those of the resistant lines.

Rhizosphere characteristics of susceptible lines.—The consistently greater bacterial populations in the rhizospheres of root rot-susceptible, as compared with root rot-resistant, lines may simply reflect a greater level of exudation and a consequent direct beneficial effect on the inoculum potential of *C. sativus*. On the other hand, the possibility that exudates may indirectly increase inoculum potential by preferentially stimulating beneficial associative microorganisms should not be overlooked (30, 33). Analyses of rhizosphere populations for bacteria having selected hydrolytic activities showed that root rot-susceptible lines had significantly greater numbers of bacteria with cellulolytic, pectinolytic, and amylolytic potentials than did rhizosphere populations from resistant lines (Table 2) (29). Whether or how these increased numbers of hydrolytic organisms in the rhizosphere of susceptible lines may be related to the ability of the pathogen to infect the plant or to produce the symptoms of disease is unknown. Subramanian (40) has suggested that pectinolytic and cellulolytic enzymes, presumably elaborated by the pathogen, are involved in the rotting of root tissues by *C. sativus*, and, indeed, pectinolytic (27) and cellulolytic (13) activity has been reported for different isolates of this fungus. Although we do not know whether the increased numbers of hydrolytic bacteria found in the rhizosphere of the susceptible lines promote pathogenesis by *C. sativus*, it is unlikely that they are secondary effects of disease development, since the test plants were grown in an essentially "disease-free" soil.

DISCUSSION.—Our results with disomic chromosome substitution lines critical for reaction to common root rot illustrate the potential of genetic model systems for simplifying the identification and elucidation of mechanisms involved in differential host responses to root pathogens. Near-isogenic lines have been used to analyze the physiological and biochemical bases of

cellular resistance and susceptibility to foliar (8, 10, 17) and root (18, 19) diseases, but, notwithstanding studies on nodulation (7, 9), no similar approach has previously been used to investigate the role of nonpathogenic microorganisms in the reaction of the host to soil-borne pathogens. Even though our studies have not shown definitively whether or how rhizosphere or laimosphere microbial populations are involved in the differential root rot reactions of the lines used, our findings do emphasize that the host genotype, as Garrett (11, p. 117) surmised, plays a major role in determining the characteristics of rhizosphere bacterial populations, presumably through its control of the quantity or quality of root exudates, or both. This point obviously needs to be stressed, since the many workers who have recognized the potential, indeed the probable, role of root exudates in directly or indirectly affecting pathogenesis by root disease fungi have not appreciated the control possible through genetic manipulation of the host (24, 31, 33, 38). Perhaps, too, this is why the host has been excluded by some from considerations of biological control (2). Others, in contrast, have included host resistance as a legitimate aspect of biological control without any distinction between cellular resistance mechanisms and resistance mediated through the agency of other living organisms (4, 36). Indeed, Sewell (34) has elegantly argued that biological control should be permitted broad coverage and has proposed the following definition: "Biological control is the induced or natural, direct or indirect limitation of a harmful organism, or its effects, by another organism or group of organisms."

In retrospect, our failure to verify any cause and effect relationship between the incidence of bacteria antibiotic in vitro to *C. sativus* and resistance to common root rot was not surprising. The antibiotic potential of these bacteria was not demonstrated in situ, and, furthermore, antagonists that may be effective in the rhizosphere or laimosphere or both need not exhibit antibiosis in vitro (39).

Extension of the previously recognized correlation between susceptibility to root disease and an intensified rhizosphere effect to the level of lines differing in only one chromosome critical for root rot reaction focuses attention on the mechanisms of susceptibility rather than of resistance. As Bawden (3) stressed, "Resistance to infection is the normal condition of plants, and susceptibility the exception." In their predilection for control mechanisms, root disease investigators should

TABLE 2. Estimates of total bacteria and of bacteria having selected hydrolytic activities in the rhizospheres of wheat lines differing in reaction to common root rot[a]

Wheat line or sample source	Reaction to root rot[b]	Total bacteria	Cellulolytic	Pectinolytic	Amylolytic
		$\times 10^6$	$\times 10^3$	$\times 10^4$	$\times 10^6$
Cadet	R	165.4 b[c]	4.7 b	4.1 d	2.4 c
Rescue	S	335.2 a	131.2 a	570.2 a	38.1 b
C-R5B	S	325.9 a	146.9 a	270.4 b	70.3 a
C-R5D	R	180.4 b	3.2 b	62.2 c	4.4 c
Nonrhizosphere		32.1 c	0.2 c	0.3 e	2.0 c

[a]Per gram of soil, oven-dry basis. Each value represents a geometric mean of three replicates. [Adapted from Neal et al. (29).]
[b]R = resistant, S = susceptible.
[c]Means in each column followed by the same letter do not differ statistically ($P = 0.01$).

not overlook the possibility that "Clarification of the events in susceptibility may be a necessary prelude to understanding resistance . . . " (8). In this regard, the fact that the susceptible line S-615 showed no greater bacterial laimosphere effect than did the resistant line S-A5B or its donor parent Apex suggests that the composition rather than the quantity of exudates might be the determining factor in disease reaction. This would also be in keeping with the preferential stimulation of bacteria having the potential for certain hydrolytic enzyme activity in the rhizosphere of susceptible lines. Furthermore, if the allele for susceptibility on chromosome 5B (20, and R. I. Larson and T. G. Atkinson, *unpublished*) does act by controlling composition of exudate, the allele for resistance probably acts through a similar mechanism. Our genetic evidence indicates that resistance as well as susceptibility is governed by active mechanisms; i.e., that resistance is not the absence of susceptibility or vice versa (20), and it does not seem reasonable that the allele governing resistance would control a characteristic or process completely unrelated to that controlled by the allele determining susceptibility.

Schroth and Hildebrand (33), Snyder (37), and others have emphasized the importance of techniques to the progress of root disease research, particularly with respect to identifying key interrelationships among the many that exist in natural field soil. Our results clearly demonstrate that resistant and susceptible lines having otherwise similar genotypes should simplify this task.

LITERATURE CITED

1. ATKINSON, T. G., J. L. NEAL, JR., and R. I. LARSON. 1974. Root rot reaction in wheat: resistance not mediated by rhizosphere or laimosphere antagonists. Phytopathology 64:97-101.
2. BAKER, R. 1968. Mechanisms of biological control of soil-borne pathogens. Annu. Rev. Phytopathol. 6:263-294.
3. BAWDEN, F. C. 1957. The role of plant hosts in microbial ecology. Symp. microbial ecology, Soc. Gen. Microbiol. 7:299-314.
4. BOSWELL, V. R. 1965. A landmark in biology, p. 3. *In* K. F. Baker and W. C. Snyder (ed.), Ecology of soil-borne plant pathogens. Univ. Calif. Press, Berkeley and Los Angeles.
5. BURRAGE, R. H., and R. D. TINLINE. 1960. Common rootrot and plant development following treatments of wheat seed with aldrin, gamma BHC, and heptachlor, with and without mercury fungicides. Can. J. Plant Sci. 40:672-679.
6. BUTLER, F. C. 1961. Root and foot rot diseases of wheat. New South Wales Dept. Agr. Sci. Bull. 77. 98 p.
7. CLARK, F. E. 1957. Nodulation responses of two near isogenic lines of soybean. Can. J. Microbiol. 3:113-123.
8. DALY, J. M. 1972. The use of near-isogenic lines in biochemical studies of the resistance of wheat to stem rust. Phytopathology 62:392-400.
9. ELKAN, G. H. 1962. Comparison of rhizosphere microorganisms of genetically related nodulating and nonnodulating soybean lines. Can. J. Microbiol. 8:79-87.
10. FLOR, H. H. 1971. Current status of the gene-for-gene concept. Annu. Rev. Phytopathol. 9:275-296.
11. GARRETT, S. D. 1956. Biology of root-infecting fungi. Cambridge Univ. Press, London. 293 p.
12. GARRETT, S. D. 1959. Biology and ecology of root-disease fungi, p. 309-316. *In* C. S. Holton et al. (ed.), Plant pathology, problems and progress 1908-1958. Univ. Wis. Press, Madison.
13. GARRETT, S. D. 1963. A comparison of cellulose-decomposing ability in five fungi causing cereal foot rots. Trans. Br. Mycol. Soc. 46:572-576.
14. GARRETT, S. D. 1965. Toward biological control of soil-borne plant pathogens, p. 4-16. *In* K. F. Baker and W. C. Snyder (ed.), Ecology of soil-borne plant pathogens. Univ. Calif. Press, Berkeley and Los Angeles.
15. HILTNER, L. 1904. Über neuere Erfahrungen und Probleme auf dem Gebiete der Bodenbakteriologie unter besonderer Berücksichtigung der Gründüngung und Brache. Arb. Deut. Landwirtsch. Ges. 98:59-78.
16. JACKSON, R. M. 1965. Antibiosis and fungistasis of soil microorganisms, p. 363-369. *In* K. F. Baker and W. C. Snyder (ed.), Ecology of soil-borne plant pathogens. Univ. Calif. Press, Berkeley and Los Angeles.
17. KIYOSAWA, S. 1971. Genetical approach to the biochemical nature of plant disease resistance. Jap. Agr. Res. Quart. 6:73-80.
18. KLARMAN, W. L., and J. W. GERDEMANN. 1963. Induced susceptibility in soybean plants genetically resistant to Phytophthora sojae. Phytopathology 53:863-864.
19. KLARMAN, W. L., and J. W. GERDEMANN. 1963. Resistance of soybeans to three Phytophthora species due to the production of a phytoalexin. Phytopathology 53:1317-1320.
20. LARSON, R. I., and T. G. ATKINSON. 1970. A cytogenetic analysis of reaction to common root rot in some hard red spring wheats. Can. J. Bot. 48:2059-2067.
21. LARSON, R. I., and M. D. MACDONALD. 1964. Cytogenetics of sawfly resistance in wheat. Can. Entomol. 94:124 (Abstr.).
22. LARSON, R. I., and M. D. MACDONALD. 1966. Cytogenetics of solid stem in common wheat. V. Lines of S-615 with whole chromosome substitutions from Apex. Can. J. Genet. Cytol. 8:64-70.
23. LEDINGHAM, R. J., T. G. ATKINSON, J. S. HORRICKS, J. T. MILLS, L. J. PIENING, and R. D. TINLINE. 1973. Wheat losses due to common root rot in the prairie provinces of Canada, 1969-71. Can. Plant Dis. Surv. 53:113-122.
24. MACURA, J. 1971. Some biological and ecological aspects of the rhizosphere effect. Folia Microbiol. 16:328-336.
25. MAGYAROSY, A., and J. G. HANCOCK. 1972. Microbial population of the laimosphere of squash (Cucurbita maxima). Plant Soil 37:187-190.
26. MARX, D. H. 1972. Ectomycorrhizae as biological deterrents to pathogenic root infections. Annu. Rev. Phytopathol. 10:429-454.
27. MUSE, R. R., H. B. COUCH, L. D. MOORE, and B. D. MUSE. 1972. Pectolytic and cellulolytic enzymes associated with Helminthosporium leaf spot on Kentucky bluegrass. Can. J. Microbiol. 18:1091-1098.
28. NEAL, J. L., JR., T. G. ATKINSON, and R. I. LARSON. 1970. Changes in the rhizosphere microflora of spring wheat induced by disomic substitution of a chromosome. Can. J. Microbiol. 16:153-158.
29. NEAL, J. L., JR., R. I. LARSON, and T. G. ATKINSON. 1973. Changes in rhizosphere populations of selected physiological groups of bacteria related to substitution of specific pairs of chromosomes in spring wheat. Plant Soil 39:209-212.
30. PARK, D. 1963. The ecology of soil-borne fungal disease. Annu. Rev. Phytopathol. 1:241-258.
31. ROVIRA, A. D. 1965. Interactions between plant roots and soil microorganisms. Annu. Rev. Microbiol. 19:241-266.
32. SADASIVAN, T. S., and C. V. SUBRAMANIAN. 1960. Interaction of pathogen, soil, other microorganisms in the soil, and host, p. 273-313. *In* J. G. Horsfall and A. E. Dimond (ed.), Plant pathology, an advanced treatise.

Vol. 2. Academic Press, New York and London.

33. SCHROTH, M. N., and D. C. HILDEBRAND. 1964. Influence of plant exudates on root-infecting fungi. Annu. Rev. Phytopathol. 2:101-132.

34. SEWELL, G. W. F. 1965. The effect of altered physical condition of soil on biological control, p. 479-493. *In* K. F. Baker and W. C. Snyder (ed.), Ecology of soil-borne plant pathogens. Univ. Calif. Press, Berkeley and Los Angeles.

35. SLYKHUIS, J. T. 1947. Studies on Fusarium culmorum blight of crested wheat and brome grass seedlings. Can. J. Res. 25(C):155-180.

36. SNYDER, W. C. 1960. Antagonism as a plant disease control principle, p. 127-136. *In* L. P. Reitz (ed.), Biological and chemical control of plant and animal pests. Amer. Assoc. Advance. Sci. Pub. 61. Washington, D.C.

37. SNYDER, W. C. 1970. Recent advances in the study of the ecology of soil-borne plant pathogens, p. 3-7. *In* T. A. Toussoun et al. (ed.), Root diseases and soil-borne pathogens. Univ. Calif. Press, Berkeley and Los Angeles.

38. STRZELCZYK, E. 1966. Effect of associated growth of bacteria from rhizosphere and non-rhizosphere soil on growth of fungi. Ann. Inst. Pasteur 111 (3 Suppl.):314-317.

39. SUBBA-ROA, N. S., and D. L. BAILEY. 1961. Rhizosphere studies in relation to varietal resistance or susceptibility of tomato to Verticillium wilt. Can. J. Bot 39:1747-1758.

40. SUBRAMANIAN, C. V. 1965. Host-pathogen interaction in foot rot of wheat. Tagber. Dtsch. Akad. Landwirtschaftswiss. 74:133-148.

41. TOUSSOUN, T. A., R. V. BEGA, and P. E. NELSON (ed.). 1970. Root diseases and soil-borne pathogens. Univ. Calif. Press, Berkeley and Los Angeles. 252 p.

42. WALKER, J. C. 1965. Host resistance as it relates to root pathogens and soil microorganisms, p. 314-319. *In* K. F. Baker and W. C. Snyder (ed.), Ecology of soil-borne plant pathogens. Univ. Calif. Press, Berkeley and Los Angeles.

Mechanisms of Control of Common Scab by Irrigation

DEREK H. LAPWOOD and M. J. ADAMS—*Plant Pathology Department, Rothamsted Experimental Station, Harpenden, Herts, England.*

INTRODUCTION.—Common scab blemishes potatoes superficially, affecting the sale of tubers rather than the yield. The lesions or scabs, first seen as small (1 mm diam) discrete spots, enlarge rapidly as the tuber swells. Labruyère (15) recognized two types, superficial and normal, and the disease is usually assessed on the proportion of tuber surface affected (21, 26). *Streptomyces scabies* (Thaxt.) Waksman and Henrici is considered the main causal organism, but other species may be involved (6). It penetrates through recently formed unsuberized tuber lenticels, perhaps stomata, but not intact skin (12). The actinomycete mycelium is fine (0.5-1 μm diam), initially confined to middle lamellae, and so difficult to detect that the early stages of infection have not been convincingly recorded. The first macroscopic symptoms are brown-spotted lenticels. The histology of recognizable lesions has been described by Jones (14) and Labruyère (15). Penetration of the tuber stimulates the formation of a wound barrier a few cells below the surface, and if the invasion is stopped, a superficial lesion is formed. If this barrier is breached, a second or third barrier may be formed below, resulting in increasingly severe forms of normal scab as the tuber grows; if the tuber stops swelling, so does the lesion.

The disease is worst on light sandy or gravelly soils, on land recently limed or plowed from long-term grass, and in the United Kingdom when June is dry. *Streptomyces scabies* probably occurs in most arable soils and sporulates freely within the dead cells of lesions. Control of the disease by seed disinfection has been tried (11), but most control measures aim at decreasing the soil population by using chemicals to kill the organism (10, 28, 31, 46) or sulfur to acidify the soil (27), or by soil amendments (47).

Wet soil prevents infection (9, 34, 38). Recently, irrigation regimes have been designed in the Netherlands for seed (15) and in the United Kingdom for ware (table use) (20) crops. In one Dutch experiment in 1967, scab on Bintje was decreased from 41% (unirrigated) to 0.2% tuber surface covered, and in the United Kingdom in 1970 on Majestic, it was decreased from 32% to 3%.

This paper describes a practical irrigation regime and how it works, outlines some ways in which moisture changes affect the soil environment, and, finally, discusses antagonism as the possible mechanism that prevents infection of young lenticels in wet soil.

CONTROL BY IRRIGATION.—*Irrigation regime.*—If the soil is dry, irrigation must start at tuber initiation (when the stolon tips first start to swell) and in ware crops must continue for at least 4-6 weeks; the soil must not be allowed to dry, or the water potential (5, 13) to fall to −0.4 bar [pF 2.6 (see 40) −0.4 atmospheres or −40 joules kgm^{-1} (see 42)].

An experiment at Rothamsted in 1967 with scab-susceptible Majestic illustrated the main problems in devising a regime for controlling scab by irrigation. A planned 4-week irrigation regime coincided with the 2nd to 5th weeks of a 6-week period predominately without rain. Irrigation was begun when plants from unsprouted seed started to form tubers and when those from sprouted seed had tubers a week or more old forming in the dry soil. Irrigation was applied to keep soil near field capacity (−0.1 bar), wetter than −0.4 bar (pF 2.6), and −0.7 bar (pF 2.9) for 4 weeks, and other plots were unirrigated. At harvest, tubers grown from sprouted seed in plots kept close to field capacity had about 1% surface area affected, with scab lesions at the stolon attachment end and a middle region free of scabs; a few tubers had scabs at the apical end. Tubers from unsprouted seed had about 1% surface area affected, but lesions were confined to the apical end. At −0.4 bar, tubers with about 2% scab showed less of a positional lesion effect, and at −0.7 bar with 3% scab, lesions were scattered over the tuber surface as on tubers from unirrigated plots with 7-10% scab. The way tubers grow has to be understood before disease incidence and distribution can be explained.

Tuber growth, soil water potential, and scab incidence.—The beginning and length of the required irrigation period is related to the growth of tubers. These form as swellings at the tip of underground stems (stolons), and as they grow, nodes (eyes) are separated from the apical bud by expansion of internode tissue between them. When first formed, tuber internodes have stomata, but as the tissue expands, the stomata are transformed to lenticels (12). Through cell division and expansion beneath them, guard cells are gradually raised to sit at the top of a small "pimple," and eventually the stoma is ruptured to expose the inner cells, thus forming a young lenticel. Most workers agree that this is the stage when infection is most likely [Fellows (12) thought but did not prove that stomata were also susceptible]. As lenticels mature, they become resistant. Therefore, each developing internode passes through a period of susceptibility and then becomes resistant. Node (eye) separation is fastest (about two per week) during the first 4 weeks after tuber initiation, and then slows so that the mature (ware) tuber may have 12 to 14 separated nodes, depending on cultivar. Internodes also expand differentially, early ones expanding so much more than those formed later that most of the mature tuber surface is composed of internodes formed during the 1st month after tuber initiation.

When soil was allowed to dry for 5, 10, or 15 days, an average of four, five, and six tuber internodes was infected, respectively (18, 19). When soil was allowed to dry for 7 days at different dates during the first 8 weeks after tuber initiation, an average of four to five internodes was scabbed after each occasion, but the later the date soil was allowed to dry, the nearer the tuber apex was the infection (Fig. 1). Because of the differential expansion of tuber internodes, the proportion of tuber surface scabbed was greatest (15%) after drying during the 3rd week and least (1.6%) after the 8th week. (The proportion with continuously irrigated control, 1.7%, gave a measure of the efficiency of the trickle irrigation system.)

It is now possible to explain the results of the experiment described in the irrigation regime section above. On tubers from sprouted seed, scab at the attachment end indicated that irrigation was not started soon enough to protect this first-formed tissue. And, on tubers from unsprouted seed, scab at the apical end (extending for four internodes toward the apex from node seven, numbering along the phyllotactic spiral from the stolon attachment) showed that irrigation for 4 weeks was insufficient to protect the later-formed parts of the tuber. Thus, irrigation should be started at tuber initiation and continued for at least 6 weeks to ensure adequate protection in prolonged dry weather. The loss in the positional effect of lesion distribution showed the gradual loss of protection as soil dried, indicating that tubers were increasingly at risk from about −0.4 bar water potential or pF 2.6.

Fig. 1-2. Fig. 1 (Top). The effect on the distribution of scabbed internodes when soil was allowed to dry (trickle irrigation lines disconnected) for 7 days at different dates during 8 weeks after tuber initiation. From left to right (with tuber apex at the top): first, no dry period; second to sixth, soil allowed to dry during 1st to 5th weeks. The second to fourth tubers, with apical lesions, had a second exposure to dry soil during the 6th to 8th weeks. Fig. 2 (Bottom). The effect of a few days of late June rain during prolonged dry weather on the distribution of scab on 25 August on tubers starting to form at different dates. From left to right (in pairs, with tuber apex at the top): tubers starting to form in late May, early June, and mid-June 1967.

Internodes infected or protected.—When soil was allowed to dry for 5 days (18), an average of four tuber internodes was scabbed, but when soil was wet for about a week, infection was prevented on only one or two internodes (17). As mentioned above, four, five, and six tuber internodes were scabbed in the experiment where soils were dried for 5, 10, or 15 days. Nodes were separating from the apical bud about one every 5 days. It was assumed, for example, in the 5-day treatment with four scabbed internodes that the internode which started to form (separated from the apical bud by at least 0.5 mm) during the dry days was infected, and that the stomatal/lenticel structures of three other internodes formed previously were still susceptible to infection; this assumed that stomata were as likely to be infected as young lenticels.

In a 1967 experiment on light sandy scabby land (17), a few days of heavy rain interrupted prolonged dry weather and kept soil wetter than −0.4 bar (below pF 2.6) for at least 5 days, protecting 1 to 1.5 internodes (Fig. 2). However, these internodes started to form not during the wet period but at least a week before the rain, when soil was dry. Since that experiment was performed, we have found that at any time early in tuber growth the internode visible nearest the apical bud [that most recently formed, defined as A minus (−) 1] has flat stomata, the second from the apex (A −2) has predominantly raised stomata, and the third (A −3) has predominantly burst or young lenticels. In dry soil (and if not infected by *S. scabies*), suberin deposition on the outer walls of the outermost and exposed cells has started on the third internode from the apex (A −3) and is complete by the fifth (A −5).

Although tubers were not checked in the experiment just described, it was known that the uninfected internode was that which was third from the apex at the time of the rain and, therefore, that it probably had young unsuberized lenticels. This internode must have passed through the stomatal stages in dry soil, suggesting that stomata are not susceptible or are unlikely to be infected by *S. scabies*. Further evidence to support this view came from Lapwood and Hering (18), who found that the four internodes that became scabbed did not include the one formed during the 5 dry days; the scab extended only to within two internodes of the apical bud at the time irrigation was restarted. If suberized lenticels resist infection and those on A −5 are suberized, and if nodes separate on the average of every 3 to 4 days, then it must take about as long for a lenticel to form and become resistant as for four nodes to separate; i.e., about 2 weeks. Subtracting the time spent in the stomatal stages (1 week), then, the young lenticel would be susceptible to *S. scabies* for about 1 week. This would explain why, if soil remains wet enough to prevent infection for about 1 week, only that internode which can reach the resistant stage (mature) in that time remains uninfected, and why the several internodes at different intermediate stages of lenticel maturation would be scabbed in soil allowed to dry for that week.

FACTORS AFFECTED BY CHANGES IN SOIL WATER POTENTIAL.—The relations between water and soil have been considered in detail by Marshall (32), water and the plant by Youngs (49), and soil water and plant disease by Cook and Papendick (5), Griffin (13), and Rotem and Palti (37). Only those possibly relevant to the infection of potato tuber lenticels will be considered here.

The soil.—In simplest terms, wetting and drying imply the filling and draining of pores in the soil. Wetting may fill all soil pores rapidly, but draining takes much longer, with the larger pores emptying first (4). Although the modern concept of water potential is perhaps more logical, the old soil moisture classification of Briggs (3), favored by Couch et al. (7), is more familiar. *Gravitational water* is water that may be removed from the soil as it dries from saturation (when all soil pores are filled with water) to field capacity (water held by soil against gravity), which approximates −0.001 to −0.1 bar (pF 0 to pF 2) (32). *Capillary water* or readily available water, is water that may be removed as the soil dries from field capacity to a permanent wilting point, −0.1 to −15.0 bars (pF 2 to pF 4.2), and is the principal water source used by plants. *Hygroscopic water* and *water vapor* (23) are beyond our range of interest here.

Common scab seems unimportant in wet soils with water potentials of −0.4 bar and above (pF 2.6 and below), but is increasingly important as soils dry in the range −0.4 to −1.0 bar (pF 2.6 to 3.0). In a saturated soil, there are few, if any, air-filled pores; oxygen diffusion is severely restricted; and the CO_2 content will rise from biological activity. As a soil dries, however, it becomes increasingly aerated.

Rain is often associated with cool cloudy weather. For example, during a wet period 8 to 20 June 1971 at Rothamsted, when soils remained wetter than −0.1 bar (pF 2.0), soil temperatures at tuber depth only exceeded 15°C once, and the average maximum and minimum temperatures were 14°C and 11°C. By contrast, during bright sunny weather from 4 to 11 June 1973, soil dried from −0.5 to −1.0 bar (pF 2.7 to 3.0), and soil temperatures exceeded 20°C for 6 to 12 hours (avg 9.6) every day.

Changes in soil moisture that alter soil aeration also affect the release of manganese, which is greater in wet soils. McGregor and Wilson (28, 29) found that manganese sulfate added to a neutral soil with low water-soluble manganese content would control scab. They suggested that low scab incidence on acid soils could be partially attributed to the concentration of soluble manganese released by them.

The potato tuber.—The first observed effect of moisture stress on tuber growth is a decrease in the rate of swelling, followed closely by a decreased rate of node separation as the stress increases. As the tuber swells, it compacts the soil around it. Large air-filled pores are likely to be lost, leading to the encirclement of the tuber by soil particles separated only by small pores.

Temperature also affects tuber growth. In the experiments where continuous trickle irrigation cooled the soil (18), a node separated about every 5 days during the 1st month after tuber initiation, whereas in nonirrigated experiments at Rothamsted, the rate was about two per week.

Changes in water potential also affect lenticel structure. In dry soil, lenticels become suberized by internode A −5, but in wet soil, following the rupture of the raised stoma and exposure of the inner cells, cell division continues and cells proliferate from the lenticel opening. When lenticels are suberized in dry soil, the meristem may form a cork barrier, but if soil becomes wet, the meristem may become active again, cutting off parenchymatous cells that burst through the cork and proliferate from the lenticel opening. As a soil dries, proliferated cells collapse, and a suberin layer is deposited and completed within 5 days. Metabolites are released into the soil from developing lenticels, but how much this process is affected by changes in soil water potential and their effect on the surrounding microflora (and *S. scabies* in particular) is unknown. However, Lawrence and Barker (22) found that tubers produced aseptically were difficult to infect unless soaked in sterile water, and that soaking increased the susceptibility of resistant Richter's Jubel. The dilution or elimination of "soluble native products" was a possible explanation.

The pathogen.—Soil water potential affects aeration, and Sanford (39) considered low concentration of O_2 important in limiting scab in wet soil. Harmsen [quoted by Labruyère (15)] found that *S. scabies* grew in soil at field capacity provided it was aerated, and that growth in sand was normal at 2.5% O_2 and could even be sustained at 0.625% O_2. In 50% CO_2, growth was about 70% of normal, so the organism should easily tolerate a rise from 0.05 to 2% CO_2 after irrigation. Williams et al. (48) inoculated unidentified *Streptomyces* spp. from sandy soil under *Pinus nigra* into sterile soil with different gas mixtures and found that decreasing O_2 in the absence of CO_2 had little effect on growth, but that increasing CO_2 above 10% (v/v) greatly decreased it. Growth was normal at $CO_2:O_2$ ratio 1:1, but drastically decreased by 3:1 and 10:1, and was greatest in humid but air-filled pores [the pores in question would have emptied at pF 1.7, see Childs (4)]. Radial growth was decreased in water-filled pores.

Labruyère (15) found little difference between total actinomycete numbers in wet or dry soils, but he found that the number of tyrosinase-producing isolates (including pathogenic *S. scabies*) was usually decreased in wet soil. The frequency of isolating tyrosinase-positive actinomycetes from tuber lenticels was high in dry soils but low in wet soils (25).

The germination speed of spores of *S. scabies* is markedly affected by temperature (41), increasing from over 48 hours at 10°C to 18 hours at 15°C, 11 hours at 20°C, and 8 hours at 25°C. In Wisconsin tank experiments, Dippenaar (9) found the most scab lesions per tuber at about 20°C, which was also most suitable for host development, although the disease occurred from 13° to 25°C. In culture, the optimum temperature for growth of the *S. scabies* strains used varied from 28° to 36°C.

Microbial populations.—Williams et al. (48) studied numbers of actinomycetes, bacteria, and fungi in soils at water potentials ranging from −0.001 to −400 bar (pF 0 to pF 5.6). All organisms were most common at −0.1 bar (pF 2) but decreased in the range −0.001 to −0.01 bar (pF 0 to 1.0) and drier than −10 bar (above pF 4.0). Between −0.2 and −0.5 bar (pF 2.3 and 2.7), the proportion of actinomycetes increased and that of bacteria decreased, and at −400 bar (pF 5.6), actinomycetes exceeded the total of bacteria and fungi. *Streptomyces* spp. mixed with sterile soil and maintained at water potentials between −0.01 and −350 bar (pF 1.0 and pF 5.5) grew much better at about −0.1 bar (pF 2.0) than in wetter or drier soils.

In wet soil, a bacterial flora predominated over fungi and actinomycetes (9), and bacterial numbers were greater than in dry soil (15), whereas numbers of tyrosinase-positive actinomycetes were greater in dry than in wet soil. Lewis (25) isolated bacteria more frequently than tyrosinase-positive actinomycetes from tuber lenticels in wet soils; in dry soil the ratios were reversed. By direct observation and culture, Lewis (24) also studied populations of actinomycetes, bacteria, and fungi in June and August on the tuber surface of cultivars differing in resistance from Ulster Chieftain (very susceptible) to King Edward (intermediate) and Seneca (resistant). The total actinomycete population (including pathogenic *S. scabies*) on the periderm surface and in lenticels decreased with increasing varietal resistance and decreased between June and August. The probability of isolating actinomycetes from a sample was less if bacteria were present. This fact plus the positive correlation between resistance and high frequencies of bacteria relative to actinomycetes indicated interactions between these organisms, but there was no evidence of antibiotic effects.

ANTAGONISM AS A CONTROL MECHANISM.—*Introduction.*—It is difficult to study common scab disease and its relation to field soils because there is no quick way to distinguish between pathogenic and nonpathogenic forms of *S. scabies* (43, 44). Recently, however, Knösel [reported by Labruyère (15)] found that pathogenic forms have greater pectic-acid transeliminase activity than do nonpathogenic forms. Pathogenic isolates can be added to sterilized soil, but in pot or glasshouse experiments it is difficult to control the water potential to ensure infection and yet maintain healthy plant growth (30). In infection studies the tuber has to be manipulated while it is still attached to the plant without interfering with further development. Furthermore, these controlled conditions give little information about natural microbial interactions in field soil or at tuber surfaces. Thus, although the mechanisms of control by irrigation in relation to presence or absence of infection or a particular distribution of lesions on tubers are now more clearly understood, information on the actual mechanisms preventing lenticel infection in wet soil remains speculative.

Suberization may account for the resistance of mature lenticels in dry soil, whereas in wet soil the lenticels proliferate but become rapidly suberized if the soil dries. It is, however, unknown whether the pathogen fails to penetrate resistant lenticels or is stopped before a recognizable symptom develops. Our results suggest that the internodes bearing flat or raised stomata (A−1 and A−2) and the lenticels on the fifth internode (A−5) are not infected. If true, this means that only young lenticels (those that have recently lost guard cells) on internodes A−3 and A−4 are susceptible. Nodes separate every 3 or 4 days during the 1st month of tuber growth, which suggests that young lenticels remain susceptible for 1 week.

The water potentials limiting infection have not been measured precisely, but the work of Labruyère (15), Langton (16), and Lapwood et al. (20) suggests that the disease is likely to occur in soils −0.4 bar (above pF 2.6) or drier and unlikely to occur in wetter soils. In practical terms: on the sandlands of Nottinghamshire, England, with 38 mm available water in the top 30 cm of soil, some scab infection was likely at soil moisture deficits greater than 20 mm or when over 50% available water was lost. On a Cambridge soil, Langton considered that the critical range occurred when between 47 and 68% of available water was lost. Labruyère (15) considered that irrigation should be applied before 20 mm of available water was lost, although the figure varied with soil type. On Rothamsted and Woburn (Bedfordshire) farms following rain, critical levels (−0.4 bar) were reached in an average of 4 days, and many lenticels could be infected if soil was allowed to dry 8 days to water potentials of −0.8 bar (pF 2.9). Baars (1) indicated that during late May or early June, when tubers start to form, evaporation averages 3 mm per day in cloudy cool weather, 3.5 mm in normal weather, and 4.5 mm per day in sunny warm weather. Hence, Labruyère advocated irrigation at least every 5 days to prevent infection in dry sunny weather. It is interesting that Williams et al. (48) found that as soil dried from about −0.5 bar (pF 2.7) the proportion of actinomycetes isolated increased and that of bacteria decreased.

Evidence for antagonism.—Of the many usages of the term antagonism, we favor that of Park (36),who used it to include "all those associations in which at least one of the interacting species is harmed." The three mechanisms of antagonism are *antibiosis*, "in which species A produces a chemical substance that is inimical to species B without species A deriving any direct benefit," *exploitation*, " in which species A inflicts harm by the direct use of species B for its own benefit," and *competition*, "which is found in the indirect rivalry of two species for some feature of the environment that is in short supply."

Sanford (39) gave evidence of antibiosis where bacteria sometimes inhibited growth of *S. scabies* in culture but not by increasing the acidity, and Daines (8) found the actinomycete susceptible to the antibiotic activity of *Trichoderma*. Orellana (35) mentions a number of actinomycetes and bacteria that inhibit growth. The effectiveness of green manuring in controlling scab has also been attributed to antibiosis, and with soya bean manuring, *Bacillus subtilis* was considered the main antagonist (45). We have found the germination of *S. scabies* spores to be inhibited by this bacterium and by metabolites from its culture medium. Millard and Taylor (33) suggested that when green manuring failed to control scab it was because certain saprophytic *Streptomyces* spp. were absent from the soil, and that when successful, green manuring encouraged growth of these species, which then suppressed *S. scabies*. Labruyère (15) searched for antagonists among soil microorganisms and found other actinomycetes to be the most active against *S. scabies*. Wet soils seem not to affect the total

actinomycete population (Labruyère), but it is not known how antagonistic species would be affected. Possibly they would prove most effective in decreasing the number of lenticels infected by *S. scabies* under favorable infection (dry) conditions rather than in preventing infection in wet soil.

There seems no evidence for *exploitation*, but *competition* between bacteria and *S. scabies* could be important. *Streptomyces scabies* will tolerate low O_2 and high CO_2 concentrations and a range of temperatures, and will grow in sterilized soil even when wet. We do not know whether lenticels can be infected in wet soil lacking other microorganisms, although Barker and Page (2) found that inoculated sterile tubers growing in glass tubes produced the most infections on areas in contact with the wall of the glass tube, where water persisted. In wet field soils with limited aeration, numbers of bacteria increase and the population of tyrosinase-positive actinomycetes, including *S. scabies*, falls. Under these conditions, the demand for O_2 will be large from bacterial activity, the rapidly swelling tuber, and the meristematic activity of the lenticels. The bacteria will include facultative anaerobes which presumably will continue to multiply, so aerobic actinomycetes are likely to suffer most.

Wet soils allow the movement of bacteria in water films, and Lewis surmised that these could reach the lenticels in large numbers before the slower growing mycelium of the actinomycete. Labruyère suggested that the leakage of metabolites might also give the bacteria a further advantage. In dry soil, bacteria are unlikely to move rapidly or in large numbers through soil pore necks less than 1.5 μm, which would be drained at a water potential of −1 bar (pF 3.0) (13). Labruyère suggested that they would be at a disadvantage and that the actinomycete would have a greater chance of reaching the lenticel first. He also speculated that the organisms arriving first would probably maintain a dominant position because the supply of metabolites from the lenticels would favor their rapid multiplication, and, in consuming them, chemotrophic attraction to other organisms would be decreased. Lewis found more actinomycetes on the tuber surface of susceptible than resistant cultivars and in dry than wet soil and concluded that the chances of infection of the susceptible cultivar or tubers in dry soil are greater because it contains a larger number of pathogenic propagules.

Conclusions.—We have not studied microbial populations in soil or on tuber surfaces, but our experiments suggest that internodes with flat or raised stomata forming in dry soil, with a presumably dominant actinomycete microflora, are not infected if, at the bursting stage (when guard cells are lost), the soil is suddenly wetted from rain or irrigation. Conversely, internodes bearing stomata in soil wetter than −0.4 bar (less than pF 2.6), with presumably bacterial-dominated populations, may be infected if, by the time they reach bursting stage, the soil has become drier than −0.4 bar (pF 2.6). Stomata become raised 100-120 μm above the surrounding tuber surface and upon bursting expose a large surface area (relative to the stomatal pore), presumably of largely uncolonized cells. In the colonization of this raised surface, the microflora on the tuber may perhaps be less important and slower to change than that in the surrounding soil.

The fact that *S. scabies* can still infect susceptible lenticels in drying soil despite earlier periods when wet soil prevented it, suggests that if antibiosis or competition does prevent infection in wet soil, these effects are soon lost. In wet soil, protected lenticels proliferate. Perhaps even the first colonists (whatever they are and by whatever means they become established) are pushed out and have no further role in preventing infection, because if the soil dries, these lenticels will become suberized and resistant to *S. scabies*. In a drying soil, conditions become increasingly favorable for infection, but even so, there must be a fairly critical condition when infection can begin, for there is usually a sharp division between scabbed and scab-free tissues.

The evidence presented suggests that with potato common scab there could be a unique host-pathogen-antagonist situation because each lenticel may be challenged, protected, or infected as it develops, and this procedure continues until lenticel formation stops.

LITERATURE CITED

1. BAARS, C. 1968. Bestrijding van schurftaantasting door beregening. Landbouwvoorlichting 25:138-142.
2. BARKER, W. G., and O. T. PAGE. 1954. The induction of scab lesions on aseptic potato tubers cultured in vitro. Science 119:286-287.
3. BRIGGS, L. J. 1897. The mechanics of soil moisture. USDA Bur. Soils Bull. 10. 24 p.
4. CHILDS, E. C. 1940. The use of soil moisture characteristics in soil studies. Soil Sci. 50:239-252.
5. COOK, R. J., and R. I. PAPENDICK. 1972. Influence of water potential of soils and plants on root disease. Annu. Rev. Phytopathol. 10:349-374.
6. CORBAZ, R. 1964. Etude des Streptomycètes provoquant la gale commune de la pomme de terre. Phytopathol. Z. 51:351-361.
7. COUCH, H. B., L. H. PURDY, and D. W. HENDERSON. 1967. Application of soil moisture principles to the study of plant disease. Va. Polytech. Inst. Bull. 4. 23 p.
8. DAINES, R. H. 1937. Antagonistic action of Trichoderma on Actinomyces scabies and Rhizoctonia solani. Amer. Potato J. 14:85-93.
9. DIPPENAAR, B. J. 1933. Environmental and control studies of the common scab disease of potatoes caused by Actinomyces scabies (Thaxt.) Guss. Sci. Bull. Dept. Agr. Union of South Africa. No. 136, 78 p.
10. EMDEN, J. H. VAN, and R. E. LABRUYÈRE. 1958. Results of some experiments on the control of common scab of potatoes by chemical treatment of the soil. Eur. Potato J. 1:14-24.
11. EMILSSON, B., and N. GUSTAFSSON. 1954. Studies of the control of common scab on the potato. Acta. Agr. Scand. 4:33-62.
12. FELLOWS, H. 1926. Relation of growth in the potato tuber to the potato scab disease. J. Agr. Res. 32:757-781.
13. GRIFFIN, D. M. 1969. Soil water in the ecology of fungi. Annu. Rev. Phytopathol. 7:289-310.
14. JONES, A. P. 1931. The histogeny of potato scab. Ann. Appl. Biol. 18:313-333.
15. LABRUYÈRE, R. E. 1971. Common scab and its control in seed-potato crops. Versl. Landbouwk. Onderz. Ned. No. 767, 71 p.
16. LANGTON, F. A. 1972. Screening potato clones for resistance to common scab (Streptomyces scabies) in the field. J. Agr. Sci. (Cambridge) 79:75-81.
17. LAPWOOD, D. H., and M. J. ADAMS. 1973. The effect of a few days of rain on the distribution of common scab (Streptomyces scabies) on young potato tubers. Ann. Appl. Biol. 73:277-283.
18. LAPWOOD, D. H., and T. F. HERING. 1968. Infection of potato tubers by common scab (Streptomyces scabies) during brief periods when soil is drying. Eur. Potato J. 11:177-187.
19. LAPWOOD, D. H., and T. F. HERING. 1970. Soil moisture and the infection of young potato tubers by Streptomyces scabies (Common scab). Potato Res. 13:296-304.
20. LAPWOOD, D. H., L. W. WELLINGS, and J. H. HAWKINS. 1973. Irrigation as a practical means to control potato common scab (Streptomyces scabies): final experiment and conclusions. Plant Pathol. 22:35-41.
21. LARGE, E. C., and J. K. HONEY. 1955. Survey of common scab of potatoes in Great Britain, 1952 and 1953. Plant Pathol. 4:1-8.
22. LAWRENCE, C. H., and W. G. BARKER. 1963. Pathogenicity of Streptomyces scabies on potato tubers cultured in vitro. Nature (London) 199:509-510.
23. LEBADEFF, A. F. 1928. The movement of ground and soil waters. Proc. 1st Int. Congr. Soil Sci. 1927:459-494.
24. LEWIS, B. G. 1962. Host-parasite relationships in the common scab disease of potato. Ph.D. thesis, Nottingham, England. 102 p.
25. LEWIS, B. G. 1970. Effects of water potential on the infection of potato tubers by Streptomyces scabies in soil. Ann. Appl. Biol. 66:83-88.
26. LOWINGS, P. H., and W. J. RIDGMAN. 1959. A spot-sampling method for the estimation of common scab on potato tubers. Plant Pathol. 8:125-126.
27. MCCREARY, C. W. R. 1967. The effect of sulphur application to the soil in the control of some tuber diseases. Proc. 4th Br. Insect. Fung. Conf. 1:303-308.
28. MCGREGOR, A. J., and G. C. S. WILSON. 1964. The effect of applications of manganese sulphate to a neutral soil upon the yield of tubers and the incidence of common scab in potatoes. Plant Soil 20:59-64.
29. MCGREGOR, A. J., and G. C. S. WILSON. 1966. The influence of manganese on the development of potato scab. Plant Soil 25:3-16.
30. MCINTOSH, A. H. 1970. A glasshouse method for testing chemicals for control of potato common scab. Potato Res. 13:241-247.
31. MCINTOSH, A. H. 1973. Glasshouse tests of chemicals for control of potato common scab. Ann. Appl. Biol. 73:189-196.
32. MARSHALL, T. J. 1959. Relations between water and soil. Tech. Comm. Bur. Soil Sci. (Harpenden). No. 50, 91 p.
33. MILLARD, W. A., and C. B. TAYLOR. 1927. Antagonism of microorganisms as the controlling factor in the inhibition of scab by green manuring. Ann. Appl. Biol. 14:202-216.
34. NOLL, A. 1939. Untersuchungen über die Biologie und Bekampfung des Kartoffelschorfes (Actinomyces). Landw. Jbr. 89:41-113.
35. ORELLANA, R. 1947. Actinomyces and bacteria antagonistic to Actinomyces scabies. Phytopathology 37:17 (Abstr.).
36. PARK, D. 1960. Antagonism—the background to soil fungi, p. 148-159. *In* D. Parkinson and J. S. Waid (ed.), The ecology of soil fungi. Liverpool Univ. Press.
37. ROTEM, J., and J. PALTI. 1969. Irrigation and plant diseases. Annu. Rev. Phytopathol. 7:267-288.
38. SANFORD, G. B. 1923. The relation of soil moisture to the development of common scab of potato. Phytopathology 13:231-236.
39. SANFORD, G. B. 1926. Some factors affecting the pathogenicity of Actinomyces scabies. Phytopathology 16:525-547.

40. SCHOFIELD, R. K. 1935. The pF of the water in soil. Trans. 3rd Int. Conf. Soil Sci. 2:37-48.

41. SHAPOVALOV, M. 1915. Effect of temperature on germination and growth of the common potato-scab organism. J. Agr. Res. 4:129-134.

42. SLATYER, R. O., and S. A. TAYLOR. 1960. Terminology in plant- and soil-water relations. Nature (London) 187:922-924.

43. VRUGGINK, H., and D. Z. MAAT. 1968. Serological recognition of Streptomyces species causing scab on potato tubers. Neth. J. Plant Pathol. 74:35-43.

44. WEINHOLD, A. R. 1970. Significance of populations of major plant pathogens in soil: bacteria including Streptomyces, p. 22-24. *In* T. A. Tousson et al. (ed), Root diseases and soil-borne pathogens. Univ. Calif. Press, Berkeley and Los Angeles.

45. WEINHOLD, A. R., and T. BOWMAN. 1965. Influence of substrate on activity of a bacterium antagonistic to Streptomyces scabies. Phytopathology 55:126 (Abstr.).

46. WEINHOLD, A. R., T. BOWMAN, and J. BISHOP. 1964. Urea formaldehyde for the control of common scab of potato. Amer. Potato J. 41:319-321.

47. WEINHOLD, A. R., J. W. OSWALD, T. BOWMAN, J. BISHOP, and D. WRIGHT. 1964. Influence of green manures and crop rotations on common scab of potato. Amer. Potato J. 41:265-273.

48. WILLIAMS, S. T., M. SHAMEEMULLAH, E. T. WATSON, and C. I. MAYFIELD. 1972. Studies on the ecology of actinomycetes in soil. VI. The influence of moisture tension on growth and survival. Soil Biol. Biochem. 4:215-225.

49. YOUNGS, E. G. 1965. Water movement in soils, p. 89-112. *In* C. E. Fogg (ed.), The state and movement of water in living organisms. 19th Symp. Soc. Expl. Biol. (Cambridge).

Effect of Bacillus and Streptomyces spp. Applied to Seed

P. R. MERRIMAN, R. D. PRICE, K. F. BAKER, J. F. KOLLMORGEN, T. PIGGOTT, AND E. H. RIDGE—*First, second, and fourth authors, Plant Pathologists, Victorian Plant Research Institute, Melbourne, Victoria; third author, Professor of Plant Pathology, University of California, Berkeley; fifth author, OIC Vegetable Research Station, Frankston, Victoria; sixth author, Soil Microbiologist, CSIRO, Division of Soils, Glen Osmond, South Australia.*

Seed inoculation with selected soil microorganisms as a method of increasing plant growth has been extensively investigated by soil microbiologists. There are many reports of the use of *Azotobacter chroococcum* and *Bacillus megatherium* var. *phosphaticum* to increase growth of agricultural crops (3, 8, 13). Seed inoculation with antagonistic microorganisms offers potential for the control of root diseases, but there are few reports of success with this method. Two exceptions are control of Fusarium diseases of flax and wheat by seed inoculation with bacteria, including *Pseudomonas* or *Achromobacter* spp. (2), and control of *Fusarium roseum* f. sp. *cerealis* on corn by seed inoculation with *Bacillus subtilis* or *Chaetomium globosum* (9).

We screened actinomycetes and *Bacillus* spp. for antagonism to *Rhizoctonia solani* Kühn, and selected isolates were tested as seed inoculants on wheat in soil infested by *R. solani*. One isolate each of *Streptomyces griseus* and *B. subtilis* was tested in the field on barley, oats, and wheat at sites where *R. solani* was associated with diseased plants. The two organisms were also tested on carrots at sites where disease was not apparent. The isolates selected for biological control of *R. solani* in pot experiments also increased plant growth; this latter effect was not related to disease control.

ISOLATION AND SCREENING FOR ANTAGONISM.—*Bacillus* spp. and actinomycetes were isolated from a clay soil, pH 8.0, at two sites, the first on undisturbed ryegrass (*Lolium rigidum*) subclover (*Trifolium subterraneum*) pasture, the second an experimental plot that had been cropped continuously with wheat for 55 years. The methods of Broadbent et al. (4) were used to screen isolates for antagonism to *R. solani*, first on agar and then in soil using peppers (*Capsicum frutescens* L. var. *grossum* Bailey) as a host. Growth of *R. solani* on agar was inhibited by 77 of 84 pasture soil isolates and 19 of 64 wheat soil isolates. Thirty-two pasture soil isolates and 10 wheat soil isolates were chosen for tests on control of *R. solani* on peppers. Disease was reduced by 10 pasture soil isolates and five wheat soil isolates.

TESTS ON WHEAT.—Seven organisms that reduced disease on peppers were applied to wheat seed and examined for control of *R. solani* in glasshouse experiments using plastic pots (11). They were identified as *S. griseus* 2-A24 from the wheat soil and *Bacillus pumilus* 1-B84, *B. subtilis* isolates 1-B3, 1-B68, 1-B77, 1-B80, and *Bacillus* sp. 1-B8ii from the pasture soil. Plants were grown in pasteurized or unpasteurized clay-sand mix (1:1) inoculated with *R. solani*. At the three-leaf stage, root discoloration associated with *R. solani* in pasteurized and unpasteurized soil was reduced by *S. griseus* and *B. subtilis* isolates 1-B3, 1-B77, and 1-B80 (Table 1).

The effect on growth and grain yield of the seven antagonists applied to wheat seed was investigated in pots of unpasteurized clay-sand mix in the presence and absence of *R. solani*. Addition of *R. solani* reduced dry weight of tops but not grain yield (Table 2). Mean dry weight of tops and mean grain yield, calculated from treatments with and without added *R. solani*, were increased by *S. griseus* 2-A24, by *B. subtilis* isolates 1-B68, 1-B77, and 1-B80, and by *B. pumilus* 1-B84. Some differing effects caused by antagonists were observed. For example, *S. griseus* increased yield in soil with and without added *R. solani*, whereas *B. subtilis* 1-B80 increased yield only in soil inoculated with *R. solani*.

FIELD TRIALS WITH CEREALS.—*Streptomyces griseus* 2-A24 and an additional isolate of *B. subtilis* A13 isolated by Broadbent et al. (4) and selected for control of *R. solani* were chosen for field trials. *Bacillus subtilis* A13 was chosen because of promise in a preliminary field trial where it had been applied as a seed inoculant to barley cv. Resibee. Growth of barley from treated seed was more vigorous than from untreated seed, and grain yield was increased by about 10%. In addition, the subsequent regrowth of pasture in the stubble following the barley crop was increased.

Methods of Merriman et al. (12) were used for application of organisms to seed, for counting propagules on seed before sowing, and for studying survival after sowing. Seed treatments were (a) *S. griseus* applied as a water suspension (5×10^4 to 3×10^5 propagules/seed), (b) *B. subtilis* applied as a water suspension (10^6 to 3×10^7 propagules/seed), (c) water, and (d) dry. Seed was treated several weeks before sowing, air dried, and stored at 5°C. Propagules per seed were counted 1 week before sowing. Organisms were tested as seed inoculants on barley, oats, and wheat at three sites where *R. solani* was one of the fungi associated with root disease in the preceding crop. Barley and oats were sown in red-brown earths, pH 6.7 to

130

6.9; wheat was sown in grey cracking clay, pH 7.5. Recommended rates of sowing and fertilizer application were followed.

There were no differences in the distribution of stunted plants among treatments. Pathogenic isolates of *R. solani* were isolated from roots of stunted plants of all three hosts. *Fusarium culmorum* and *F. graminearum* were isolated from barley and oats; *Pythium irregulare* from barley, oats, and wheat; and *P. debaryanum* and *P. mamillatum* from barley (12). *Streptomyces griseus*

increased grain yield, dry weight of foliage, tiller number of oats (Fig. 1), and tiller number of wheat. *Bacillus subtilis* increased grain yield of oats, tiller number of oats (Fig. 1), and tiller number of wheat (Table 3). On barley, *S. griseus* advanced head emergence by 10 days.

Survival of the seed inoculants on barley, oats, and wheat was studied 5 weeks after sowing, when plants were at the two-leaf stage. Isolations were made from washed pericarps or roots from composite plant samples collected from all replicates in each treatment. The results showed

TABLE 1. The effect of antagonists of *Rhizoctonia solani* applied to wheat seed on symptoms caused by *R. solani* in pasteurized and unpasteurized soil

Seed treatment[a]	Pasteurized soil			Unpasteurized soil		
	Plant height	Root length	Root length discolored	Plant height	Root length	Root length discolored
	mm	*mm*	*mm*	*mm*	*mm*	*mm*
Nil	319	116	0	298	112	0
Nil + *R. solani*	130	69	17	229	90	26
2-A24 + *R. solani*	233	91	8	174	85	9
1-B80 + *R. solani*	194	105	5	191	97	7
1-B77 + *R. solani*	207	86	5	232	115	12
1-B68 + *R. solani*	253	96	14	222	83	9
1-B3 + *R. solani*	167	73	5	165	80	9
1-B8ii + *R. solani*	168	52	6	217	103	17
1-B84 + *R. solani*	152	69	10	124	65	15
P = 0.05	67	32	6	78	31	10
P = 0.01	90	43	8	104	41	13

[a]Isolate number: A = actinomycete, B = bacterium; prefix 1 from pasture soil, prefix 2 from wheat soil.

TABLE 2. The effect of inoculation of wheat seed with antagonists of *Rhizoctonia solani* on grain yield and dry matter production in unpasteurized soil

Seed treatment[a]	Grain weight, g			Dry weight of tops, g		
	−*R. solani*	+*R. solani*	Mean	−*R. solani*	+*R. solani*	Mean
Nil	21.93	20.98	21.45	51.6	46.1	48.9
2-A24	26.43	27.68	27.06	60.4	61.4	60.9
1-B80	21.92	27.41	24.66	52.8	62.2	57.5
1-B77	25.04	22.85	23.94	55.4	54.7	55.0
1-B68	24.96	24.23	24.60	57.9	56.6	57.2
1-B3	24.86	19.93	22.39	59.6	51.1	55.3
1-B8ii	22.92	20.77	21.84	55.8	50.0	52.9
1-B84	25.63	22.37	23.99	60.8	54.0	57.4
P = 0.05		3.06	2.16		5.2	3.7
P = 0.01		4.03	2.85		6.8	4.8

[a]Isolate number: A = actinomycete, B = bacterium; prefix 1 from pasture soil, prefix 2 from wheat soil.

TABLE 3. Effect of inoculating seed with *Streptomyces griseus* (2-A24) or *Bacillus subtilis* (A13) on grain yield, dry matter, and tiller production of oats and wheat

Seed treatment	Grain yield, kg		Dry matter, g/7 m row		Tiller number	
	Oats	Wheat	Oats	Wheat	Oats	Wheat
S. griseus	23.2	108.4	658.5	839.0	88.7	111.5
B. subtilis	22.4	103.4	585.8	806.8	76.2	110.5
Water	16.0	100.6	527.8	720.7	59.3	83.3
Dry	16.9	101.3	521.8	630.7	64.8	75.3
P = 0.05	2.8	21.7	90.7	175.8	12.6	14.4
P = 0.01	3.8	29.3	122.7	237.9	17.0	19.5

that organisms resembling *S. griseus* and *B. subtilis* were present on pericarps at levels between 10^5 and 10^6 propagules/pericarp. Counts from pericarps of controls were low. Evidence for colonization of roots by seed inoculants was inconclusive; this aspect requires further investigation.

FIELD TRIALS WITH CARROTS.—The effect of *S. griseus* 2-A24 and *B. subtilis* A13 on yield of carrots cv. Royal Star was investigated (12). The seed treatments were (a) *S. griseus* applied either as a water suspension or in combination with a bentonite-sand pellet (10^6 propagules/seed), (b) *B. subtilis* applied as a water suspension or with a pellet (between 10^6 and 2×10^6 propagules/seed), and (c) water. Application of a pellet enables precision sowing of small seeds like carrots. The trial was sown at a vegetable research station in sandy soil, pH 6.2, where severe outbreaks of carrot root diseases had not previously occurred. Recommended rates of sowing and application of fertilizer were followed.

The yield was increased by *B. subtilis* with pellet and water and by *S. griseus* with water (Table 4), particularly the yield of very large and large grade carrots.

DISCUSSION.—*Bacillus* and *Streptomyces* spp. selected for biological control of *R. solani* also promote plant growth. An inadvertent selection for growth-promoting organisms apparently was made during the screening for antagonism. The mechanism of this growth response is not understood, but it does not appear to be specific to particular soil types or plant groups, because growth was affected in different soils on plants belonging to the Gramineae and Umbelliferae.

The evidence from experiments with pasteurized soil indicates control of *R. solani*, but the results with unpasteurized soil indicate that plant growth responses may be due to factors other than control of root diseases. There are several possible explanations for the plant growth response. *Azotobacter chroococcum*, *Bacillus* and *Streptomyces* spp., and other soil bacteria produce gibberellin-like compounds that increase plant growth (6, 7, 10). Growth responses may also result from increased nutrient availability to the plant. For example, evidence suggests that *B. megatherium* var. *phosphaticum* affects growth by increasing the availability of phosphorus (1, 13). The seed inoculants may also benefit plant growth by detoxification of organic substances in soil that affect growth adversely (15), or by exclusion from the rhizosphere of microorganisms that produce growth-inhibiting compounds (5). Nitrogen fixation may also contribute to increased growth, but this is regarded as unlikely. *Bacillus subtilis* and *S. griseus* are not known as nitrogen-fixing organisms, and there are probably insufficient energy sources in the rhizosphere to support active nitrogen fixation (13).

Seed inoculation with bacteria like *A. chroococcum* and *B. megatherium* var. *phosphaticum* is extensively practiced in Russia and has been tested experimentally in other countries, including Australia and the United Kingdom. Experiments in those countries demonstrated that seed inoculation could increase growth of cereals and vegetables but results were variable, and in many

Fig. 1. Growth differences among oat plants (cultivar Avon) sampled from field plots and grown from seed treated with *A)* water, *B)* a cell suspension of *Streptomyces griseus* 2-A24 in water, and *C)* a cell suspension of *Bacillus subtilis* A13 in water.

TABLE 4. Effect on the yield of carrots of inoculating seed with *Streptomyces griseus* (2-A24) or *Bacillus subtilis* (A13) applied either in combination with a pellet or as a water suspension

| Seed treatment | Total | Yield, t/ha[a] | | | |
		Very large	Large	Medium	Small
B. subtilis pellet	83.75	18.00	29.25	21.88	14.63
S. griseus pellet	53.25	2.75	9.63	17.00	24.03
B. subtilis	63.38	7.63	17.13	21.25	17.38
S. griseus	62.75	6.88	20.63	19.88	15.38
Pellet	56.75	2.88	11.75	20.00	22.13
Water	53.75	5.00	13.38	18.00	17.38
$P = 0.05$	7.83	3.73	5.15	5.50	5.23
$P = 0.01$	10.53	5.00	5.90	7.38	7.00

[a]Very large, 5.72 × 20.32 cm; large, 4.45 × 20.32 cm; medium, 3.18 × 15.24 cm; small, 1.91 × 13.70 cm.

instances there was no effect on growth (7, 13, 14). In this investigation, the effects of *B. subtilis* and *S. griseus* on grain yield of cereals were variable, but the effects of the inoculants on vegetative growth appeared to be more consistent. Tiller number of oats and wheat, dry weight of foliage of oats, and root weight of carrots were parameters of vegetative growth increased by the inoculants. *Bacillus subtilis* and *S. griseus* may be better applied to seed of crops grown for foliage and root production than to seed of grain crops.

The effect of the seed inoculants on different crops and their consistency in promoting growth under field conditions need further investigation before commercial application can be considered. An evaluation is also required of the role of *B. subtilis* and *S. griseus* as agents for biological control of root disease and as "bacterial fertilizers." The properties of antagonism and growth promotion may be related or distinct, and some understanding of the mechanisms may assist in improving techniques for selecting and applying organisms beneficial to plant growth.

ACKNOWLEDGEMENT.—The authors wish to thank the Australian Wheat Industry Research Council for a grant supporting the work on cereals, and W. Birkenhead, S. Isaacs, and L. Williams for technical assistance.

LITERATURE CITED

1. BAJPAI, P. D., and W. V. B. SUNDARA RAO. 1971. Phosphate solubilising bacteria. Part III. Soil inoculation with phosphorus solubilising bacteria. Soil Sci. Plant Nutr. 17:46-53.
2. BERESOVA, J. F., and A. N. NAUMOVA. 1939. A bacterial method for the control of fungus diseases of agricultural plants. Mikrobiologiya 8:186-205.
3. BOWEN, G. D., and A. D. ROVIRA. 1968. Influence of microorganisms on the growth and metabolism of plant roots, p. 199-201. *In* W. J. Whittington (ed.), Root growth. Proc. Univ. Nottingham Easter School in Agr. Sci., Butterworths, London.
4. BROADBENT, P., K. F. BAKER, and Y. WATERWORTH. 1971. Bacteria and actinomycete antagonists to fungal root pathogens in Australian soils. Austral. J. Biol. Sci. 24:925-944.
5. BROWN, M. E. 1972. Plant growth substances produced by microorganisms of soil and rhizosphere. J. Appl. Bacteriol. 35:443-451.
6. BROWN, M. E., and S. K. BURLINGHAM. 1968. Production of plant growth substances by Azotobacter chroococcum. J. Gen. Microbiol. 53:135-144.
7. BROWN, M. E., S. K. BURLINGHAM, and R. M. JACKSON. 1964. Studies on Azotobacter species in soil. III. Effects of artificial inoculation on crop yields. Plant Soil 20:194-214.
8. BROWN, M. E., R. M. JACKSON, and S. K. BURLINGHAM. 1968. Growth and effects of bacteria introduced into soil, p. 531-551. *In* T. R. G. Gray and D. Parkinson (ed.), Ecology of soil bacteria. Liverpool Univ. Press.
9. CHANG, I. P., and T. KOMMEDAHL. 1968. Biological control of seedling blight of corn by coating kernels with antagonistic microorganisms. Phytopathology 58:1395-1401.
10. KATZNELSON, H., and S. E. COLE. 1965. Production of gibberellin-like substances by bacteria and actinomycetes. Can. J. Microbiol. 11:733-741.
11. MERRIMAN, P. R., R. D. PRICE, and K. F. BAKER. 1974. The effect of inoculation of seed with antagonists of Rhizoctonia solani on the growth of wheat. Austral. J. Agr. Res. 25:213-218.
12. MERRIMAN, P. R., R. D. PRICE, J. F. KOLLMORGEN, T. PIGGOTT, and E. H. RIDGE. 1974. Effect of seed inoculation with Bacillus subtilis and Streptomyces griseus on the growth of cereals and carrots. Austral. J. Agr. Res. 25:219-226.
13. MISHUSTIN, E. N., and A. N. NAUMOVA. 1962. Bacterial fertilizers, their effectiveness and mode of action. Mikrobiologiya 31:543-555.
14. RIDGE, E. H., and A. D. ROVIRA. 1968. Microbial inoculation of wheat. Trans. 9th Int. Cong. Soil Sci. (III):473-481. Adelaide, Australia.
15. SCHREINER, O., and E. C. SHOREY. 1909. The isolation of harmful organic substances from soils. USDA Bur. Soils Bull. 53 p.

PART VI

ELUCIDATION AND EXPLOITATION OF NATURALLY OCCURRING BIOLOGICAL CONTROL

Elucidation and Exploitation of Naturally Occurring Biological Control: An Introduction

KENNETH F. BAKER, *Department of Plant Pathology, University of California, Berkeley.*

The papers and discussion of this session of the symposium constitute, by my tally, the sixth major conference on biological control of plant pathogens. The five preceding conferences were:

1963, Berkeley, California: First International Symposium on Factors Determining the Behavior of Plant Pathogens in Soil. (Published as *Ecology of Soil-Borne Plant Pathogens*.)

1968, London, England: Second International Symposium on Factors Determining the Behavior of Plant Pathogens in Soil. (Published as *Root Diseases and Soil-Borne Pathogens*.)

1971, Philadelphia, Pennsylvania: 63rd Annual Meeting of The American Phytopathological Society, "Biological Control of Soil-Borne Pathogens—Mission Impossible?" (Published In Soil Biology and Biochemistry, Vol. 5, 1973.)

1972, Burnley, Victoria, Australia: International Symposium on Biological Control of Root Pathogens. (Proceedings not published.)

1973, Lausanne, Switzerland: International Symposium, "Perspectives de Lutte Biologique Contre les Champignons Parasites des Plantes Cultivees et les Pourritures des Tissues Ligneux." (Proceedings published, 1973.)

An International Symposium in the First Intersectional Congress of the International Association of Microbiological Societies will be held in Tokyo, Japan, in 1974. The proceedings of the symposium will be published.

Although 5 years elapsed between the first two of the symposia, six conferences have been held in the last 4 years. This is clear evidence of the increasing interest and volume of work in biological control of plant pathogens. The appearance in 1974 of the first book devoted wholly to this subject, *Biological Control of Plant Pathogens*, by K. F. Baker and R. J. Cook, is further evidence of this increased interest.

Naturally occurring biological control is recognized by situations in which the plant pathogen cannot persist, establish, or produce disease in a given field soil, or where a substrate of near sterility is invaded by an antagonist before the pathogen arrives. The organizing committee of this session of the symposium selected examples of these four types for the papers to be presented.

A) The pathogen is present in field soil, but the disease produced decreases with continuing monoculture, exemplified by take-all disease of wheat, discussed by P. J. Shipton.

B) Although the pathogen is repeatedly introduced to field soil, it does not survive in it. This is illustrated by Fusarium wilt pathogens in numerous soils, and by the nonpersistence of *Fusarium* sp. in pine duff, discussed by T. A. Toussoun.

C) The pathogen is present in field soil but does not produce disease. The suppression of *Phytophthora cinnamomi* in certain soils in eastern Australia, discussed by Patricia Broadbent, is of this type.

D) The pathogen arrives on a medium of near sterility after an antagonist has already established itself there. This is exemplified by planting *Peniophora gigantea* on freshly cut stumps to prevent invasion by *Fomes annosus*, discussed by John Rishbeth.

The participants of this session have demonstrated that success in studies of biological control is most likely to be attained with persistence. Rishbeth continued his studies for about 15 years, Toussoun for nearly 10, Shipton for 7, and Broadbent for 5. The field of biological control has been ill-served by many brief, inconclusive, and prematurely terminated studies. The participants demonstrate the benefit from persistence in this difficult field, and the level of success attained by them is evidence that biological control is entering an era of successful application.

Take-All Decline During Cereal Monoculture

PETER J. SHIPTON—*Plant Pathology Division, North of Scotland College of Agriculture, School of Agriculture, Aberdeen, Scotland.*

As long ago as 1852, the symptoms of take-all were recognized in South Australia. Take-all has since become a classic soil-borne disease meriting the extensive research conducted over the last century (6, 22). It is remarkable, however, that only in the last 10 years has the spontaneous decline of take-all during cereal monoculture been appreciated and investigated. The widespread demonstration of this phenomenon could be of great potential importance if it can be exploited as an effective natural biological method of control for an intractable disease for which neither resistance (6, 33) nor chemical control (28) is available and where extended crop rotations are not feasible. Although the name *Ophiobolus graminis* Sacc. has been long established, it is taxonomically indefensible, and the fungus is now correctly known as *Gaeumannomyces graminis* (Sacc.) Arx and Olivier var. *tritici* J. Walker (42).

The pathogen is distributed worldwide on more than 160 species in the Gramineae, among which wheat is particularly susceptible.

MONOCULTURE AND TAKE-ALL.—Garrett (11) stated that "In general theory, any pathogen specialised to a particular crop should continue to cause fairly widespread and severe disease in that crop for so long as monoculture continues." Nevertheless, spontaneous reductions in take-all, sometimes accompanied by an improvement in yield, in a continuous succession of susceptible cereal crops (such as wheat and barley) have been reported from time to time since the end of the last century. Progressive but gradual disappearances of take-all in wheat in South Australia were observed regularly over a 25-year period after reclamation (24), and it was suggested then (1) that continued cultivation gradually increased the antagonistic soil bacteria that provided a progressive check to the fungus. Later, in the United States, patches of severe take-all in Kansas enlarged with successive crops of wheat and then decreased in size and finally disappeared, although the soil still contained the fungus (9). In Switzerland, spring barley has been grown in monoculture over several centuries in the mountainous areas, and Zogg (43) has described how yields initially fell sharply but increased progressively over a prolonged period of monoculture. This he attributed to a fading-out of an epidemic of take-all because of specific qualitative and quantitative changes in the soil microflora.

Glynne and her co-workers at Rothamsted have collected the majority of the basic critical observations on take-all and yields of winter wheat under monoculture. The decline of take-all was first measured at the Woburn Experimental Farm in 1933; take-all appeared in the third consecutive crop and continued to increase to the fifth crop when less than 35% of plants of wheat or 10% of barley had been infected the year before, diminishing only where these levels of infection had been exceeded in the previous crop (13). Autointoxication or dying-out of the pathogen were suggested as possible explanations.

The significance of this reduction was not investigated at the time, so the consistent changes in take-all and their practical significance were appreciated first by Slope in 1962 (34) after the fourth successive crop of winter wheat had shown 8% less take-all and 625 kg/hectare greater grain yield than either the second or third crop over 3 consecutive years of trials. When the trial was concluded, similar observations had been recorded, with some seasonal variation, for 7 successive years. As a result, the term take-all decline (TAD) was coined by Slope and Cox (35) to describe the phenomenon of the spontaneous reduction in disease and increase in yield with extended monoculture.

They concluded that "Although the differences are small they are important because the yields of the second and third wheat crops cancel out much of the benefit gained in the first after one not susceptible to take-all" (8).

EVIDENCE FOR THE WIDESPREAD OCCURRENCE OF TAKE-ALL DECLINE.—Concurrent investigations during the last 10 years have demonstrated the widespread existence of TAD in Britain, particularly its role in the well-established commercial practice of continuous barley growing.

A wealth of data has accumulated on the Experimental Husbandry Farms of the Ministry of Agriculture on the behavior of take-all and yields in spring barley and wheat, as well as winter wheat, under different soil and climatic conditions. For the first time, clear evidence was found for TAD in spring barley and spring wheat. This was best documented at the Experimental Farm in Hampshire (30), where detailed comparisons of its expression in both cereals could be made simultaneously (Fig. 1).

With each successive crop, take-all increased, reaching maximum intensity (with greatest incidence and severity) in wheat about the 4th or 5th year, 1 year earlier than in barley, which was less severely attacked under comparable conditions. In both cereals, the disease subsequently declined in severity and only to a lesser extent in incidence.

Nitrogen application determined the level of yield at all stages in the cropping sequence. It did not influence take-all until the peak of the disease (Fig. 2), when there was a reduction in severity and a small reduction in incidence in proportion to the rate of nitrogen, which was complementary to that produced by TAD (30) and subject to seasonal variation.

Yields inversely reflected the concurrent alterations in disease intensity (Fig. 3). Lowest yields coincided with the peak of take-all, with a smaller reduction in barley than in wheat. After the onset of TAD, the yields of barley improved to a level almost equivalent to those of the 1st year crops, although the yield recovery of spring wheat was less complete.

In surveys in the mid-1960's, farm yields were collected from 579 barley fields in the south of England that had grown up to nine successive crops (29) (Fig. 3).

Even with allowance for error, the similarity of field yields to experimental results is remarkable. Simultaneous disease assessments in samples of these crops also showed that take-all was lower in crops with more than 7 years of monoculture than in those following four to six barley crops (29). Continuous cultivation of barley and sometimes wheat is recognized as economic, especially on heavy clays or chalky loams, but it is recognized as a greater risk on light sandy soils, where losses of 30% or more may occur (21).

Take-all decline is recognized in Switzerland, Holland, France, West Germany, Yugoslavia, and Denmark (12, 15, 18, 23, 41, 43). The similarity of TAD throughout Europe is all the more remarkable in view of the range of climatic and agronomic conditions, soil types, and chemical and physical factors in these countries.

It is clear that the development and decline of take-all follows a consistent pattern, characteristic of the species of cereal, apparently requiring only the continuous cultivation of susceptible crops for its initiation and development. However, the occurrence of a peak of disease at a particular stage is a regular though not invariable precursor to the onset of TAD (21). Once established, TAD permits a recovery in yield, due more to a decline in disease severity than incidence, and it persists indefinitely at a fairly consistent level of intensity as long as monoculture continues. Such variation as does occur is attributable mainly to differences in soil type and previous cropping history. For example, in Denmark (15), TAD develops more slowly in light sandy soils and at low pH values, with the least recovery in yield. Chernozem soils in Yugoslavia (40, 41) show TAD most quickly and extensively, with black marsh soils markedly less and pseudogley soils least favorable to its development. Clay loam and peat soils previously cultivated with mixed crops for several years in the polders of Holland (12) encourage TAD more than do sandy loam or recently drained peat soils.

In the United States, a phenomenon similar, if not identical, to TAD exists in the dryland areas of Washington State (31). When water for irrigation becomes available, devastating attacks of take-all develop on land newly reclaimed from the native

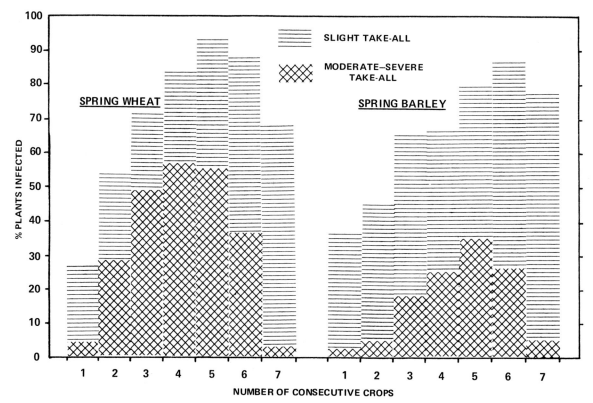

Fig. 1. Duration of monoculture and incidence and severity of take-all infection of spring wheat and spring barley, Hampshire, England, 1963-1969. Plants with slight take-all: 1-25% roots infected at growth stage 10.5.1; moderate-severe take-all: 25-100% roots infected. (After Shipton, 1972.)

sagebrush/bunch grass desert, but when water is applied to dryland fields in the same area that have a history of decades of wheat-fallow cultivation, take-all remains absent or negligible.

EXPLANATIONS FOR THE DECLINE OF TAKE-ALL.—Although TAD is a widespread phenomenon showing superficially remarkably similar symptoms wherever it occurs, it is unknown whether its causes or mechanisms are identical in all instances: indeed, no particular occurrence is yet fully understood, since cause and effect relationships are difficult to unravel. This may be due to the fact that "Because TAD is part of the ecology of wheat fields, it is almost certainly accompanied by disturbances, some of which it causes, some of which influence it" (5). It has been concluded from field evidence, however, that the development and decline of take-all are associated with monoculture of cereals and with the concurrent development and influence of factor(s) inimical to the pathogenicity and survival of *G. graminis*. Suggestions have been made that TAD follows changes in the virulence of the pathogen (3), accumulations of toxic products from the fungus or host (12, 13), alterations in nutrient availability (10), or changes in physical and chemical conditions of the soil (5). Certain of these hypotheses may be applicable in particular situations, but, on theoretical or experimental

grounds, most can be dismissed as probably inapplicable to the widespread phenomenon.

One intriguing possibility remains unresolved. Recently, virus or virus-like particles have been obtained from the mycelium of isolates of *G. graminis*. In France (17, 19), these were associated with reduced pathogenicity of the fungus and with TAD, although these associations could not be confirmed at Rothamsted (27).

MICROBIOLOGICAL MECHANISMS FOR DECLINE.—Theoretical and experimental considerations suggest that the most likely explanation for TAD is that it is brought about by a microbiological interaction between the pathogen and other microorganisms in the soil involving some form of generalized antagonistic or inhibitory action in the widest possible sense, such as antibiosis, competition, or hyperparasitism associated with the increase in fungal inoculum during monoculture (5, 12, 26, 41). *Gaeumannomyces graminis* is particularly sensitive to many microorganisms (including some hyperparasites) (7, 16, 32, 43), and, since they occur widely, almost any natural nonsterile soil shows a degree of activity against *G. graminis*. The expression of take-all in a host crop growing in any arable soil is a reflection partly of cultural factors and partly of the effect of the dynamic microbiological balance in soil that limits the activity of each of its components, including the pathogen. Gerlagh

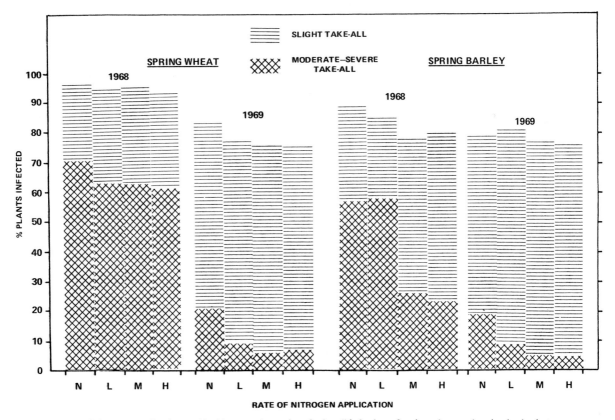

Fig. 2. Rate of nitrogen application and incidence and severity of take-all infection of spring wheat and spring barley in two years, 1968 and 1969, in crops showing take-all decline after 4 or more years of monoculture, Hampshire, England. Nitrogen applied at nil (N), low (L), medium (M), or high (H) rates. (After Shipton, 1972.)

(12) postulated that the more pronounced diminution of take-all occurring in TAD soil was due to an additional characteristic source of potent antagonism specific to *G. graminis* that was in turn induced by monoculture or by the fungus itself (Fig. 4).

For clarity and brevity, the term "antagonism" will be used hereafter without qualification to embrace the range of microbial activity characteristic and *specific* to take-all decline, in contrast to the generalized inhibition with nonspecific effect present in all biologically active soils. Hence, soils in which TAD occurs will be described as decline (TAD) or antagonistic and vice versa.

BIOLOGICAL PROPERTIES OF DECLINE SOILS.—The decline of take-all under experimental conditions has been achieved by the inoculation of natural soils with *G. graminis* and the repeated sowing of wheat in successive cycles (12, 38). Experimental conditions probably encourage a more exaggerated expression of decline (12), but even soils known to support heavy attacks of take-all in the field can develop marked antagonism. As in the field, TAD regularly followed severe attacks of the disease, whereas the extent or speed of development of antagonism depended on soil type and previous history (12, 40).

The widespread occurrence of antagonistic soils and their association with TAD in commercial crops has been demonstrated in England (20, 29). Soils from fields with a history of cereal cropping were contrasted with those from adjacent fields not cropped with cereals or from adjacent uncultivated areas; a consistently greater degree of antagonism occurred in soils with a history of more than five consecutive susceptible cereal crops than in soils with fewer or no cereal crops.

It is notable that antagonism can be induced only with live, virulent inoculum of *G. graminis*; no antagonism results when dead or avirulent inoculum is used (12, 40).

Furthermore, antagonism is not conditional on the presence of host roots, although it is enhanced by them. Similarly, other pathogens such as *Fusarium culmorum* and *Helminthosporium sativum* did not induce antagonism to *G. graminis,* although these fungi are known to induce antagonism to themselves under experimental conditions (12, 38). In his experiments, Gerlagh (12) stressed the specificity of reaction between the fungus and its antagonists.

Elimination of antagonism from soil.—It is possible to deduce the nature of the antagonism that soils contain by treating them with heat or chemicals. Decline soils can be rendered almost totally nonantagonistic by the use of broad-spectrum biocides or nematicides (12, 31), although not by the use of antibiotics. Results of pasteurization (2) of antagonistic soils both in Holland and the United States showed remarkable agreement (Fig. 5) in the elimination of antagonism from a range of soils by temperatures between 40° and 60° C and hence by an increase in take-all (12, 31). This is circumstantial evidence indicating that the antagonism is due to non-spore-forming bacteria and certain saprophytic fungi, rather than to actinomycetes or spore-forming bacteria, most of which have greater heat tolerance (2).

The results of dilution of antagonistic soil with varying proportions of nonantagonistic soil can be explained most satisfactorily by invoking microbiological activity in decline soils (12). Mixtures of soils inoculated with the pathogen (12) showed progressively more antagonism and greater yield in direct proportion to the amount of antagonistic soil; sometimes as little as 5% or even 1% w/w antagonistic soil halved the severity of infection (31), which is consistent with an antibiotic effect.

Transfer of antagonism.—Microorganisms dominant in one soil usually fail to express themselves in a dissimilar soil. But successful inoculations of a given soil with

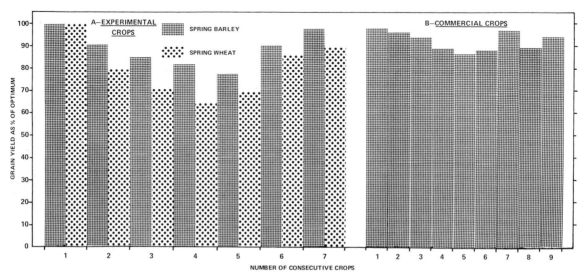

Fig. 3. Duration of monoculture and relative grain yield of spring barley and spring wheat expressed as a percentage of the optimum yields obtained (usually in the first crops). *A)* Data from two experimental crops, Hampshire, 1963-1969; *B)* Data from a total of 579 commercial field crops of spring barley, Hampshire and Berkshire, England, 1965-1968 (*personal data*).

antagonists can sometimes be achieved by using another physically and chemically compatible soil. In Washington State (31), antagonism was restored to antagonistic soils fumigated with methyl bromide and inoculated with *G. graminis* by recontamination with the identical untreated natural antagonistic soil at rates down to 1% w/w (Fig. 6).

Plants sown in soil fumigated and then inoculated with

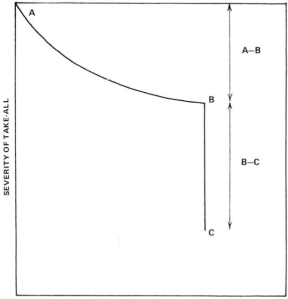

Fig. 4. Hypothetical relationship between number of microorganisms in soil and severity of take-all attack, on arbitrary scales. The amelioration of disease (A-B) is produced by "generalized" antagonism present in all soils, the sharp attenuation (B-C) by "specific" antagonism present only in take-all decline soils. (After Gerlagh, 1968.)

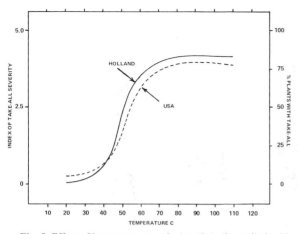

Fig. 5. Effect of heat treatment of take-all decline soils for 30 minutes at a range of temperatures on residual antagonism measured by severity of take-all infection in bioassays of wheat seedlings inoculated with *Gaeumannomyces graminis*, Holland and United States. (After Gerlagh, 1968; Shipton et al., 1973.)

G. graminis were severely attacked by take-all, and many died before tillering, but in soil recontaminated and inoculated, plants were only somewhat more heavily affected than those in untreated inoculated soil. The effect was out of proportion to the rate of recontamination and could only be explained by the multiplication of the antagonistic microflora in its parent soil.

Other experiments showed that it is possible to transfer antagonism into another soil under contrasting climatic and soil conditions. The influence of three antagonistic soils from the arid alkaline Columbia Basin could be detected after incorporation into a fumigated field soil west of the Cascade Mountains. The latter soil had an acid reaction and four times the annual rainfall.

Recently, Pope and Jackson (26) have reported that extended protection against subsequent infection was given to the roots of wheat seedlings exposed briefly to nonsterile antagonistic soil or its extracts and transplanted into inoculated antagonistic soil, even though the active principle was present in amounts as little as 1 mg soil/plant. It was concluded that the specific antagonism in decline soils was bacterial in origin, whereas the generalized antagonism in nonantagonistic soils was largely fungal (25). In Gerlagh's experiments (12), the antagonistic factor in sterile extracts of soils survived heating to 60° or 100° C, temperatures known to eliminate antagonism in whole soil.

Inoculum density in decline soils.—A reduction in the number of infective units in soil during monoculture could result in a lower rate of infectivity and a decline of take-all. Several separate estimations of infectivity or inoculum density have confirmed that this could be a contributing factor (14, 26, 40).

MICROORGANISMS IN DECLINE SOILS AND THEIR INFLUENCE ON TAKE-ALL.—Attempts to

define the species or type of microorganisms in decline soils responsible for antagonism to *G. graminis* have had only limited success, although their mode of action is now at least partially understood (41).

In view of the range of conditions under which TAD occurs, it seems improbable that a single species or even a small group of microorganisms could be the sole cause of decline in all cases. It is possible that, whereas the phenomena appear universally similar, the causes may be peculiar to individual situations. It is not surprising that investigations comparing the broad composition of the microflora of decline and nondecline soils have been mostly unsuccessful or inconclusive (7, 12) and that population differences, although sometimes marked, could not be related to decline. Pope (25), however, did find that total populations of bacteria and actinomycetes in bulk TAD soils (30) exceeded those in nondecline soils by a considerable margin, thereby supporting his deductions from the results of ultrafiltration of decline soil extracts that antagonism was principally bacterial in origin (25).

Comparisons of the rhizosphere microflora of wheat in decline and nondecline soils by Vojinović (39, 40, 41) showed that the total number of organisms isolated from senescent wheat roots after harvest differed significantly, but not when bulk soils were compared. Furthermore, the rhizosphere of plants in the crop showing symptoms either of the severest level of disease or of the decline of take-all was richest in total microbial populations (excluding fungi), being particularly rich in antagonists to *G. graminis*. These decline soils or their extracts inhibited infection of wheat roots and reduced the saprophytic colonization and survival of the fungus in buried wheat straw. Most actinomycetes and bacteria caused pronounced inhibition of in vitro growth of the fungus and were isolated three times more often from decline than from nondecline soils. The most inhibitory isolates were of certain actinomycetes that greatly reduced infection in sterile conditions, but which were less effective in natural soil.

Severely infested host debris was degraded by a microflora different from that associated with debris of healthy or lightly infected plants. Infested debris almost doubled the population of bacterial antagonists, and a spore-forming, motile, gram-negative *Bacillus* sp. formed over 90% of the isolations. Subsequently, the infectivity of nonsterile infested debris was found to be almost negligible compared to that of sterile infested debris. Vojinović (40) concluded that the soil microflora in decline soils both reduced the inoculum potential and provided an enhanced source of antagonism to the pathogen, since its antagonistic activity against the parasitic phase of the fungus could be relatively more important than that against the saprophytic phase because of the greater contact between the pathogen and the host under cereal monoculture. Antagonism in extracts of the rhizosphere of inoculated, heavily infected plants appeared to be enhanced specifically by live mycelium. Since this extract was more potent than extracts of the rhizosphere of plants in decline soils, it may be that the microflora developing on the roots of

infected plants itself diminishes with the decline of the pathogen.

Hyphae of *G. graminis* respond to exudates of wheat roots by growing through nondecline field soils to the root surface, where the mode of infection may vary according to the nutrient status of the source of inoculum (4). In decline soils, however, both the tropic response and preinfection morphology of the hyphae of the fungus are modified by the microflora (26, 41).

According to Pope and Jackson (26), the rate of hyphal growth is reduced in decline soil: there is no response to roots beyond 5 mm, whereas in nondecline soil there is response up to 8 mm. This difference was used to explain the relative degree of infectivity of the two soils in theoretical terms based on the size of the zone around the root within which the hyphae could respond. The reduced response of *G. graminis* might be caused by antibiotics or by a decreased uptake of the specific root exudates because of enhanced microbial metabolism in the rhizosphere.

The much reduced branching of runner hyphae and an absence of fine infection hyphae seen in the rhizosphere in decline soils was attributed by Vojinović (41) to exhaustion of the medium by the abundant microflora or to direct decomposition of the hyphae by lysis.

Role of nutrients in decline.—A new approach to the search for the mechanism of TAD has been made by the group at Rothamsted (5) who have attempted to establish associations rather than cause and effect relationships between chemical and biological factors and TAD. In bulk soil that had supported maximum disease (usually in the second wheat crop), total nitrogen fell to its lowest level, whereas in the rhizosphere, although nitrate-nitrogen (NO_3-N) was also at a minimum, ammonium-

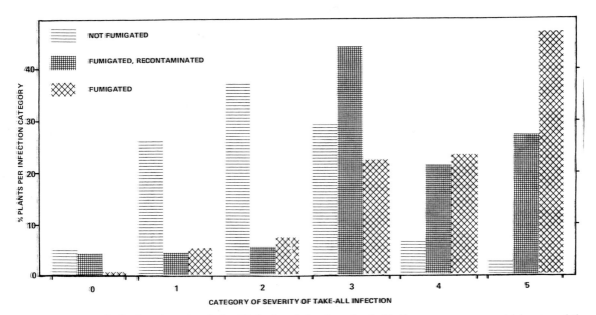

Fig. 6. Frequency distribution of severity of take-all infection of wheat inoculated with *Gaeumannomyces graminis* in antagonistic field soil with plots fumigated, fumigated and recontaminated with nonfumigated antagonistic soil at about 0.5% w/w, and not fumigated (control). Take-all severity assessed 4 months after sowing, with categories 0 = healthy to 5 = very severe infection, Washington State, United States. (After Shipton et al., 1973.)

nitrogen (NH_4^+-N) was at a maximum. Sterilization of this soil nevertheless released the lowest quantities of NH_4^+-N that were still sufficient to overwhelm the original differences. Likewise, the greatest degree of inhibition of the fungus was produced by soil that had shown maximum take-all and minimum populations of ammonifying and denitrifying bacteria in the rhizoplane.

Some changes appeared to be associated consistently with TAD and with levels of nitrogen and numbers of bacteria usually at a minimum when disease was maximal, whereas the greatest expression of many factors (e.g., numbers of microorganisms, etc.) occurred when TAD had become established. Evidently, monoculture produces many simultaneous changes, and TAD does not occur in isolation.

Effect of nonhost crops on decline.—Nonhost crops can produce a substantial effect on the rate of initiation of TAD and, especially when used as "break" crops, on the status of TAD in succeeding cereal crops, depending on the species and duration of the break crop (12, 40). For example, an antagonistic soil fallowed or cultivated with potatoes or peas for 1 year apparently retains its antagonism, but a single crop of oats or sugarbeet will virtually eliminate it. Combinations of other break crops, however, produce a reduction in antagonism and in pathogen inoculum in proportion to the number of crops grown. Further cultivation of wheat or barley rapidly increases take-all to levels considerably greater than in other cereal crops under monoculture. This has been attributed to the elimination or suppression of antagonism (12).

The benefit of increased yield in the first crop of wheat or barley after the break may later be partially nullified, since yields in the second and third successive crops often fall drastically below yields of crops in monoculture because of the increase in take-all (8). At Rothamsted (36), yields over the three consecutive crops after the break averaged only 4,040 kg/hectare with 36% take-all, compared to 3,850 kg/hectare from continuous wheat with 45% take-all, showing that the result of a break may be only marginally favorable (and sometimes even negative), especially if the break is one of oats or beans. The use of break crops cannot be justified, particularly under such circumstances, as their economic value is often below that of the cereal crop grown in monoculture.

CURRENT THEORIES ON MECHANISM OF DECLINE.

—Research into the mechanism of TAD has led to the development of four theories, complementary in many respects, that share the common concept of microbial antagonism to *G. graminis*.

A) Gerlagh (12) postulates that it is the dominance of fungal inoculum alone that induces a highly specific microbial antagonism against itself. Antibiotics produced by a balanced microflora antagonize the fungus in its parasitic and saprophytic phases.

B) Pope and Jackson (26) suggest that the reduction of infectivity is the result of inhibition of the tropic response of the fungal hyphae to the host roots consequent upon the development of a bacterial microflora in the rhizosphere and possibly caused by antibiotic action.

C) Vojinović (41) also considers that a specific antagonistic microflora develops in the rhizosphere after infection of the host in response to the parasitic mycelium. Severe attacks of take-all increase the quantity of infested debris and its associated antagonistic microflora, especially during host senescence. In turn, the enhanced antagonism decreases the survival of inoculum and its infectivity by inhibition of hyphal branching.

D) The Rothamsted group (5) suggest that TAD operates through changes in the soil microflora in response to disease progress and that these limit the disease by modifying the root environment nutritionally, perhaps by a shift in the NO_3^-/NH_4^+-N ratio in the rhizosphere.

FUTURE DEVELOPMENTS AND EXPLOITATION OF DECLINE.

—Current theories on TAD have much in common, and a comprehensive explanation of the mechanisms involved should eventually be possible. The establishment of consistent relationships among the many factors operating in TAD could provide useful evidence, but cause and effect may be difficult to distinguish. The role of the nutrition of host, pathogen, and microflora in the rhizosphere merits more attention, as do the changes in chemical and physical factors such as pH (37) that occur during monoculture. Identification of characteristic activity groupings of organisms may prove more rewarding than taxonomic tabulations.

For the farmer, TAD has been recognized as a practical asset, provided optimum standards of husbandry are maintained, and there are obvious attractions in a naturally occurring biological control of disease for which there are no other effective alternatives in intensive cereal cultivation.

An understanding of TAD and its practical exploitation may eventually be regarded as a milestone in the progress of the control of soil-borne diseases by biological methods.

LITERATURE CITED

1. ANONYMOUS. 1934. Take-all of wheat, p. 72-73. Waite Agr. Res. Inst. Rept. 1925-1932.
2. BAKER, K. F. 1962. Principles of heat treatment of soil and planting material. J. Austral. Inst. Agr. Sci. 28:118-126.
3. BROADFOOT, W. C. 1933. Studies on foot and root rot of wheat. II. Cultural relationships on solid media of certain microorganisms in association with Ophiobolus graminis Sacc. Can. J. Res. 8:545-552.
4. BROWN, MARGARET E., and D. HORNBY. 1971. Behaviour of Ophiobolus graminis on slides buried in soil in presence or absence of wheat seedlings. Trans. Br. Mycol. Soc. 56:95-103.
5. BROWN, MARGARET E., D. HORNBY, and VIVIENNE PEARSON. 1973. Microbial populations and nitrogen in soil growing consecutive cereal crops infected with take-all. J. Soil Sci. 24:296-310.
6. BUTLER, F. C. 1961. Root and foot rot diseases of wheat. Sci. Dept. Agr., New South Wales., Bull. 77. 98 p.
7. EHLE, H. 1966. Einfluss einer Superinfektion mit antagonistichen Actinomyceten auf die Ophiobolose in Gefässversuch. Z. PflKrankh. PflPath. Pflschutz 73:321-326.
8. ETHERIDGE, JUDITH, and D. B. SLOPE. 1967. Continuous wheat growing and the decline of take-all, p.

123. Rothamsted Exp. Stn. Rept. for 1966.

9. FELLOWS, H., and C. H. FICKE. 1934. Cereal and forage crop disease investigations, p. 94-97. Kans. Agr. Exp. Stn. Rept., 1932-1934.

10. GARRETT, S. D. 1948. Soil conditions and the take-all disease of wheat. IX. Interaction between host plant nutrition, disease escape, and disease resistance. Ann. Appl. Biol. 35:14-17.

11. GARRETT, S. D. 1970. Pathogenic root infecting fungi. Cambridge Univ. Press, Cambridge. 294 p.

12. GERLAGH, M. 1968. Introduction of Ophiobolus graminis in new polders and its decline. Netherlands J. Plant Pathol. 74. Suppl. 2. 97 p.

13. GLYNNE, MARY D. 1935. Incidence of take-all on wheat and barley on experimental plots at Woburn. Ann. Appl. Biol. 22:225-235.

14. HORNBY, D. 1969. Methods of investigating populations of the take-all fungus (Ophiobolus graminis) in soil. Ann. Appl. Biol. 64:503-513.

15. JENSEN, A. 1973. Investigations on diseases in cereal monocultures in Denmark. Proc. Conf. on Role of Pathogenic Organisms in Crop Rotation Problems of Cereals and Other Arable Crops, Versailles, France.

16. LAL, A. 1939. Interaction of soil microorganisms with Ophiobolus graminis Sacc., the fungus causing the take-all disease of wheat. Ann. Appl. Biol. 26:247-261.

17. LAPIERRE, H., J. -M. LEMAIRE, B. JOUAN, and G. MOLIN. 1970. Mise en évidence de particules virales associées à une perte de pathogénicité chez le Piétin-échaudage des céréales, Ophiobolus graminis Sacc. Comptes rendus hebd. Acad. Sci. Paris. 271:1833-1836.

18. LEMAIRE, J. -M., and M. COPPENET. 1968. Influence de la succession céréalière sur les fluctuations de la gravité du Piétin-echaudage (Ophiobolus graminis Sacc.). Ann. Épiphytol. 19:589-599.

19. LEMAIRE, J. -M., H. LAPIERRE, B. JOUAN, and G. BERTRAND. 1970. Découverte de particules virales chez certaines souches d'Ophiobolus graminis, agent du Piétin-échaudage des Céréales, conséquences agronomiques prévisibles. Comptes rendus hebd. Séances Acad. Agr. Fr. 56:1134-1138.

20. LESTER, E., and P. J. SHIPTON. 1967. A technique for studying inhibition of the parasitic activity of Ophiobolus graminis (Sacc.) Sacc. in field soils. Plant Pathol. 16:121-123.

21. MUNDY, E. J. 1969. Twelve years of continuous barley cropping. Exp. Husb. 18:91-101.

22. NILSSON, H. E. 1969. Studies of root and foot rot diseases of cereals and grasses. I. On resistance to Ophiobolus graminis Sacc. Lantbr-Hogsk. Annlr. 35:275-807.

23. OBST, A., and R. DIERECKS. 1970. Crop rotation and soil pathogens. Trends in Germany. Eur. Plant Protection Org. Conf. on Soil-Borne Pathogens, Arnheim, Belgium.

24. PERKINS, A. J. 1917. Eyre's Peninsula. Its agricultural development and the work of the Department of Agriculture. J. Dept. Agr. S. Austral. 20:684-690.

25. POPE, A. M. S. 1972. The decline phenomenon in take-all disease of wheat. Ph.D. thesis, Univ. Surrey, Guildford, England. 161 p.

26. POPE, A. M. S., and R. M. JACKSON. 1973. Effects of wheatfield soil on inocula of Gaeumannomyces graminis (Sacc.) Arx and Olivier var. tritici J. Walker in relation to take-all decline. Soil Biol. Biochem. 5:881-890.

27. RAWLINSON, C. J., D. HORNBY, VIVIENNE PEARSON, and J. M. CARPENTER. 1973. Virus-like particles in the take-all fungus, Gaeumannomyces graminis. Ann. Appl. Biol. 74:197-209.

28. SALT, G. A. 1971. Soil fumigation and root-rots of wheat, p. 138-146. Rothamsted Exp. Stn. Rept. for 1970.

29. SHIPTON, P. J. 1967. Take-all decline, p. 96-107. Proc. 4th Br. Insect. Fung. Conf., Vol. 1.

30. SHIPTON, P. J. 1972. Take-all in spring sown cereals under continuous cultivation: disease progress and decline in relation to crop succession and nitrogen. Ann. Appl. Biol. 71:33-46.

31. SHIPTON, P. J., R. J. COOK, and J. W. SITTON. 1973. Occurrence and transfer of a biological factor in soil that suppresses take-all of wheat in Eastern Washington. Phytopathology 63:511-517.

32. SIEGLE, H. 1961. Über Mischinfektionen mit Ophiobolus graminis und Didymella exitialis. Phytopathol. Z. 42:305-348.

33. SIMMONDS, P. M. 1953. Root rots of cereals, II. Bot. Rev. 19:131-146.

34. SLOPE, D. B. 1963. Cereal bean experiment, p. 116-117. Rothamsted Exp. Stn. Rept. for 1962.

35. SLOPE, D. B., and JUDITH COX. 1964. Continuous wheat growing and the decline of take-all, p. 108. Rothamsted Exp. Stn. Rept. for 1963.

36. SLOPE, D. B., and JUDITH ETHERIDGE. 1971. Grain yield and incidence of take-all (Ophiobolus graminis Sacc.) in wheat grown in different crop sequences. Ann. Appl. Biol. 67:13-22.

37. SMILEY, R. W., and R. J. COOK. 1971. Influence of nitrogen fertilizers on rhizosphere pH and take-all of wheat caused by Ophiobolus graminis. Phytopathology 61:911-912 (Abstr.).

38. TYNER, L. E. 1940. The effect of crop debris on the pathogenicity of cereal root rotting fungi. Can. J. Res. 18:289-306.

39. VOJINOVIĆ, Ž. D. 1970. Jedna metoda za odredivanje gustine populacije antagonista prema O. graminis u. zemljištu. Zaštita bilja, br. 110/111:365-372.

40. VOJINOVIĆ, Ž. D. 1972. Antagonists from soil and rhizosphere to phytopathogens. Inst. Soil Sci., Beograd, Yugoslavia. Final Tech. Rept. 130 p.

41. VOJINOVIĆ, Ž. D. 1973. The influence of microorganisms following Ophiobolus graminis Sacc. on its further pathogenicity, p. 91-101. Eur. and Mediterranean Plant Protection Org. Bull. 9.

42. WALKER, J. 1972. Type studies on Gaeumannomyces graminis and related fungi. Trans. Br. Mycol. Soc. 58:427-457.

43. ZOGG, H. 1951. Studien über die Pathogenität von Erregergemischen bei Getreidefusskrankheiten. Phytopathol. Z. 18:1-54.

Fusarium-Suppressive Soils

T. A. TOUSSOUN—*Institute for Fungus Research, San Francisco, California, and Fusarium Research Center, Department of Plant Pathology, The Pennsylvania State University, University Park.*

HISTORICAL REVIEW.—*The first experimental proofs.*—Investigations on Fusarium-suppressive soils started more than 50 years ago. L. Knudson, on behalf of the United Fruit Company, began his studies in 1922 on the correlation between soil types, soil pH, and rate of spread of Fusarium wilt of bananas in Guatemala; these studies were continued by N. J. Volk. This research, largely unpublished, has been summarized by Stover (31). Volk characterized "resistant" soils as those in which the spread of the disease was so slow that the productive life of plantations of susceptible bananas extended for more than 20 years. He termed "nonresistant" those soils in which the same banana varieties succumbed to the disease at such a rate that the plantations had to be abandoned in 10 years or less. He also noted the existence of soils of intermediate effect, which he termed "semiresistant." Reinking (18) and Reinking and Manns (20) found a correlation between resistance of soils and population of the pathogen. They studied instances where areas of severe disease bordered on areas of slight disease. Although of different soil types, these areas had almost identical pH, were deposited by the same river, and were under identical meteorological conditions. The *Fusarium* population was assayed in the immediate vicinity of diseased plants, and in each case these workers found a clear-cut relation of low population with low disease incidence. High disease incidence coincided with high soil populations. They found that resistance appeared to be determined by soil type rather than by pH. The pathogen was found in greater abundance in light-textured than in heavier textured soils.

It appears doubtful today that these workers were able to differentiate the pathogen from the saprophytic clones of *Fusarium oxysporum* solely on the basis of morphological criteria, and their pathogen population data are therefore open to question. Nevertheless, these studies should be considered landmarks in this area, and I would like to quote Reinking's conclusions (19): "In the adjoining areas examined the diseased area did not gradually radiate out from the 100% diseased area into the less diseased area, but there was a distinct line of demarcation between the two areas in respect to disease that corresponded with the change in the soil type. A line could be drawn between severe disease and no or slight disease following the change in soil . . . . In all extreme differences in infection on adjoining areas, it was found that the disease was most severe on the sandy soil and less severe on the clay soil regardless of pH concentration." He goes on to report: "The banana disease investigations also showed that in each case examined the number per gram of soil of the parasite causing the disease was always greatest in abundance in the badly diseased areas as compared to the slightly diseased areas even though the sampled areas were separated only by some 20 to 30 feet in distance. It was found that the organism could easily establish itself in the sandy soil types where large numbers were always isolated. It apparently had difficulty in establishing itself in clay soil types . . . . Apparently in heavily infected soil, root infections are so numerous that the plant is unable to ward off attack . . . . The effect of the different soil types upon the possible production or non-production of resistance within the plant to attack by the parasite is not of such great importance." Reinking (19) also determined that the pathogen could be isolated from nonresistant soils 10 years after susceptible bananas had last been planted, but could not be easily isolated from resistant soils after a like period. He concluded that since the fungus appeared able to exist saprophytically for varying periods of time, depending on the resistance of soils, length of crop rotations should be dependent on soil type. Stover (31) has summarized an unpublished experiment of Reinking and Manns in two virgin woodland areas, one with resistant and one with nonresistant soils, as determined by chemical and other analyses. These were planted to susceptible banana varieties, the propagative material having been obtained from disease-free areas in resistant soil. The plantation on nonresistant soil was destroyed by disease within 3 years. The plantation on resistant soil was only 5-16% diseased after 8 years.

Another landmark is the work of Walker and Snyder (37) on the relation of soil type to Fusarium wilt of pea. At a time when the pea industry was being increasingly jeopardized by the disease, they noted that the disease progressed little from year to year in some soils and that certain Wisconsin red clay soils had remained disease-free. They established experimental plots on the clay soil and on sandy loam brought in from areas where the disease was severe. Wilt developed on the loam, but they were unable to establish the pathogen in the clay soil. Three successive plantings of susceptible peas were made in greenhouse pot tests. Wilt incidence increased rapidly to 60% when field-infested sandy loam was mixed with the noninfested sandy loam in a 1:10 ratio. It reached 7.5% in the clay soil similarly infested and did not exceed this value, indicating that this soil prevented the establishment of the pathogen.

Earlier evidence dating to start of this century.—The work cited above gave early experimental proof of the existence of Fusarium-suppressive soils. Evidence of the probable existence of such soils dates back even earlier. Atkinson (1), who first described Fusarium wilt of cotton in 1892, noted that Fusarium wilt was especially prevalent in Alabama and Arkansas in light sandy soils. Orton (14, 15) stated in 1907 and in 1908 that cotton wilt was most prevalent in sandy soils and more rarely found in soils containing considerable clay, even when these bordered infested fields in sandy soil. He found the disease to be less severe in low spots in a field and surmised that this was caused by these areas being wetter and richer in organic matter. Lewis (9) observed in 1911 that the nature of the soil seemed to determine the distribution of the disease in Georgia, and he noted that, except for two instances, wilt had never been found on the clay soil of that state. He concluded: "The fungus causing the wilt disease of cotton may be looked upon as a weed indigenous to some soils, and like them thriving best on certain soils." Gilbert (6) reported in 1914 that Fusarium wilt occurred in every cotton-producing state from North Carolina to Texas and that a line could be drawn between the sandy and alluvial soils of the Coastal Plain, where wilt is prevalent, and the clay soils of the Piedmont Plateau, where the disease is rare. Young (40), in a comprehensive review of the subject in 1928, reported that it was common knowledge among workers on Fusarium wilt of cotton in the United States that the disease is destructive in sandy soils but is of very slight importance in clay soils. And he observed that, since all studies pointed to a relationship between "large amounts of vigorously growing inoculum" in the soil and wilt severity, the pathogen must be able to grow vigorously in

soils that are warm, well aerated, and contain sufficient organic matter to support good saprophytic growth of the pathogen. Bolley and Manns (2) noted that Fusarium wilt of flax was more important on light sandy soils that are well adapted to flax-raising than on the heavier soils of the Red River Valley of North Dakota. Taubenhaus et al. (32), in their 1928 study on the role of pH on cotton root rot and cotton wilt, reported: "Fusarium wilt of cotton, in common with Fusarium wilt of many other plants has been considered more destructive on sandy soils than on heavier soils."

End results of research of the early period.—This short review of early work clearly indicates an awareness that the distribution of Fusarium wilt diseases was influenced by soil type, and that this influence was probably exerted upon the saprophytic development of the pathogen in the soil. In addition, 40 years ago, there was experimental basis for these assertions and proof of the existence of Fusarium-suppressive soils. Also, there was a developing interest in other soil-borne diseases and in biological control. For example, Henry (7) showed, in in vitro experiments, that the normal microflora of Edmonton black loam suppressed *Helminthosporium sativum* and *Fusarium graminearum*. He concluded that the antagonism of the soil microflora deserved continuing study. However, further studies along these lines proved to be largely disappointing, and interest gradually dwindled. A number of factors accounted for the loss of interest. The problems were a good deal more complicated than originally envisaged and required more detailed knowledge than was readily forthcoming on the

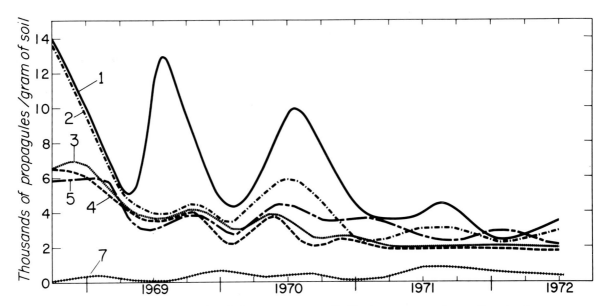

Fig. 1. Fluctuations of *Fusarium* populations in forest and nursery soils. Numbers on curves refer to soil types and treatments as follows: *1)* nursery soil left at nursery undisturbed, kept uncovered, *2)* nursery soil left at nursery, undisturbed, covered with 15-cm layer pine needles, *3)* nursery soil taken to pine forest, kept uncovered, *4)* nursery soil taken to pine forest, covered with 15-cm layer pine needles, *5)* nursery soil taken to laboratory and kept air dry (control), and *7)* forest soil taken to nursery, left uncovered. The following forest soils contained no detectable *Fusarium* and were not graphed: *6)* forest soil left at pine forest, undisturbed, covered with 15-cm layer pine needles, and *8)* forest soil taken to nursery and covered with 15-cm layer pine needles. Soils 3, 4, and 6 were replicated. (Courtesy of R. S. Smith, Jr., Pacific Southwest Forest and Range Experiment Station, U.S. Forest Service, Berkeley, California.)

kinds of organisms in the soil, their relationship, their activities, etc. Simplistic hopes that control should be obtained through fertilization, liming of the soil to change the pH, etc., were not fulfilled. Meanwhile, the need to stem the ravages of these diseases grew continuously, and a quick practical control was needed if the cultivation of many important crops was not to be jeopardized. The solution came with the development of resistant varieties, at least insofar as Fusarium wilts were concerned. The pressure was relieved, but, unfortunately, much of the burgeoning information on Fusarium-suppressive soils and biological control of soil-borne diseases was placed in limbo. It might well have been completely forgotten were it not for Garrett's treatises (4, 5) and for the fact that the United Fruit Company, because of the inherent difficulty in breeding resistant banana varieties and for other reasons, persevered in its original research.

This research was centered on the chemical and physical characterization of resistant (which they now termed "longlife") and nonresistant ("shortlife") soils, and included pH, available calcium, magnesium, phosphorus, potassium, total soluble salts, texture, drainage, cation exchange capacity, organic matter, etc. These studies, largely unpublished, have been reviewed by Stover (31). They enabled the productive life of banana soils to be predicted with a fair degree of accuracy. A major goal, however, that of finding the element(s) responsible, along with the concomitant hope that the application of the element(s) would enable the "life" of the soils to be extended at will, was not fulfilled. This line of research culminated in the work of Stotzky and his associates (25-30). Basing their investigation on the facts that, of all the soil components studied, available potassium was the most consistent indicator of productive life of soils but that adding potassium to a soil did not change its life, they searched for more fundamental soil characteristics. With few exceptions, they found that montmorillonoid-type clays were present in all longlife and "intermediate life" soils and were absent in shortlife soils. Further work in the laboratory showed that these clays stimulated bacterial activity, which might act as a check on the development of the pathogen, suppressing it.

Despite these developments, the United Fruit Company was finally forced to use resistant banana varieties. It is not unreasonable to point to this, the latest in a series of disappointments, as one of the major reasons for the renewed disillusionment about the prospects of practical biological control of soil-borne plant diseases. Nevertheless, the existence of suppressive soils cannot be denied, nor should their study be stifled. Indeed, thanks in part to the development of new techniques for the study of fungi in situ in the soil, new knowledge on the existence and nature of Fusarium-suppressive soils has come to light.

INVESTIGATIONS DURING THE LAST TWO DECADES.—Wensley and McKeen (38) studied the relationship of the muskmelon wilt *Fusarium* population to disease incidence in wilt-suppressive and wilt-conducive Canadian soils. They found consistently higher incidence of wilt in the conducive soil, roughly three times higher on the average. This relationship corresponded

with a generally higher pathogen population in the conducive soil when they assayed the sites of dead and dying plants. However, random sampling did not show a consistent population difference between the two soils. The authors concluded that factors besides population size of the pathogen influenced the wilt potential of these two soils. Komada and Ezuka (8) infested 10 soils collected in Japan with the radish, cucumber, and tomato wilt fusaria and followed the fate of these populations for 2 years in pots outdoors. Populations decreased rapidly in seven soils and remained essentially unchanged in two soils. They determined also that the nonpathogenic *F. oxysporum* population, though low, did not change markedly. R. J. Cook (*personal communication*) has found that *F. roseum* 'Culmorum' is destructive to wheat in the sandy soils of the low rainfall area of eastern Washington State. It is rare in the higher organic matter soils of the Palouse area, where rainfall is higher. The difference is apparently quite marked, and the population of the pathogen appeared to be capable of existing indefinitely in the sandy soils of the low rainfall area. Cook studied an exceptional natural outbreak of the disease in the Palouse area and found that the population declined to negligible levels in 2 years. He also experimentally infested plots in this area and recorded a similar population decline, even though a wheat crop was grown in the interim.

Bean root rot in Washington State.—One of the more intriguing investigations on Fusarium-suppressive soils has been that of Burke (3). Fusarium root rot of beans is a major disease in newly irrigated lands of the Columbia River Basin. However, certain fields long cultivated to beans remain free of the disease. Both soils are fine sandy loams, the major distinction being that the conducive soil is loessial and the suppressive soil contains limy lacustrine constituents. When both soils were artificially infested with diseased bean root inoculum, less root rot developed in the suppressive soil, particularly at low inoculum levels. Autoclaving destroyed the suppressive effect, suggesting that it was of biological origin. However, no differences in the microflora of the two soils could be found, and attempts to transfer the suppressing qualities from one soil to the other were inconclusive. Of added interest was the observation that the suppressive effect was not evident in virgin soils of the suppressive type. This, Burke suggested, indicated that the factor(s) responsible was not native to the suppressive soil but developed as a result of continuous cropping to beans. He suggested that the suppression could be due to a combination of physical and microbiological factors, none of which is sufficient to produce the effect alone. Burke followed the fate of macroconidia of the pathogen in water agar films on buried glass slides. Vegetative growth was suppressed in the conducive soil, and numerous large chlamydospores were formed. In the suppressive soil, on the other hand, hyphal development was more extensive, but there were ultimately fewer and smaller chlamydospores. Thus there appeared to be an inverse relationship between vegetative growth and chlamydospore formation. Consequently, chlamydospores formed in the suppressive soil were fewer in number and less robust in terms of size, wall thickness, and reserve food material. Similar differences between suppressive and conducive soils were noted by Burke in Colorado, and he has recently mentioned (*personal communication*) that after 20 years, the suppressive soil used in his original experiments is still suppressive.

Wilt in California.—Fusarium-suppressive soils may be more prevalent than is generally realized. The picture is probably obscured, as it is in other areas, by the widespread use of resistant varieties. Fusarium wilt of pea is a case in point. Snyder (*personal communication*) has observed the disease in California since 1933 in only a few restricted localities on the coast. Oswald (16) believed that soil type restricted the distribution of *F. roseum* 'Culmorum' in California. Smith (Nash) et al. (24) noted that cotton wilt was serious in only a few fields of the San Joaquin Valley and that after its initial rapid spread in the lighter, more acid soils, further spread slowed down. They also found that pathogen population appeared to be inversely correlated with the percentage of clay in the soil, all of which recalls the earlier observations on cotton wilt. The Salinas Valley (Monterey County) offers another example. Here, despite long and intensive cultivation of a wide variety of crops, Fusarium wilt diseases do not occur. They do occur in the adjacent Castroville area. Smith (Nash) and Snyder (22) investigated these two soil types with Fusarium wilt of sweet potato. In greenhouse tests, they found the disease to be less severe in the wilt-suppressive soils of Salinas. Chlamydospores were allowed to form naturally from macroconidia added to the wilt-conducive and wilt-suppressive soils, as well as to a noncultivated parent soil of the suppressive (Salinas) type. The inoculum was adjusted to 50, 500, and 5,000 chlamydospores/g soil. Wilt was most severe at all inoculum levels in the conducive soil. The parent nonagricultural soil was intermediate, and here, as was the case in Burke's studies, cultivation apparently improved the suppressive effect. A population count was made at the 500 chlamydospore/g soil level at the end of 8 weeks, when two series of pathogenicity tests were run. No change was found in the wilt-suppressive soil. It had doubled in the conducive soil and in the noncultivated parent soil of the suppressive type. Once again, results were similar to Burke's observations.

Saprophytes vs. pathogens.—Smith (Nash) and Snyder (23) later compared the behavior, in wilt-suppressive and wilt-conducive soils, of chlamydospores of the sweet potato, banana, and tomato wilt fusaria, as well as the behavior of three saprophytic clones of *F. oxysporum* isolated from the wilt-suppressive soil. Each saprophyte was chosen for its close resemblance in culture to one of the pathogens. The saprophytes were native to the suppressive soil. Chlamydospores were allowed to develop from conidia, and their germination was observed when carbon and nitrogen nutrients were added. Germination of the pathogens was lower in the wilt-suppressive soil, and the germ tubes were strikingly shorter and appeared to be stunted. The difference in percentage germination between the two soils decreased with increased concentration of added nutrients, but the differences in germ tube length remained. Chlamydospore germination in the pathogen was less

than that of saprophytes in the wilt-suppressive soil. Of equal interest was the observation that in both soils, germ tube length of the saprophytes was considerably greater than that of the pathogens, thalli were formed, and daughter chlamydospores were produced more rapidly. Finally, certain bacteria were found to increase more rapidly in the wilt-suppressive soil when nutrients were added.

The difference in behavior of saprophytes and pathogens reflected the fact that the suppressive soil acted principally against the pathogens: it was wilt-pathogen-suppressive; it was not suppressive to the saprophytes native to it. Such a phenomenon should not surprise anyone familiar with the Fusarium disease situation in the Salinas Valley, where Fusarium wilts are absent but other Fusarium diseases like bean root rot are prevalent. Wensley and McKeen (38) also noted differences between saprophytic *F. oxysporum* and the muskmelon wilt pathogen, and Komada and Ezuka (8) found that soils were not suppressive to the native saprophytic *F. oxysporum*, only to the pathogenic forms of this species that they studied.

Ecological studies.—Such differences may be important in the elucidation of Fusarium-suppressive soils and constitute one of the reasons that population studies should be intensified to gain an understanding of the distribution of fusaria in various soil types. Nash and Snyder (13) undertook an ecological study of this kind in Monterey County when they compared *Fusarium* populations in cultivated and noncultivated areas of both wilt-suppressive and wilt-conducive soil types. The cultivated soils of both types had quantitatively and qualitatively higher *Fusarium* populations, with the exception of the wilt-inducing fusaria, the sole example of which caused radish wilt, which occurred only in very low populations under the best of circumstances in the wilt-conducive soil prior to the introduction of resistant radish varieties. Apparently, cultivation exerts a qualitative and quantitative additive effect on the *Fusarium* population, there being no clear distinction in this respect between the two soil types.

Lim and Chew (10) also found that cultivation quantitatively and qualitatively increased the *Fusarium* population in Singapore soils. They found no fusaria in six of the noncultivated soils, three of which were from forest preserves. Windels and Kommedahl (39) found that fusaria were more prevalent in the cultivated soils of Minnesota than in noncultivated soils. Meyer (11) found *F. oxysporum* to be prevalent in cultivated soils in Belgium and to reach 80-90% of the total fungus flora in the rhizospheres of cucumber, melon, pea, and tomato. *Fusarium oxysporum* was generally absent from forest soils. Miller et al. (12), in a survey of the principal geologic regions and soil types of Georgia, noted that fusaria were quite common in cultivated soils and absent in forest soils. Park (17) found *F. oxysporum* in high proportions in cultivated soils in England and in half the uncultivated grassland soils he surveyed. He found none in woodland and heath soils and comments that he never found records of *F. oxysporum* in any of the lists of isolations from woodland and heath soils he examined. He believed that *F. oxysporum* has a wide tolerance to soil type, since he found it in soils ranging from heavy clay to sand, and to soil moisture content, since he found it in soils ranging from inundated salt marsh to sand dunes. Park thought it was the plant cover, especially the root system of plants, that affects the ecology of *F. oxysporum*. In New Zealand, Thornton (33-35) arrived at similar conclusions with respect to the ecology of *F. oxysporum*, *F. roseum* 'Culmorum,' and *F. solani* in forest and pasture soils. He believes that the Gramineae selectively encourage these fusaria. The density of grass roots is very high and their living, dying, and dead roots may be contributing the necessary nutrients as exudates and organic residues. These studies deal with several species treated as a unit, and it would be most worthwhile to re-examine the whole situation on a subspecific clonal level.

We may find that ecological differences exist among saprophytic fusaria, as they have been shown to exist between saprophytic and pathogenic forms with respect to suppressive soils. A side benefit would be the development of a nomenclatural system for these saprophytes. Nevertheless, these studies point out the possible role of plant cover in the ecology of the fusaria, a role that could have an important bearing on the proper management of *Fusarium*-suppressive soils. For example, Smith (Nash) et al. (24) found that populations of the cotton wilt pathogen increased after a barley crop but decreased after cowpeas. The latter, presumably, should be the preferred crop in a rotation designed to enhance or preserve the suppressive qualities of a soil.

Fusarium in California conifer forests.—The work of R. S. Smith, Jr., of the Pacific Southwest Forest and Range Experiment Station, U.S. Forest Service, has further elucidated the interrelationship of plant cover and soil type. Smith (21) noted that seedlings of *Pinus lambertiana* Dougl. grown in forest nurseries carried a high population of *F. oxysporum* in their rhizosphere, whereas direct-seeded native seedlings did not. He therefore followed the fate of the *Fusarium* populations of nursery seedlings transplanted into the forest. At the time of transplanting, 90% of the root samples had one or more propagules of *F. oxysporum*. Within 1 year, a rapid decline had occurred, and none could be recovered after 3 years. This finding led to an extensive survey of *Fusarium* populations (36) in conifer forests of California. No fusaria were found in pine stands, but they could be isolated from contiguous grasslands and occasionally from brushland as *F. oxysporum* and *F. roseum*. Attention was focused on the pine needle residue extract. In forest soils, a high proportion of the population of chlamydospores germinated, accompanied by stunting and destruction of germ tubes before daughter chlamydospores could be formed. Thus, it seemed that the combination of pine needle extract and forest soil contained the factor(s) responsible for the general absence of fusaria from the pine stands in California. This problem was investigated by Smith (*personal communication*), and I am indebted to him for his unpublished data.

Soil plots were established in a pine nursery at Placerville and in a pine stand in Stanislaus County. These plots were 0.6 × 0.9 m and were 13 cm deep. The experiment was started in the late summer of 1968. Assays were made every month for the first 2.5 years, every 3 months for the next year, and every 6 months thereafter. At the start of the experiment, the forest soil had no fusaria. The nursey soil had a population ranging from 14,000 propagules/g in summer to 7,000 propagules/g in winter. It consisted mainly of three broad groups of *F. oxysporum* and occasional clones of *F. roseum* and *F. solani*.

The results (Fig. 1) show that the population in soil 2 declined during the first winter and stayed at this low level thereafter, whereas that in soil 1 declined every winter and rose every summer. The rise in population coincided with the development of annuals that were not present in soil 2 because of the 15-cm layer of pine needles. The population fluctuation of soil 1 decreased in amplitude with time. A possible explanation may be that prior to the establishment of the plots, the nursery soil was fertilized and otherwise cultivated; these activities ceased on the plots after they were established, and annuals developed unhindered. As time progressed, however, annuals gradually gave way to perennials, which became a major part of the plant cover. In other words, what was observed was the reversion of a cultivated soil to its noncultivated or virgin state and the concomitant change brought to the *Fusarium* population. This phenomenon recalls the work cited earlier which showed that cultivation increased the population of fusaria.

The annual plant effect could also be seen in soils 3 and 4. No annuals grew in these soils because of the pine needle litter, shade, and other factors in the pine forest. The initial drop-off in population in soils 3, 4, and 5 at the time the plots were established probably was due to the drying of the soils when they were moved. The population decline in all three was of the same magnitude, and thereafter the population behaved similarly in all three. Since soil 5 was a control, kept air dry in the laboratory, we may conclude that the fluctuations were due to variables inherent in the population count technique. The population in soils 3 and 4 gradually declined over the 5 years, presumably because of a lack of herbaceous rootlets to parasitize (or lack of other food), and this decline therefore represented loss of viability, as with the laboratory control, soil 5. This decline was much slower than that recorded by Smith (21) for *F. oxysporum* in the rhizosphere of pine seedlings transplanted to the forest. This difference may have resulted because, in the present study, the fusaria were transported in a large quantity of their native soil, which acted as a buffer against the suppressive environment. It may also reflect the difference between pathogens or parasites and saprophytes in *F. oxysporum*. The sampling technique did not detect fusaria in soil 8, which, because of the pine needle cover, had no annuals. The *Fusarium* population in soil 7 was very low and stayed at this low level consistently, despite the presence of some annuals. Thus, in the years that have elapsed, the *Fusarium* invasion of these forest soils from the surrounding nursery soil has been meager at best. The suppressive effects of these soils continue to be exerted, apparently aided by the inherently slow movement of these fungi in undisturbed soil.

CONCLUSION.—It is apparent from this brief survey that two distinct types of *Fusarium*-suppressive soils exist. In the classic type, the suppressive effect is exerted on pathogenic forms, as illustrated by the studies of Smith (Nash) and Snyder and others. In the second type, as illustrated by forest soils, all fusaria in the soils are affected. Obviously, the mechanisms and factors involved in the two types are different, and their elucidation should become the concern of plant pathologists.

R. S. Smith's experiment demonstrates the interplay of plant cover and soil type on the ecology of fusaria. On the one hand, annuals apparently support the development of fusaria through exudates, residues, etc. On the other hand, pine trees, through their residues, suberized roots, specialized rhizosphere organisms, and their competitive effects (i.e., shading, etc.) on other plants have so influenced the chemical and biological composition of the soil as to render it suppressive. The salient point of this demonstration, and one that could be applied to biological control of soil fungi in general, is the slowness with which such changes occur. Such processes cannot be made to yield quick results. It also points up the necessity of carefully husbanding the *Fusarium*-suppressive soils that now exist.

LITERATURE CITED

1. ATKINSON, G. F. 1892. Some diseases of cotton. Ala. Agr. Exp. Stn. Bull. 41. 65 p.
2. BOLLEY, H. L., and T. F. MANNS. 1932. Fungi of flax seed and of flax-sick soil. N.D. Agr. Exp. Stn. Tech. Bull. 259. 57 p.
3. BURKE, D. W. 1965. Fusarium root rot of beans and the behavior of the pathogen in different soils. Phytopathology 55:1122-1126.
4. GARRETT, S. D. 1938. Soil conditions and the root-infecting fungi. Biol. Rev. 13:159-185.
5. GARRETT, S. D. 1944. Root disease fungi. Chronica Botanica Co., Waltham, Mass. 177 p.
6. GILBERT, W. W. 1914. Cotton wilt and root rot. USDA Farmer's Bull. 625. 21 p.
7. HENRY, A. W. 1931. The natural microflora of the soil in relation to the root-rot problem of wheat. Can. J. Res. (C) 4:69-77.
8. KOMADA, H., and A. EZUKA. 1970. Ecological study of Fusarium diseases of vegetable crops. I. Survival of pathogenic Fusaria in different soil types, p. 1-6. [English summary.] Res. Progr. Rept. Tokai-Kiuki Nat. Agr. Exp. Stn. 6.
9. LEWIS, A. C. 1911. Wilt disease of cotton in Georgia and its control. Cotton anthracnose: preliminary report. Ga. State Bd. Entomol. Bull. 34. 31 p.
10. LIM, G., and C. H. CHEW. 1970. Fusarium in Singapore soils. Plant Soil 33:673-677.
11. MEYER, J. A. 1967. Recherches sur les Fusarioses. II. Ecologie et pathogenie du Fusarium oxysporum. Ann. Epiphyties 18:241-247.
12. MILLER, J. H., J. E. GIDDENS, and A. A. FOSTER. 1957. A survey of the fungi of forest and cultivated soils of Georgia. Mycologia 49:779-808.
13. NASH, SHIRLEY M., and W. C. SNYDER. 1965. Quantitative and qualitative comparisons of Fusarium populations in cultivated fields and noncultivated parent soils. Can. J. Bot. 43:939-945.
14. ORTON, W. A. 1907. Sea Island cotton, its culture, improvement and diseases. USDA Farmer's Bull. 302. 48 p.
15. ORTON, W. A. 1908. Cotton wilt. USDA Farmer's Bull. 333. 24 p.
16. OSWALD, J. W. 1942. Taxonomy and pathogenicity of fungi associated with root rot of cereals in California, with special reference to the Fusaria and their variants. Ph.D. thesis, Univ. Calif., Berkeley. 129 p.
17. PARK, D. 1963. The presence of Fusarium oxysporum in soils. Trans. Br. Mycol. Soc. 46:444-448.
18. REINKING, O. A. 1934. Parasitic and other fusaria counted in Costa Rica and Panama soils. Zent. Bakt. 90:6-17.
19. REINKING, O. A. 1935. Soil and Fusarium diseases. Zent. Bakt. Parasitenk. Infekt. 91:243-255.
20. REINKING, O. A., and M. M. MANNS. 1933. Parasitic and other fusaria counted in tropical soils. Zeitschr. Parasitenk. 6:23-75.
21. SMITH, R. S., JR. 1967. Decline of Fusarium oxysporum in the roots of Pinus lambertiana seedlings transplanted into forest soils. Phytopathology 57:1265.
22. SMITH, SHIRLEY N., and W. C. SNYDER. 1971. Relationship of inoculum density and soil types to severity of Fusarium wilt of sweet potato. Phytopathology 61:1049-1051.
23. SMITH, SHIRLEY N., and W. C. SNYDER. 1972. Germination of Fusarium oxysporum chlamydospores in soils favorable and unfavorable to wilt establishment. Phytopathology 62:273-277.
24. SMITH, S. N., W. C. SNYDER, and F. MOYNIHAN. 1970. Population of Fusarium oxysporum f. vasinfectum in field soil in relation to cotton wilt, p. 69-70. Proc. Beltwide Cotton Prod. Res. Conf.
25. STOTZKY, G. 1966. Influence of clay minerals on microorganisms. II. Effects of various clay species, homoionic clays and other particles on bacteria. Can. J. Microbiol. 12:831-848.
26. STOTZKY, G. 1966. Influence of clay minerals on microorganisms. III. Effect of particle size, cation exchange capacity, and surface area on bacteria. Can. J. Microbiol. 12:1235-1246.
27. STOTZKY, G., J. E. DAWSON, R. T. MARTIN, and C. H. H. TER KUILE. 1961. Soil mineralogy as factor in spread of Fusarium wilt of banana. Science 133:1483-1484.
28. STOTZKY, G., and R. T. MARTIN. 1963. Soil mineralogy in relation to the spread of Fusarium wilt of banana in Central America. Plant Soil 18:317-337.
29. STOTZKY, G., and L. T. REM. 1966. Influence of clay minerals on microorganisms. I. Montmorillonite and kaolinite on bacteria. Can. J. Microbiol. 12:547-563.
30. STOTZKY, G., and L. T. REM. 1967. Influence of clay minerals on microorganisms. IV. Montmorillonite and kaolinite on fungi. Can. J. Microbiol. 13:1535-1550.
31. STOVER, R. H. 1962. Fusarial wilt (Panama disease) of bananas and other Musa species. Commonwealth Mycol. Inst., Phytopathol. Paper 4. 117 p.
32. TAUBENHAUS, J. J., W. N. EZEKIEL, and D. T. KILLOUGH. 1928. Relation of cotton root rot and Fusarium wilt to the acidity and the alkalinity of the soil. Tex. Agr. Exp. Stn. Bull. 389. 19 p.
33. THORNTON, R. H. 1960. Growth of fungi in some forest and grassland soils, p. 84-91. *In* D. Parkinson and J. S. Waid (ed.), Ecology of soil fungi. Liverpool Univ. Press, Liverpool.
34. THORNTON, R. H. 1960. Fungi of some forest and pasture soils. N. Z. J. Agr. Res. 3:699-711.
35. THORNTON, R. H. 1965. Studies of fungi in pasture soils. I. Fungi associated with live roots. N. Z. J. Agr. Res. 8:417-449.
36. TOUSSOUN, T. A., W. MENZINGER, and R. S. SMITH, JR. 1969. Role of conifer litter in ecology of Fusarium: stimulation of germination in soil. Phytopathology 59:1396-1399.
37. WALKER, J. C., and W. C. SNYDER. 1933. Pea wilt and root rots. Wis. Agr. Exp. Stn. Bull. 424. 16 p.

38. WENSLEY, R. H., and C. D. MCKEEN. 1963. Populations of Fusarium oxysporum f. melonis and their relation to the wilt potential of two soils. Can. J. Microbiol. 9:237-249.

39. WINDELS, CAROL E., and T. KOMMEDAHL. 1971. Comparisons of Fusarium spp. and populations in cultivated and noncultivated soils. Phytopathology 61:1026 (Abstr.).

40. YOUNG, V. H. 1928. Cotton wilt studies I. Relation of soil temperature to the development of cotton wilt. Ark. Agr. Exp. Stn. Bull. 226. 50 p.

Soils Suppressive to Phytophthora Root Rot in Eastern Australia

PATRICIA BROADBENT and KENNETH F. BAKER—*Biological and Chemical Research Institute, New South Wales Department of Agriculture, Rydalmere, Australia, and Department of Plant Pathology, University of California, Berkeley.*

LOSSES DUE TO PHYTOPHTHORA ROOT ROT.—*Phytophthora cinnamomi* Rands is a devastating pathogen in Australia in horticultural crops like avocado (*Persea americana* Miller) and pineapple (*Ananas comosus* L.) (29), in ornamental nurseries on a wide range of plants (28), and in native eucalyptus forests. In 1972 in New South Wales, surveys showed that 10% of avocado trees had been killed by Phytophthora root rot (D. Leigh, *personal communication*).

Phytophthora cinnamomi was first reported in association with the death of native vegetation in Australia in 1956 (13), but it was not until 1965 that the tremendous losses in jarrah forests in Western Australia were attributed to Phytophthora root rot (23). In southwestern Australia, jarrah dieback has devastated 80,000 hectares or 5% of the commercially productive eucalyptus forest (18). The disease occurs in patches, often bounded sharply by healthy forest. First symptoms are chlorosis and rapid death of many plants of the understory and shrub layers, followed by chlorosis, twig dieback, and death of the dominant *Eucalyptus marginata* Sm. Aerial photographs show a strong association between the occurrence of dieback and roadways and natural water drainages (21), indicating that *P. cinnamomi* is an introduced fungus in this part of Australia.

Disease within an open sclerophyll forest dominated by *Eucalyptus* species in the Brisbane Ranges in Victoria has also been associated with *P. cinnamomi* spread during roadbuilding activities and through drainage channels (22, 30, 32). Dominants, codominants, intermediates, and suppressed trees are affected in the dieback in the mixed hardwood forests of Eastern Victoria (15). Species of the Macrantherae section of eucalyptus show resistance to dieback, whereas those of the Renantheroideae (stringy bark) species are susceptible. Lightly affected stands usually show poor crown development and dieback in a few of the major limbs. In more severe cases, a considerable part of the crown is lost, there are many dead limbs, and diseased trees closely resemble overmature trees. The natural regeneration is stunted and spindly. *Phytophthora cinnamomi* has also invaded dry sclerophyll woodland, heaths, and swamps in the National Park at Wilson's Promontory, Victoria (31).

Phytophthora cinnamomi has been associated with the death of native plants like *Banksia aemula* R. Br. following heavy rains in the low sclerophyll shrub woodland ("wallum") of the coastal lowlands of southeastern Queensland (20). Coaldrake (11) considered topographic control of perched water tables the major factor governing plant distribution in the coastal lowlands (wallum) and thought there was no steady ecotonal advance toward a stable climax. Pegg and Alcorn (20) pointed out that *P. cinnamomi* may be indigenous in this region and that its presence in this ecologically unstable region between heath and low sclerophyll shrub woodland may be one of the factors governing ecotonal oscillation.

Pratt et al. (26) have suggestd that *P. cinnamomi* may be indigenous to native forest areas in southeastern New South Wales. Observations made by Broadbent and Fraser in 1972 (*unpublished data*) in an isolated area of the northern Budawang Ranges of New South Wales led to similar conclusions. *Phytophthora cinnamomi* could be readily isolated from soil in wet areas where drainage was impeded by indurated sandstone and the vegetative cover was sedges and grasses. The summer-autumn period prior to the field study had been extremely wet, and the margins of these wet areas had extended to encompass sites where *Pultenea brunioides, Kunzia parvifolia* and *K.* aff. *capitata, Baeckia imbricata, Leptospermum scoparium* var. *rotundifolium, Mirbelia pungens, Bossiaea piamensis, Aotus ericoides, Platysace lanceolata, Acacia hamiltoniana,* and *Casuarina distyla* were growing. The increased soil wetness apparently favored the fungus, and disease occurred. *Phytophthora cinnamomi* was readily isolated from the roots of *Pultenea brunioides, Dillwynia sericea,* and *Acacia hamiltoniana.* This is a special area because of an unusual rock formation that ensures areas of permanently poor drainage that are subject to seepage. The recolonization of the intermittently saturated areas by susceptible shrubs suggests an intermittent or varying situation. *Eucalyptus sieberi,* which is very susceptible to root rot, was restricted to the deeper well-drained soils of the ridges. Pratt et al. (26) reported that the confined habitats of the fungus in the southern Budawang Ranges and its association with a resistant or tolerant vegetation in undisturbed inaccessible habitats suggest an equilibrium established over a long period of time.

Additonal evidence suggesting that *P. cinnamomi* may be indigenous to Eastern Australia and Papua includes: (a) the isolation of the A_1 mating strain from the high mountains of Papua (27), (b) the occurrence of both A_1 and A_2 mating types and the not uncommon isolation of the A_1 type in Eastern Australia (25), (c) the fact that trees and shrubs in vulnerable situations in Eastern Australia show a degree of resistance to *P. cinnamomi* compared with those in Western Australia (1), and (d) the occurrence of soils highly suppressive to *P. cinnamomi* root rot in southeastern Queensland (6).

The importance of soil disturbance in the development of Phytophthora root rot in native forests has been stressed by Pratt and Heather (24), but factors affecting the physiology of the host may be equally important. Observations in the Gosford area of New South Wales suggest that psyllids (*Glycaspis* spp.) feeding on blue gums (*Eucalyptus saligna* Sm.) in the valley floor reduce the sapwood content of trees (D. T. Hartigan, *personal communication*), making them more prone to injury from Phytophthora root rot.

FOREST SOILS AND ROOT ROT OCCURRENCE.—Soils in southern Australia, where root rot is severe, are generally infertile.

In Western Australia, most of the jarrah soils are lateritic podzolics, infertile, moderately acid(pH 5.5-6.5), and low in exchangeable cations. Texture, soil depth, and drainage vary, but shallow sands, silts, and gravelly clays over indurated laterite or dense clays predominate (21).

The soils in the Brisbane Ranges of Victoria are shallow lateritic gravelly grey to yellow-brown sands over mottled red and yellow sandy clays. There is an impervious mottled clay pan at depths of 0.6 to 2 m. The soils become waterlogged over long wet periods in winter and spring when perched water tables may develop (32). In the eucalyptus forests of eastern Victoria, dieback is worst on relatively infertile sites where drainage is impeded (15). The soils of Wilson's Promontory are derived from siliceous sands that are deficient in nitrogen and phosphorus and subject both to waterlogging and drying out (31).

The wallum soils of southern Queensland are acid and low in organic matter; total nitrogen is generally below 0.1%, and the total cation exchange capacity is low (11). The root rot soils of the Budawang Mountains are of similarly low fertility (Broadbent and Fraser, *unpublished data*).

Possibility of soils suppressive to Phytophthora root rot in northeastern Australia.—Apparently, then, infertile forest soils in southern Australia are conducive to Phytophthora root rot when excessive moisture is present, but suppressive soils might exist farther north in eastern Australia. As pointed out by Newhook and Podger (18), there have been few reports of *P. cinnamomi* in susceptible rainforest vegetation in Queensland, despite adequate opportunities for the spread of *P. cinnamomi* in drainage water from infected crops like avocado and pineapple. Little is yet known about the distribution of *P. cinnamomi* in the subtropics.

In surveys of avocado groves in northern New South Wales and southeastern Queensland in 1971 and 1972, two groves were found at Mt. Tamborine in which root rot was not a problem, even though climatic factors were favorable for disease development (6). Since the two fields were not free from the pathogen, their relative freedom from root rot was thought to have resulted from the effects of soil properties unfavorable to the disease.

DESCRIPTION OF AN AVOCADO GROVE WITH SOIL SUPPRESSIVE TO PHYTOPHTHORA CINNAMOMI ROOT ROT.—The grove studied consisted of 280 Fuerte avocado trees planted in 1940 and subsequently. Cover crops were planted in the newly cleared land for 2 years before avocados were planted; the mature cover crops were disked in along with fowl manure, dolomite, and superphosphate. An area of clean cultivation was maintained around the young trees; the cut cover crop was piled around, but was not in contact with the trees. This practice was continued until the trees were 5 years old, after which the fallen avocado leaves maintained organic matter on the soil under the trees. In general, the following practices have been maintained since. Fowl manure has been applied twice a year in March and September at 5 tons/hectare per application. Synthetic fertilizer (12.5% N, 12.5% P_2O_5, 20% K_2O) has been applied in March-April and October-November at 0.45 kg/tree per year of growth. Dolomite has been applied at 2.5 tons/hectare whenever pH has fallen below 6. Summer cover crops of *Dolichos lab-lab* L. (22 kg seed/hectare) and maize (*Zea mays* L.) (44 kg/hectare) have been sown from October to December following the first summer storms. They have been disked in before maturity. Winter cover crops of New Zealand blue lupine (*Lupinus angustifolius* L.) (44 kg/hectare) have been grown and disked in about May, when they are in flower.

PRELIMINARY APPRAISAL OF ANTAGONISTIC POTENTIAL OF SOIL TO PHYTOPHTHORA ROOT ROT.—Tests were conducted with a susceptible host growing in small rectangular plastic trays of soil to measure the ability of *P. cinnamomi* to cause root rot in suppressive and conducive soils. Fifty-four soils from many areas in Australia were tested for their antagonistic potential to Phytophthora root rot. These soils were treated with aerated steam at 60° C for 30 minutes or with flowing steam at 100° C for 30 minutes, or they were untreated. Small punnets were filled with treated soils seeded with avocado, jacaranda (*Jacaranda acutifolia* Humb. & Bonpl.), or *Eucalyptus marginata* Sm., or *E. sieberi* F. Muell, and placed in a growth cabinet running on a 27° C day/21° C night 12-hour cycle. Cornmeal-sand inoculum (1 g/punnet) or autoclaved succulent lucerne (alfalfa) stems 2 cm long on which *P. cinnamomi* had grown were mixed with soil in half the punnets in each treatment.

In soils that were conducive to root rot in the field, seedlings developed root rot in the three treatments where *P. cinnamomi* was added and often in untreated soil in the absence of additional inoculum. In the soil suppressive to root rot in the field, jacaranda and eucalyptus seedlings grew well in inoculated untreated soil and soil steamed at 60° C for 30 minutes but developed root rot in soil treated at 100° C and similarly inoculated. Plants grew well in suppressive soil after all three treatments when *P. cinnamomi* was not added. Microbial factors operating toward suppression in untreated soil survived treatment at 60° C, but not at 100° C for 30 minutes (8).

SPORANGIUM AND CHLAMYDOSPORE PRODUCTION BY PHYTOPHTHORA CINNAMOMI IN VARIOUS SOILS AND SOIL

EXTRACTS.—Sporangium production of *P. citrophthora* and *P. cinnamomi* is much reduced in the suppressive soil or soil leachate (6). For example, Coulter counts of zoospores of *P. citrophthora* from leachates of two conducive soils were over 4,000/ml each, whereas there were only from 290 to 774 zoospores/ml in leachates of four suppressive soils. The reduction of sporangium formation and zoospores may be sufficient to prevent effective invasion of the host without eliminating the pathogen from the soil under most environmental conditions. Failure of *P. cinnamomi* to form many sporangia in the suppressive soil did not result in greater chlamydospore production.

MICROORGANISMS IN SOIL EXTRACT.—Infection by zoospores occurs only when certain bacteria that stimulate sporangium production are present in soil solution (16), and sporangium production is often correlated with the inhibition of mycelial growth of *P. cinnamomi* caused by bacteria. Living soil microorganisms, particularly *Pseudomonas* spp. (2, 3, 10, 14, 17) and *Chromobacterium violaceum* (Schroeter) Bergonzini (33), have been implicated in the production of sporangia by the fungus in soil extracts.

The suppressive soil and soil extract contained more microorganisms, particularly *Bacillus* spp. and actinomycetes, than any other soil tested. It did not have as many *Chromobacterium* as some Western Australia forest soils (6). In soil suppressive to root rot, *Pseudomonas* numbers were generally as high or higher than in conducive soil. No single isolates of *Pseudomonas* or *Chromobacterium* from suppressive or conducive soils were found to be stimulatory to sporangium formation (7).

With the abundance of diverse microorganisms present in the suppressive soil and the apparent nonspecificity of microbial stimulation of sporangium formation, it is reasonable to assume that this soil did not lack stimulatory microorganisms. Under conditions of waterlogging, addition of large amounts of *Phytophthora* inoculum, or addition of an organic amendment (subterranean clover meal), the suppressive properties of the soil were reduced. It seemed probable that, under most field conditions, the organisms in the soil were suppressing the stimulatory bacteria or decomposing the stimulatory material.

Influence of microorganisms on sporangium formation.—Leachates from soils suppressive and conducive to sporangium formation were passed through Millipore filters (0.22 μ); the filter pads were washed in 9 ml sterile water, and 1 ml of the suspension formed the inoculum (Table 1). Sterile soil leachates did not stimulate sporangium formation, but sporangia were formed in abundance when a small amount of inoculum of conducive soil 30 was added. Stimulation is therefore associated with the presence of microorganisms. Inoculum of the suppressive soil provided much less stimulation to sporangial formation than did inoculum of the conducive soil (Table 1) and gave more sporangial breakdown, regardless of whether the soil extract was suppressive or conducive before sterilization.

GROWTH AND SURVIVAL OF PHYTOPHTHORA CINNAMOMI IN STEAM-AIR TREATED SOILS.—The mycelial growth and survival of *P. cinnamomi* were studied in soils suppressive and conducive to root rot. The soils were steam-air treated at 49°, 60°, and 100° C for 30 minutes or were left untreated. Moist soil was placed in glass vials over the fungus growing on a weak nutrient base. Sterile lucerne

TABLE 1. Effect of sterile leachates and microbial inoculum from soils conducive (C) and suppressive (S) to Phytophthora root rot upon zoospore production and sporangium breakdown (6)

Sterile leachate, 9 ml	Microbial inoculum, 1 ml	Effect upon numbers of zoospores/ml leachate and upon sporangium breakdown			
		P. citrophthora		*P. cinnamomi*	
		Zoospores	Sporangia	Zoospores	Sporangia
Soil 30, C	Soil 30, C	215	66/318[a]	340	9/110[a]
Soil 30, C	Soil 29, S	74	7/173	95	5/21
Soil 29, S	Soil 29, S	71	35/66	56	few[b]
Soil 29, S	Soil 30, C	964	21/354	156	many[b]

[a]Sporangia broken down/total sporangia.
[b]Chlamydospores formed.

TABLE 2. Chemical analyses of soils suppressive (S) and conducive (C) to Phytophthora root rot of avocado in the Mt. Tamborine area (6)

Soil	pH	Conduct., mmho/cm	Soluble P in ppm	Exchangeable m.e. %				Bulk density	Total nitrogen %
				Ca	Mg	Na	K		
10, C	5.3	0.63	4.6	9.7	3.0	.2	1.72	0.8	0.465
12, S	5.5	0.26	4.9	22.9	6.0	.8	.47	0.91	0.673
19, C	4.8	0.47	77.0	8.4	1.2	.1	.76	0.91	0.474
25, S	6.5	0.07	4.9	21.2	5.1	.2	1.74	0.98	0.617

TABLE 3. Chemical analyses of red basaltic rainforest and avocado soils and the presence of Phytophthora root rot

Soil	Vegetation[a]	Root rot status	pH	Conduct., mmho/cm	Soluble P in ppm	Exchangeable m.e. %				Total N %	Mn ppm	NH₄-N ppm	NO₃-N ppm	Clay[b]
						Ca	Mg	Na	K					
Alstonville	A	+	4.7	.30	14.4	2.3	1.3	1.3	1.28	.555	16	35	35	K
Rous Mill	RF	–	5.3	.19	9.1	9.5	2.7	1.5	1.16	.668	55	66	37	K
Wilson's Creek	A	+	4.5	.11	9.1	1.5	0.35	0.05	0.29	.347		0.2	27.2	K
	RF	–	5.9	.19	17.5	22.6	13.4	0.8	1.21	.349	9	30	19	K + M
Mt. Tamborine	A	+	4.8	.47	77.0	8.4	1.2	0.1	0.76	.474				V + K
	RF	–	5.3	.6	5.6	20.4	5.4	1.5	1.21	.857	55	132	34	
Mt. Tamborine	A	–	6.5	.07	4.9	21.2	5.1	0.2	1.74	.617	27	10.1	10.0	I + K
	RF	–	5.5	.26	5.6	19.0	2.9	1.0	0.4	.673	34	54	39	K
Mt. Tamborine	A	–	5.5	.26	4.9	22.9	6.0	0.8	0.46	.673	12	91	38	V + K
	RF	–	6.1	.31	2.5	26.6	8.8	1.2	1.05	.761		58	28	
Mt. Irvine	C	+	5.5	.13	2.7	5.7	1.6	0.2	1.36	.531	6.0	15.1	17.6	
	RF	–	6.0	.24	2.5	19.2	5.6	0.1	1.89	.585	47.6	59.4	0.83	V + K

[a] A = avocado, RF = rainforest, C = cultivated.
[b] K = kaolinite, M = montmorillonite, V = chloritized vermiculite, I = illite.

(*Medicago sativum* L.) stems were used to bait *P. cinnamomi* from the soil surface after 1, 2, and 6 weeks. It was more difficult to bait *P. cinnamomi* from the surface of untreated suppressive than conducive soils. The fungus grew readily through all soils treated at 49°, 60°, and 100°C. Isolation frequency of *P. cinnamomi* had declined by the 6th week in suppressive soil 29 treated at 60°C for 30 minutes. This confirmed previous experiments suggesting that antagonists associated with suppression of *P. cinnamomi* survived treatment at 60°C for 30 minutes. In other soils considered suppressive to root rot, isolation frequency at 6 weeks was high in soils treated at 60°C, but low in soils treated at 49°C. Apparently 49°C is near the survival threshold for antagonistic microorganisms in these soils.

CORRELATION OF SOIL ANALYSES AND PHYTOPHTHORA ROOT ROT IN RED BASALTIC SOILS.

—Borst (5) reported that the application of bulky organic mulches or sources of exchangeable calcium lowered the losses from root rot. These two properties are characteristic of the suppressive Mt. Tamborine soil 29 (Table 2). A survey was carried out in New South Wales to determine the correlation between root rot incidence and soil analyses. In general, red basaltic soils in which root rot losses are minimal have a high content of organic matter, sometimes with a humus layer 9 inches deep, high exchangeable calcium levels (20-27 m.e. %), with the calcium apparently tied in the organic cycle, a pH of 5.5-7.0, high levels of ammonium and nitrate nitrogen, and high biological activity. These trends are reflected in Table 2, which gives the chemical analyses of soils suppressive and conducive to root rot in the Mt. Tamborine area of southeastern Queensland. Soils in which root rot losses are minimal have calcium, nitrogen, and organic matter levels comparable to those in nearby rainforest soils (Table 3). Similar correlations have been made in Queensland by K. G. Pegg (*personal communication*). No association could be found between the clay mineral composition of the soil and its root rot status (Table 3).

The krasnozems of the north coast of New South Wales and southeastern Queensland carry a tropical rainforest vegetation. For a luxuriant rainforest vegetation to flourish on soils poorly supplied with certain plant nutrients, there must be a rapid turnover of nutrients. The saprophytic fungi and bacteria prominent in rainforests achieve this under the warm moist conditions by rapidly decomposing fallen leaf litter and softwood trees. Higher contents of organic carbon, nitrogen, and exchangeable cations in upper horizons of the profile suggest a closed cycle in which the cation requirements of the virgin vegetation are supplied by leaf litter. The replacement of this vegetation with crops appears to have interrupted the cycle and replaced it with an exploitive system (12, 19).

We postulate that drastic microbiological changes must accompany these physical and chemical changes, making the environment more favorable for the establishment and development of *Phytophthora* spp. We think that by applying fowl manure, dolomite, and synthetic fertilizers and by incorporating cover crops of *Dolichos lablab*, maize, and New Zealand blue lupine, the farmer has maintained the physical, chemical, and

microbiological properties of the suppressive soil 29 similar to those in the nearby virgin rainforest (soil 25). Higher organic matter levels provide a more complex biological antagonism to host pathogens, as well as improve soil structure and moisture-holding capacity.

The maintenance of high calcium levels in soils and the addition of calcareous amendments may affect root rot by improving soil structure, but there also appears to be a complex relationship between soil drainage and aeration, calcium supply, and the resistance of the host plant to fungal attack (4, 9).

Newhook and Podger (18) have suggested that if *P. cinnamomi* proves to be widespread in the subtropics without producing symptoms, it could be due to a lack of stress of plants whose growth rhythms are "in phase" with the distribution of rainfall. We think an ecological balance may have evolved in the rainforests of southern Queensland among the various hosts, the pathogen, the antagonists, the stimulators and inhibitors of sporangium formation, and the abiotic environment.

Cultural practice and Phytophthora root rot.—The forest soils in which Phytophthora root rot is a problem are of low fertility. The red basaltic soils under rainforest vegetation are of high fertility and are naturally suppressive to Phytophthora root rot. When these are cleared and planted to avocados, the fertility levels fall and the soils become conducive to the development of Phytophthora root rot.

To reduce the losses from Phytophthora root rot in avocados on red basaltic soils, it is imperative that clean nursery stock be used and that the chemical, physical, and microbiological properties of the rainforest soils be maintained. The latter can be achieved by the use of summer and winter cover crops disked in before maturity, the application of dolomite and lime to maintain pH and calcium levels, the application of fowl manure, and the maintenance of a balanced fertilizer program.

ACKNOWLEDGEMENTS.—The authors gratefully acknowledge the information and slides provided by K. Pegg, F. Podger, G. Weste, and G. Marks, and the constructive comment, interest, and vigorous support given to the work by L. R. Fraser. Soil analyses were done by J. Bull and M. Maguire.

LITERATURE CITED

1. ANONYMOUS. 1971. Root rot disease of native forests. Rural Res. CSIRO (Australia) 74:2-8.
2. AYERS, W. A. 1971. Induction of sporangia in Phytophthora cinnamomi by a substance from bacteria and soil. Can. J. Microbiol. 17:1517-1523.
3. AYERS, W. A., and G. A. ZENTMYER. 1971. Effect of soil solution and two pseudomonads on sporangium production by Phytophthora cinnamomi. Phytopathology 61:1188-1193.
4. BELLANY, G., W. A. HEATHER, and B. H. PRATT. 1971. The effect of calcium chloride treatment in reducing the rate of spread of Phytophthora cinnamomi within roots of Lupinus angustifolius, p. 3(e), 13-15. Proc. Austral. Plant Pathol. Conf., Hobart.
5. BORST, G. 1970. Selection, management of avocado soils. Calif. Citrograph 55:263-265.
6. BROADBENT, P., and K. F. BAKER. 1974. Behavior of

Phytophthora cinnamomi in soils suppressive and conducive to root rot. Austral. J. Agr. Res. 25:121-137.

7. BROADBENT, P., and K. F. BAKER. 1974. Association of bacteria with sporangium formation and breakdown of sporangia in Phytophthora spp. Austral. J. Agr. Res. 25:139-145.

8. BROADBENT, P., K. F. BAKER, and Y. WATERWORTH. 1971. Bacteria and actinomycetes antagonistic to fungal root pathogens in Australian soils. Austral. J. Biol. Sci. 24:925-944.

9. CHAPMAN, H. D. 1965. Chemical factors of the soil as they affect soil microorganisms, p. 120-141. *In* K. F. Baker and W. C. Snyder (ed.), Ecology of Soil-Borne Plant Pathogens. Univ. Calif. Press, Berkeley and Los Angeles.

10. CHEE, K. H., and F. J. NEWHOOK. 1966. Relationship of microorganisms to sporulation of Phytophthora cinnamomi Rands. N. Z. J. Agr. Res. 9:32-43.

11. COALDRAKE, J. E. 1961. The ecosystem of the coastal lowlands ("wallum") of southern Queensland. CSIRO (Melbourne) Bull. 283. 138 p.

12. COLWELL, J. D. 1958. Observations on the pedology and fertility of some krasnozems in northern New South Wales. J. Soil Sci. 9:46-57.

13. FRASER, L. R. 1956. Phytophthora cinnamomi attacking native plants. Austral. Plant Dis. Rec. 8:12.

14. MANNING, W. J., and D. F. CROSSAN. 1966. Effects of a particular soil bacterium on sporangial production in Phytophthora cinnamomi in liquid culture. Phytopathology 56:235-237.

15. MARKS, G. C., F. Y. KASSABY, and S. T. REYNOLDS. 1972. Die-back in the mixed hardwood forests of eastern Victoria; a preliminary report. Austral. J. Bot. 20:141-154.

16. MARX, D. H., and W. C. BRYAN. 1970. The influence of soil bacteria on the mode of infection of pine roots by Phytophthora cinnamomi, p. 171-172. *In* T. A. Toussoun, et al. (ed.), Root Diseases and Soil-Borne Pathogens. Univ. Calif. Press, Berkeley and Los Angeles.

17. MARX, D. H., and F. A. HAASIS. 1965. The induction of aseptic sporangial formation in Phytophthora cinnamomi by metabolic diffusates of soil microorganisms. Nature (London) 206:673-674.

18. NEWHOOK, F. J., and F. D. PODGER. 1972. The role of Phytophthora cinnamomi in Australian and New Zealand forests. Annu. Rev. Phytopathol. 10:299-326.

19. NICHOLLS, K. D., and B. M. TUCKER. 1956. Pedology and chemistry of the basaltic soils of the Lismore district, New South Wales. Austral. CSIRO Soil Publ. 7. 153 p.

20. PEGG, K., and J. L. ALCORN. 1972. Phytophthora cinnamomi in indigenous flora in southern Queensland. Search 3:257.

21. PODGER, F. D. 1972. Phytophthora cinnamomi, a cause of lethal disease in indigenous plant communities in Western Australia. Phytopathology 62:972-981.

22. PODGER, F. D., and D. H. ASHTON. 1970. Phytophthora cinnamomi in dying vegetation on the Brisbane Ranges, Victoria. Austral. For. Res. 4:33-36.

23. PODGER, F. D., R. F. DOEPEL, and G. A. ZENTMYER. 1965. Association of Phytophthora cinnamomi with a disease of Eucalyptus marginata forest in Western Australia. Plant Dis. Reptr. 49:943-947.

24. PRATT, B. H., and W. A. HEATHER. 1971. Association of environmental factors with Phytophthora cinnamomi disease in native forest, p. 3(e), 7-8. Proc. Austral. Plant Pathol. Conf., Hobart.

25. PRATT, B. H., W. A. HEATHER, and C. J. SHEPHERD. 1972. Transcontinental occurrence of A_1 and A_2 strains of Phytophthora cinnamomi in Australia. Austral. J. Biol. Sci. 25:1099-1100.

26. PRATT, B. H., W. A. HEATHER, and C. J. SHEPHERD. 1973. Recovery of Phytophthora cinnamomi from native vegetation in a remote area of New South Wales. Trans. Br. Mycol. Soc. 60:197-204.

27. SHAW, D. E., E. G. CARTLEDGE, and D. J. STAMPS. 1972. First records of Phytophthora cinnamomi in Papua New Guinea. Papua New Guinea Agr. J. 23:46-48.

28. TITZE, J. F., and C. R. PALZER. 1969. Host list of Phytophthora cinnamomi Rands with special reference to Western Australia. Dept. Nat. Dev., For. Timber Bull. Tech. Note 1. 58 p. 1970 addendum, 6 p.

29. VEITCH, R., and J. H. SIMMONDS. 1929. Pests and diseases of Queensland fruits and vegetables. Queensland Dept. Stock Agr., Brisbane. 198 p.

30. WESTE, G., D. COOKE, and P. TAYLOR. 1973. The invasion of native forest by Phytophthora cinnamomi. II. Post-infection vegetation patterns, regeneration, decline in inoculum, and attempted control. Austral. J. Bot. 21:13-29.

31. WESTE, G., and C. LAW. 1973. The invasion of native forest by Phytophthora cinnamomi. III. Threat to the National Park, Wilson's Promontory, Victoria. Austral. J. Bot. 21:31-51.

32. WESTE, G., and P. TAYLOR. 1971. The invasion of native forest by Phytophthora cinnamomi. I. Brisbane Ranges, Victoria. Austral. J. Bot. 19:281-294.

33. ZENTMYER, G. A. 1965. Bacterial stimulation of sporangium production in Phytophthora cinnamomi. Science 150:1178-1179.

Stump Inoculation: A Biological Control of Fomes annosus

J. RISHBETH—*Botany School, University of Cambridge, England.*

Fomes annosus is a pathogen of conifers that is widely distributed in the Northern Hemisphere and is considered to be the most serious cause of disease in British forests. The fungus is essentially root-inhabiting, but its mycelium has a very limited ability to grow through soil or survive in it. *Fomes annosus* breaks down both cellulose and lignin, and the extensive root decay that commonly develops may result in killing, wind-throw, loss of increment, or heart rot. In most regions, characteristic polypore fructifications are formed at or near ground level, and these liberate spores under a wide range of conditions.

The fungus is often present from the outset in young conifer stands growing on former woodland or forest sites, as it can survive for many years in stumps. In general, however, it is absent at first from similar stands established on treeless sites like pasture or arable land. Whereas trees on the former type of site often become infected as soon as their roots make contact with those of stumps containing *F. annosus*, those on the latter are generally affected by root disease only after stumps have been created, most commonly by thinning. For reasons that are not fully understood, killing attacks are often more severe on this latter type of site despite the later start. Trees may be infected directly by spores colonizing wounds, but infection via stumps is more common. In the latter process the mycelium passes from stump roots onto living ones that are in contact with or have actually united with them, and the fungus then often spreads radially from tree to tree in the same manner. Root infection of standing trees from free mycelium or spores in the soil can probably be discounted for practical purposes, at least under the conditions in Britain.

Stumps may be colonized by *F. annosus* either at the cut surface by air-borne spores, or from roots already infected at the time of thinning, or, less commonly, through roots not infected until after thinning (6, 9). Similar but more extensive opportunities for stump infection occur during clear felling at the end of a rotation. Chiefly because of the close correlation between stump and root infection, disease caused by *F. annosus* tends to be minimal in natural, unexploited forests and maximal in intensively managed ones. Many attempts have been made to limit stump infection by *F. annosus*. But once the fungus has become well established in stumps and has infected roots of adjacent trees, it is difficult or impossible to control its further spread. Almost certainly, some degree of biological control over *F. annosus* is exerted naturally at root surfaces, particularly in certain soils, but the chance of exploiting this possibility seems remote at present. Pine stumps provide highly selective substrates, in the sense that few fungi can colonize them initially, and are also very favorable for growth of *F. annosus*. Some of the most spectacular and destructive outbreaks of root disease have occurred in pine plantations, especially those on light-textured soils. Consequently, much of the work on stump protection has been carried out in them.

STUMP TREATMENTS.—At first, cut surfaces were treated with materials that hardly affected the underlying tissues and tended to exclude all types of fungi (Table 1). In these experiments, the proportion of stumps colonized by *F. annosus* and the number of trees subsequently attacked were closely related (16). This relationship held whether such stump infection varied naturally with the time of year at which trees were thinned or was influenced by stump treatment. Similarly, Ross (23) has recently reported a good correlation between the number of stumps of slash pine (*Pinus elliottii*) infected by *F. annosus* and losses in the stand 5 years later. Although the earlier experiments demonstrated the effectiveness of stump treatment, the methods used were disadvantageous, mainly because they retarded stump colonization by other fungi. Stumps tended to remain alive longer, and *F. annosus* could readily invade them from roots already infected or could invade them following mechanical damage. It is difficult to extract stumps after thinning, and injury to adjacent root systems is too great for the method to be practicable at this stage.

More recent methods of treatment were superior because stump tissues were killed and their selectivity thus destroyed. Potentially useful substances like disodium octaborate, urea, and ammonium sulfamate vary in their toxicity to stump fungi and in the rate at which they break down or are leached from stump tissues (17). *Fomes annosus* is generally prevented from colonizing the stump surface, either because of this toxicity, or through competition from saprophytic fungi that become established as a result of the treatment, or by a combination of these effects. However, by no means all substances controlling *F. annosus* at the surface of a stump can prevent its entry from infected roots. Effectiveness appears to depend mainly upon the type of microflora that develops and in particular whether another wood-rotting basidiomycete, *Peniophora gigantea*, is favored by the treatment. It seemed logical, therefore, to test the effect of inoculating pine stumps with this fungus.

INOCULATION OF PINE STUMPS.—*Peniophora gigantea* was already known to replace *F. annosus* in pine

TABLE 1. Proportion of stumps infected by *Fomes annosus* in two plantations of Scots pine (*Pinus sylvestris*) in East Anglia, England, and number of trees subsequently attacked

Plantation A (stumps not treated)

	Month of thinning					
	Feb	Mar	May	June	Sept	Nov
Stumps with *F. annosus* 1 year after cutting, %	52	24	0	0	28	0
Number of trees affected 7 years after cutting[a]	60	43	7	4	31	7

Plantation B

	Stump treatment			
	None	Creosoted	Painted	Extracted
Stumps with *F. annosus* 1 year after cutting, %	36	4	3	
Number of trees affected 7 years after cutting[b]	30	5	5	5

[a]In one plot of 0.4 ha.
[b]Mean of three plots each of 0.9 ha.

stumps under natural conditions (14), but early attempts to exploit this were only partially successful. Further observations on such replacement were made by Meredith (10, 11), who showed that *F. annosus* is unlikely to become dominant in stumps also colonized by *P. gigantea* unless the balance of inoculum at the cut surface is greatly in its favor. With most mixed inoculations, whether natural or artificial, *F. annosus* first enters the stump but is later progressively replaced by *P. gigantea*. This antagonistic action of *P. gigantea* probably results from hyphal interference; i.e., it is a short-range antibiotic effect (7).

By means of an improved inoculation method, it was shown that *P. gigantea* was the most successful competitor of *F. annosus* at the cut surface of Scots pine stumps (19). In other trials with stumps of both Scots pine and Corsican pine (*Pinus nigra* var. *maritima*), *P. gigantea* gave good control at all times of the year. Moreover, it soon grew into lateral roots and rapidly decomposed stump tissues. It replaced *F. annosus* in roots colonized by this fungus after felling, but generally failed to do so in roots invaded by the parasite before felling; these roots were resinous as a result of the host response to infection. Asexual spores (oidia) of *P. gigantea* formed in culture on malt agar provided satisfactory inoculum. A dosage of 1×10^4 viable spores controlled any naturally occurring infection by *F. annosus* at the surface of stumps having a wood diameter of 16 cm. Contamination of the suspension is rarely a problem because the stump tissues are so selective (19).

Air-borne spores of *P. gigantea*, like those of *F. annosus*, are common and widely distributed in Britain (18). The incidence of natural stump infection by *P. gigantea* is somewhat erratic, however. In some localities, spores of *P. gigantea* may be absent, whereas those of *F. annosus* are present: a good deal is likely to depend on the history of the area. Even in places where *P. gigantea* is abundant, periods of dry weather tend to restrict spore production, since its sporophores desiccate more rapidly than those of *F. annosus*. During the hot dry summer of 1959, for instance, an unusually low ratio of *P. gigantea* to *F. annosus* spores (1:30) was detected at Cambridge during a period when northeasterly winds were transporting them from Thetford forest, about 40 km away. That natural stump infection may reflect such a situation is suggested by data obtained during 1955. In

one locality in East Anglia, *P. gigantea* became dominant in pine stumps created during a moderately wet period but was almost absent from those cut during a subsequent dry spell. Over this latter period, *F. annosus* became dominant instead (16). Therefore, under conditions in Britain, it seems necessary to inoculate every stump with *P. gigantea* in order to protect pine stumps adequately against infection by *F. annosus*. From recent observations on deposition of *P. gigantea* spores in the southeastern United States (23), it seems very doubtful whether natural inoculum of the fungus would give consistent protection in this area either.

If stump inoculation is to be practicable, inoculum must be produced on a relatively large scale. Spore suspensions from cultures grown on wood or agar were satisfactory for experimental work, but a much more compact source of inoculum seemed essential for routine use in forests. Dehydrated tablets were prepared; they contained about 1×10^7 viable spores and would dissolve in water to give a suspension sufficient for treating about 100 stumps. Their storage life, about 2 months at 22° C, represented the period over which the number of spores exceeded 3×10^6; this figure was based on the number of stumps to be treated, the dosage per stump, 1×10^4, and a threefold safety factor (19). A nontoxic dye could be incorporated with the suspension for marking treated stumps.

The first experiments demonstrating the effectiveness of stump inoculation were completed by 1960. By this time, the Forestry Commission in Britain was committed to a policy of stump protection against *F. annosus* for all conifers (20). A large proportion of plantations had been established on nonwoodland sites and were therefore subject to little or no root infection initially. It was hoped that stump treatment applied during the critical early thinnings would restrict entry of *F. annosus* and the extent to which it subsequently increased. Even in partially affected stands, stump treatment might well limit the number of new infection foci.

The advantages of stump inoculation are several. The spore suspension is not toxic to mammals, and there is no risk of side effects from accumulation of residues. Decay of stumps is rapid in contrast to some other methods that inhibit it. Moreover, with stump inoculation, *F. annosus* is controlled as well at the cut surface and probably better in stump roots. *Peniophora gigantea* often produces sporophores 6 to 12 months after inoculation, thus increasing the number of its spores in the air. It has not been recorded as attacking healthy trees and is widely distributed, so *P. gigantea* can be used in many parts of the world. Even though some chemical treatments favor biological control by promoting growth of fungal competitors, this secondary effect ultimately depends on the presence of such fungi, whereas control by stump inoculation is independent of the air spora or other sources of inoculum. The cost of the treatment is comparable with that of chemical methods, varying from £1.25 to £3.75 per ha. The chief disadvantages arise because the material used is living: it progressively loses viability during storage (the higher the temperature, the more rapid the loss). The method might be less satisfactory than a chemical one in extremes of hot or cold weather; e.g., through death of *P. gigantea* or its failure to grow sufficiently in desiccated tissues. However, this does not appear to be a serious problem in Britain.

When trials set up by the Forestry Commission Research Branch confirmed the effectiveness of stump inoculation, it was decided that the advantages of the method outweighed the disadvantages, and it was introduced in some parts of East Anglia in 1962. Since then, the use of *P. gigantea* has been extended to all suitable pine forests managed by the commission, an area totaling about 62,000 ha (24). Inoculum is prepared commercially in fluid form and is distributed in sachets. It has a storage life of about 4 months, provided the storage temperature does not exceed 20°C. Sachets are tested in a Forestry Commission laboratory before a batch is released, and later again during storage. The contents of a sachet are mixed well with 5 liters of water, and dye is added. Directly after felling, the spore suspension is applied to the stump surface from a spouted polyethylene container and is then distributed by brush. Spot checks are made on suspensions being used in the forest, and stump samples are collected periodically to determine the extent of colonization by *P. gigantea*. There is no formal proof at present that stump inoculation during thinning reduces the amount of killing by *F. annosus* in pines, although from the evidence presented in Table 1, it would certainly be expected to. It is hoped that an experiment set up by the Forestry Commission Research Branch will provide information about this point within the next few years.

A collection of *P. gigantea* isolates is maintained at Cambridge, and the one being used on a large scale can be changed periodically. Prior testing is desirable because effectiveness in colonizing stumps varies with the isolate. In a preliminary trial with eight isolates, for instance, the mean area of cross section occupied by *P. gigantea* in the upper portion of Scots pine stumps 8 months after standard inoculation varied from 67 to only 18%. Results obtained with the four most promising isolates, from an experiment in which stumps of Corsican pine were each inoculated with 1×10^4 spores, are given in Table 2. Although the extent of colonization by *P. gigantea* varied, good control over natural *F. annosus* infection was obtained throughout. Results from such experiments and from tests on fresh pine disks in the laboratory were poorly correlated, so field trials are essential, at least for the present.

TABLE 2. Natural colonization of Corsican pine (*Pinus nigra* var. *maritima*) stumps by *Fomes annosus* after inoculation with various isolates of *Peniophora gigantea*

Peniophora isolate	Area of stump section (%) colonized after 8 months by	
	F. annosus[a]	P. gigantea[a]
None	20	55[b]
A	1	87
B	0	81
C	1	70
D	0	66

[a]Mean of 12 replicates.
[b]Natural infections.

TABLE 3. Influence of supplementary treatments upon colonization of conifer stumps inoculated with *Peniophora gigantea*

Supplementary treatment	Area (%) of outer wood in stump cross section colonized by *Peniophora* after 1 1/2-2 years in		
	Sitka spruce[a]	Douglas fir[a]	European larch[a]
None	3	20	34
5% ammonium sulfate	3	73***	21
5% ammonium sulfamate	35**[b]	100***	32
10% sodium thiosulfate	1	49*	62

[a]Mean of 10 replicates.
[b]*, **, and *** indicate significance at the 5%, 1%, and 0.1% levels, respectively.

In some circumstances, stump inoculation with *P. gigantea* will probably prove useful after clear felling of pines. This might be the case, for example, where the plantation has little or no root infection but where a risk of stump colonization by air-borne spores of *F. annosus* exists. Stump inoculation is of doubtful value where a plantation is heavily infected at the time of felling, however. The Forestry Commission Research Branch has set up experiments in East Anglia after clear felling first-rotation stands of Scots pine. After carrying out various treatments, the sites were replanted with Corsican pine (5). After 12 years the mean percentages of trees killed in the various plots of one experiment were: no treatment (control) 29.5; stumps inoculated with *P. gigantea*, 24.5; stumps treated with ammonium sulfamate, 31.5; stumps treated with urea, disodium octaborate, or creosote, 35.9, 35.2, and 36.7, respectively; and stumps removed, 4.1. Inoculation was significantly better than treatment with substances other than ammonium sulfamate, which generally favors colonization by *P. gigantea*. In a similar experiment at another site, representative mean percentages were: no treatment, 36.4; inoculation, 24.4; stumps removed, 5.9. Therefore, although stump inoculation can be expected to reduce killing slightly under these conditions, it may not do so to an acceptable extent. This failure is almost certainly due to the inability, mentioned above, of *P. gigantea* to replace *F. annosus* in very resinous roots. At sites where stump extraction is feasible and economical, this method is superior to inoculation. On the basis of earlier observations (16), the low level of initial infection following stump extraction would probably not lead to a serious increase of *F.*

annosus during the rotation. Therefore, provided stump protection was adequate during thinning, extraction might be unnecessary after the next clear felling.

DEVELOPMENTS IN OTHER COUNTRIES.—Apparently, stump inoculation has been used on a forest scale only in Britain. In many countries, no form of stump treatment against *F. annosus* has been introduced because damage by the fungus is not considered sufficiently serious, or because the method is thought to be uneconomical or incapable of dealing adequately with the intractable problem of long-established root disease. In a few countries, stumps are protected by chemical means, as indeed species other than pines are in Britain; the use of borax in the southern United States is an example. However, some interest in the use of *P. gigantea* has been expressed elsewhere. In the USSR, inoculation of freshly cut pine stumps with this fungus protected them from infection by *F. annosus* and reduced the number of potential infection sources in plantations (12). Delatour (3) reported an experiment in France in which reasonably good control of infection was similarly obtained in stumps of maritime pine (*Pinus pinaster*). Driver and Ginns (4) found that inoculating stumps of slash pine with *P. gigantea* controlled colonization by *F. annosus* in southern Georgia, although infection was in any case prevented by high temperatures during summer. In this region, it seems probable that stump inoculation will be combined with mechanical harvesting of pines in the near future.

It has also been suggested that development of *P. gigantea* might be encouraged in plantations without necessarily inoculating every stump (15). Since earlier observations had shown spores of *P. gigantea* to be present in the air at all seasons in Georgia, Boyce (2) thought it might be practicable to cut scattered understory pines several months before regular thinning to achieve a natural increase of its sporophores. Sufficient spores of *P. gigantea* might then be produced to minimize stump colonization by *F. annosus* at the time of thinning, except during extended dry periods. Blakeslee and Stambaugh (1), who confirmed that low humidity sharply curtails spore release by *P. gigantea*, also suggested that it may be possible to synchronize fruiting of this fungus with thinning operations. Further experiments (involving monthly thinning for example) should show whether such dependence on local sources of *P. gigantea* spores is justified in particular areas. The intriguing possibility exists that in some places climatic conditions so favor consistent spore release by the fungus that control over infection of pine stumps by *F. annosus* is virtually automatic.

STUMP INOCULATION FOR SPECIES OTHER THAN PINE.—The extension of stump inoculation to other types of trees has not yet progressed beyond an experimental stage. In East Anglia, nine basidiomycetes were tested for their ability to compete with *F. annosus* in stumps of Norway spruce (*Picea abies*). Stumps were inoculated with spores of *F. annosus* and with spores or mycelial fragments of each of the other fungi (21). After 16 months the mean area of outer wood colonized by *F. annosus* in cross sections taken from the tops of stumps ranged from 20 to 40% for the control and for stumps inoculated with ineffective fungi such as *Hypholoma fasciculare*; it ranged from 7 to 12% for those inoculated with partially effective species such as *Trechispora brinkmanni*, but was nil where *Polyporus adustus* or *Peniophora gigantea* had been used. This experiment was performed in April. Kallio (8) reported that natural infection of Norway spruce stumps by air-borne spores of *F. annosus* was controlled in Finland at all times of the year by *P. gigantea*, but was not consistently controlled by four other fungi. When he deliberately inoculated stumps with *F. annosus*, *P. gigantea* again gave virtually complete control in April and subsequently until September, but was not so effective at other times of the year. Such experiments repeated at intervals throughout the season are especially valuable. The prospects for inoculation of Norway spruce stumps are therefore reasonably good, should the need arise for such a method. Direct inoculation of standing spruce with a competing fungus (13) has been suggested as a possibility for limiting heart rot by *F. annosus*, but this method seems more problematical, although it is a remarkably challenging idea.

A method of biological control for stumps of Sitka spruce (*Picea sitchensis*) would be very useful in Britain, but the prospects are less than encouraging. In an experiment similar to that outlined for Norway spruce, only two fungi of 13 controlled *F. annosus* infection in inoculated stumps. Even these species, *P. gigantea* and *Stereum sanguinolentum*, had colonized only 17 and 24%, respectively, of the outer wood after 16 months, and the central wood had not been colonized at all. The major weakness of this situation is that, under certain circumstances, later natural infection of the cut surface by *F. annosus* might not be controlled, nor might its entry from infected roots. As with pines, isolates of *P. gigantea* vary considerably in their ability to colonize these stumps, but so far none has been found that grows consistently well in them. Some improvement may be obtained with a supplementary treatment, and this is also true for stumps of some other conifers. Table 3 records the results of inoculating stumps of three kinds of conifer with *P. gigantea* and then treating them in different ways. Treatment with 5% ammonium sulfamate improved colonization of stumps of Sitka spruce and Douglas fir (*Pseudotsuga menziesii*), whereas with those of European larch (*Larix decidua*) it did not. Without any supplementary treatment the fungus hardly colonized stumps of Sitka spruce and colonized those of Douglas fir rather poorly. Throughout a number of such experiments, there has been a strong tendency, as might be expected, for *F. annosus* to be controlled better where growth of *P. gigantea* was more extensive.

Whether *P. gigantea*, though convenient, is the best fungus to use with such conifers is doubtful. The search for suitable competitors is complicated by the fact that laboratory screening (on freshly cut wood disks, for example) is unreliable except perhaps as a means of eliminating very ineffective fungi, hence the emphasis on stump experiments. Ideally it would be desirable to use fungi for which no supplementary treatment is necessary, but supplements may well be essential, at least with Sitka spruce. For such a method to be practicable, treatment and inoculation would have to be combined. For instance, if *P. gigantea* spores are suspended in 5% ammonium sulfamate, their viability is unaffected over a period of 8 hours. It might be argued that treatment of stumps with 5% ammonium sulfamate alone would give the desired effect, but, as with pines, this would only apply when sufficient natural inoculum of an effective competitor was present.

It is conceivable that biological control of *F. annosus* will be needed for stumps of broad-leaved trees like birch, because in some areas such stumps are serious sources of infection. If so, a supplementary chemical method would usually be required to kill stump tissues and promote growth of potential competitors. Observations so far suggest that in this respect ammonium sulfamate is generally more effective than 2,4,5-trichlorophenoxyacetic acid. In one experiment with birch (*Betula verrucosa*), good colonization by *Polystictus versicolor* and *Polyporus adustus* was obtained in stumps inoculated with these fungi and then treated with 40% ammonium sulfamate (22). The extent to which these fungi suppress infection by *F. annosus* has not yet been determined, however.

In conclusion, it appears that creating a stump offers a potentially useful opportunity for influencing the fungal colonization of the woody tissues above and below soil level. Inoculation of stumps is much more likely to succeed than is inoculation of soil, with its large resident microflora, because of the initial absence or scarcity of microorganisms in the fresh stump and because much of the tissue is living and therefore selective. Even dead wood provides a poor substrate for most microorganisms. The established use of *P. gigantea* for protecting pine stumps shows that this method of biological control is practicable and suggests that in appropriate circumstances it could be extended to other fungus-tree stump combinations for control of *F. annosus*.

ACKNOWLEDGEMENT.—The author is indebted to B. J. W. Greig of the Forestry Commission Research Branch for providing information about experiments still in progress.

LITERATURE CITED

1. BLAKESLEE, G. M., and W. J. STAMBAUGH. 1973. The influence of environment upon the physiology of Peniophora gigantea on Pinus taeda. Proc. Fourth Int. Conf. on Fomes annosus, Athens, Georgia.

2. BOYCE, J. S. 1966. Sporulation by Peniophora gigantea with reference to control of Annosus root rot. Forest Sci. 12:2-7.

3. DELATOUR, C. 1972. Essai de traitement d'un peuplement de pins maritimes contre le Fomes annosus à l'aide du Peniophora gigantea. Rev. Forest. Francaise 24:201-202.

4. DRIVER, C. H., and J. H. GINNS. 1969. Ecology of slash pine stumps: fungal colonization and infection by Fomes annosus. Forest Sci. 15:2-10.

5. GREIG, B. J. W., and D. A. BURDEKIN. 1970. Control and eradication of Fomes annosus in Great Britain, p. 21-32. Proc. Third Int. Conf. on Fomes annosus, IUFRO Sect. 24, Aarhus, Denmark.

6. HODGES, C. S. 1964. The effect of competition by Peniophora gigantea on the growth of Fomes annosus in stumps and roots. Phytopathology 54:623 (Abstr.).

7. IKEDIUGWU, F. E. O., C. DENNIS, and J. WEBSTER. 1970. Hyphal interference by Peniophora gigantea against Heterobasidion annosum. Trans. Br. Mycol. Soc. 54:307-309.

8. KALLIO, T. 1971. Protection of spruce stumps against Fomes annosus (Fr.) Cooke by some wood-inhabiting fungi. Acta Forest. Suecica 117:1-20.

9. KUHLMAN, E. G. 1969. Numbers of conidia necessary for stump root infection by Fomes annosus. Phytopathology 59:1168-1169.

10. MEREDITH, D. S. 1959. The infection of pine stumps by Fomes annosus and other fungi. Ann. Bot. (London) 23:455-476.

11. MEREDITH, D. S. 1960. Further observations on fungi inhabiting pine stumps. Ann. Bot. (London) 24:63-78.

12. NEGRUTSKII, S. F. 1963. On the use of antagonistic fungi for combating the fungus Fomes annosus. Microbiology (Moscow) 32:632-635.

13. RICARD, J. L. 1970. Biological control of Fomes annosus in Norway spruce (Picea abies) with immunizing commensals. Stud. Forest. Suecica 84:1-50.

14. RISHBETH, J. 1950. Observations on the biology of Fomes annosus, with particular reference to East Anglian pine plantations. I. The outbreaks of disease and ecological status of the fungus. Ann. Bot. (London) 55:365-383.

15. RISHBETH, J. 1952. Control of Fomes annosus Fr. Forestry 25:41-50.

16. RISHBETH, J. 1957. Some further observations on Fomes annosus Fr. Forestry 30:69-89.

17. RISHBETH, J. 1959. Stump protection against Fomes annosus. II. Treatment with substances other than creosote. Ann. Appl. Biol. 47:529-541.

18. RISHBETH, J. 1959. Dispersal of Fomes annosus Fr. and Peniophora gigantea (Fr.) Massee. Trans. Br. Mycol. Soc. 42:243-260.

19. RISHBETH, J. 1963. Stump protection against Fomes annosus. III. Inoculation with Peniophora gigantea. Ann. Appl. Biol. 52:63-77.

20. RISHBETH, J. 1967. Control measures against Fomes annosus in Great Britain, p. 299-306. XIV IUFRO Congress, Sect. 24, Munich, Germany.

21. RISHBETH, J. 1970. The possibility of stump inoculation for conifers other than pines, p. 110-120. Proc. Third Int. Conf. on Fomes annosus, IUFRO Sect. 24, Aarhus.

22. RISHBETH, J. 1972. Biological control for root diseases of trees, p. 29-35. *In* V. J. Nordin (ed.), Biological control of forest diseases. XV IUFRO Congress, subject Group 2, Gainesville, Florida, 1971.

23. ROSS, E. W. 1973. Fomes annosus in the southeastern United States: relation of environmental and biotic factors to stump colonization and losses in the residual stand. USDA Forest Serv. Tech. Bull. 1459. 26 p.

24. WEBB, P. J. 1973. An alternative to chemical stump protection against Fomes annosus in state and private forestry. Scot. Forest. 27:24-25.

PART VII

RESISTANCE IN PLANTS TO ROOT-INFECTING FUNGI

Breeding for Resistance to Soil-Borne Pathogens

J. C. WALKER—*Retired Professor of Plant Pathology, University of Wisconsin, Madison.*

About the turn of the 20th century, vascular fusaria were rapidly increasing in importance in several crops in the United States. Following the pioneer work of Orton and Bolley on cotton and flax, respectively, the use of resistance as a major control for these diseases came rapidly to the forefront. Perhaps no other group of diseases has been so successfully reduced to a minor place in crop production by this method. Another vascular disease, bacterial wilt or Granville wilt of tobacco, was successfully controlled a little later.

Pioneer work in the control of a cortical root rot, also started early in this century, is that with Thielaviopsis disease of tobacco initiated by James Johnson. Black shank of tobacco, incited by *Phytophthora*, has also been reduced by this means.

It is not my purpose to go into an extensive review of the numerous successful and unsuccessful programs for controlling soil-borne pathogens by development of resistant varieties. Suffice it to say this means of control is still one of the important ones and in some cases, as with vascular fusaria, it is the only practical one.

There is nothing unique about the ups and downs of resistance programs in this area. In some cases, monogenic, vertical, or specific resistance has been advantageous, especially when the organism involved is quite stable in pathogenicity, e.g., cabbage yellows, pea wilt, tomato wilt. In other cases, multigenic resistance is the only one available and is a major means of control, e.g., with bean root rot, tobacco root rots, and some vascular wilts.

What I have to say now is based largely on my own experiences during a 50-year period. In some instances, the methods of assay for resistance have been improved in precision and speed. Examples of this are cabbage yellows (*Fusarium oxysporum* f. *conglutinans*) and tomato wilt (*F. oxysporum* f. *lycopersici*). By close control of root temperature, plants can be tested in the seedling stage, and large populations can be screened in a short period of time. This is particularly valuable where multiple resistance (i.e., resistance to several diseases) is the object. In cabbage, for instance, yellows resistance has now been combined with resistance to four other diseases. In tomato, several resistances have been combined with wilt resistance.

In cabbage yellows, this refined technique distinguishes between the higher more stable monogenic vertical resistance and the less stable multigenic horizontal resistance. These cannot be distinguished satisfactorily in field tests. The same may be said for tomato wilt.

Speedy indoor precision tests are not always satisfactory. For one thing, they tend to be extreme and sort out only highly resistant individuals; many usable resistant individuals are lost. In general, I believe it is fair to state that horizontal multigenic resistance is not as stable under the range of environmental conditions to which the crop is ordinarily exposed as are many cases of monogenic resistance. For this reason, they may be better sought for and assayed by field tests than by highly controlled indoor methods. As I have said on numerous occasions, if we had only the commonly occurring multigenic resistance to cabbage yellows to deal with, we would probably be using field resistance tests primarily. Incidentally, in the latest multiple resistant cabbage variety, released by Williams et al., the selection and assaying for resistance to the four other diseases (mosaic, internal tipburn, Rhizoctonia head rot, and powdery mildew) have been done entirely in the field.

In other words, the method of assay must evolve along with an increase in the understanding of the disease concerned. In the search for new sources of resistance, the field test should normally be applied first and over as wide a range of samples of the host (and possibly its close relatives) as possible. This has been done in numerous cultivars like tomato, pea, and potato and in numerous field crops. As assay methods are refined, they are not necessarily improved in all respects. In all likelihood, they are improved for the isolation of one type of resistance, but they reduce or eliminate the possibility of detecting one or more other types.

An interesting experience I may cite has to do with the assay of young onion seedlings for resistance to pink root (*Pyrenochaeta terrestris*). If seed is sown in sand culture to which inoculum has been added, the disease progresses rapidly at the desirable temperature. However, if the temperature is too high and/or the inoculum is too heavy, all seedlings in a population showing a percentage of field-resistant plants will succumb entirely. By reducing the temperature and inoculum to suboptimum levels, a rapid assay equivalent to field resistance tests was devised whereby thousands of seedlings were assayed every 30 days. A third variable, which was traced to seed vitality, was found later. If seeds were matured under adverse conditions so they were shriveled rather than plump, many resistant plants acted as susceptibles. This is just another example of the need for continuous attention to the assay method, whether it be in the field or indoors.

One of the most perplexing problems in breeding for resistance is the variability of the pathogen in regard to its selective pathogenicity. Some root pathogens are just as ornery to the breeder in this regard as are the true rusts. In my 50 years of studying cabbage diseases, the yellows organism has behaved admirably, and up to now only one pathogenic race that attacks cabbage has been defined. On the other hand, the club root organism (*Plasmodiophora brassicae*) has proved to be just the opposite. Just as a promising selection gave excellent performance in several locations, another race appeared right in my own backyard. Furthermore, I had no barberry bush or south wind to blame. I do not know what occurred. It may have been a recent mutation or it may have been a very rare strain that remained in the background until I provided the right substrate upon which this obligate parasite multiplied rapidly. Obviously, this possibility must be kept uppermost in dealing with any disease, both in the early stage of screening for resistant plants and in the final stages of transferring resistance to desirable cultivars. Where the organism is extremely variable in selective pathogenicity, success in the development of resistant varieties is likely to be spotty and often may be impossible.

Despite some discouragement in its study, resistance still holds a properly important place in the field of root rot diseases. Although high resistance may not always be attainable, even moderate resistance may help make other measures more effective. For this reason alone, it must retain a prominent position. Another major reason for continued attention is that breeding of crop plants is going on with increased velocity. The demands of greater uniformity in maturity naturally result in the narrowing of gene composition. This is especially true in the synthesis of inbreds for hybrid varieties. Our well-known catastrophic experiences with oat blight and southern leaf blight of corn have served to warn pathologists of their continuous responsibility to plant improvement. We must be ever on the alert for the detection of undescribed pathogenic races. And we must keep ever before us that sorting and rearranging of genes by the breeder, for other purposes may lead to entirely new combinations of susceptibility to currently known races.

In conclusion, I should like to quote a statement I have published elsewhere: "While it would be an ideal situation if we could look forward to having, eventually, disease-resistant varieties in all crop plants, it is undoubtedly too much to expect in view of the variability of some pathogens. The need for resistant varieties is greatest in the cases of diseases which do not lend themselves to other remedial measures. To be effective, resistance to a given disease must be combined with the currently desirable agronomic or horticultural characteristics of standard varieties and with resistance which may have been established for one or more other diseases. Continual adjustment is therefore needed to meet changing crop requirements as well as changes in pathogenicity of the causal organism. Disease resistance, to be applied as a control program, therefore, is a continuous process which is most effective when coordinated with the general improvement of the crop concerned. The responsibility of plant pathology in plant improvement is, indeed, a large and important one. It should emphasize the continuous study of the pathogens concerned and their variability; their reaction to environment; the nature of host resistance, its heredity, and its reaction to environment; and the improvement of testing and screening techniques, whereby resistance can be most readily and effectively integrated with a general plant-improvement program."

Sources and Nature of Verticillium Wilt Resistance in Some Major Crops

STEPHEN WILHELM, *Department of Plant Pathology, University of California, Berkeley.*

Rather than attempt to present a compendium of resistance sources and to review the reported biochemical mechanisms of resistance, I shall discuss the sources and nature of resistance in a general way and attempt to synthesize areas of knowledge and provide guidelines useful to breeders of wilt-resistant crops. There are also pitfalls I hope to expose. I shall speak largely from firsthand experience and from years of observations. Of necessity, I must overlook some areas of research and some crops and draw heavily upon California experience.

HISTORICAL BACKGROUND.—Verticillium wilt, caused by *Verticillium albo-atrum* Rke. and Berth. (Fig. 1) (I use the designation to include both microsclerotial and dark mycelial forms), is one of the major hazards of dicotyledonous crops of temperate world agriculture. Important floricultural crops like roses, chrysanthemums, strawflowers, and other daisy types also are affected. The disease achieves maximum destructiveness in semiarid regions of irrigated agriculture where rain falls primarily in winter and spring, and summer and fall are dry. The disease intensifies under monoculture of major crop hosts like cotton, mint, peppers, potatoes, safflower, tobacco, and tomatoes and on weeds of the genera *Xanthium*, *Solanum*, *Physalis*, and *Senecio*.

The fungus is spread by contaminated seed (13), within one-seeded dry fruits (23, 26), by fruits of weeds and spinach (12, 20), by vegetative cuttings and tubers (9, 11), by scions and buds (22), by wind and surface ground water (10), and by the soil itself, which may contain up to 100 or more microsclerotia per gram (1). Besides being infected through both foliage and roots, the potato (*Solanum tuberosum* L.) (34), through its tubers, probably has served as the primary vehicle of the spread and establishment of the *Verticillium* fungus throughout the temperate world. For example, first identifications of Verticillium wilt of cotton in California in 1927 (27) and of tobacco in New Zealand in 1959 (36) were associated with previous potato cultivation. Not uncommonly, wild lands, virgin areas, and lands long in nonirrigated, nonhost orchards like apple and pear show evidence of *Verticillium* infestation when brought under irrigated crop cultivation for the first time (40). The fungus may be indigenous to certain wild land soils, but I believe a more likely possibility is that it was introduced by crops like potatoes cultivated previously nearby or even at great distances. Perpetuation of inoculum by weeds like species of *Xanthium* during and long after either abandonment of the potato agriculture or a change to other crops implements spread to distant areas by the infected burred

fruit caught, for instance, in the fur of animals (12). There are records of nearly a century ago in California of unilateral dying of branches of young apricot trees and of brown blight (Verticillium wilt) of strawberries. These serve to establish the likelihood of widespread distribution of *Verticillium* in major agricultural areas of California simultaneously with the first attempts to develop agriculture.

EPIDEMIOLOGY.—The epidemiology of *Verticillium* follows a pattern. Initial outbreaks as reported in cotton (27), mint (14), hops (18), and tobacco (8) are typically mild and local. Attacks subsequently become severe and widespread, and the crop is either discontinued, ameliorated through resistance, or rotations are established. The severe outbreaks are brought on by strains of the fungus more virulent than the original, by a greater inoculum potential, or by both. In the Soviet Union, within 15 years after its release in 1940 and extensive cultivation in Uzbekistan, the wilt-resistant cotton variety 108F began to succumb to wilt. In a few more years the resistance was lost, and the 108F variety became the standard of susceptibility to which new experimental cotton lines were compared (30). A similar course of events has reportedly occurred in one area of cotton production in California since the release in 1948 of the wilt-resistant Acala 4-42 (25). It also happened in the wilt-resistant, mechanically harvested processing tomato lines designated by the prefix VF and developed by G. C. Hanna of the Department of Vegetable Crops at

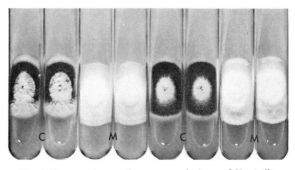

Fig. 1. Young cultures of two cotton isolates of *Verticillium albo-atrum* initiated from single conidia. Conidial type colonies *(C)* begin as white, appressed mycelial growth and ultimately become black when microsclerotia have formed; *albo-atrum* refers to this phenomenon. White colonies or mycelial types *(M)* arise from the *C* type. They produce few or no microsclerotia, are pathogenic, and do not revert to the *C* type.

the University of California, Davis. The resistance of these tomatoes was derived from a Utah breeding line, Loran Blood, which in turn was derived from a single accession of the wild Peruvian cherry tomato, *Lycopersicon pimpinellifolium* (24). Thus, although we lack knowledge of a conventional sexual stage in *Verticillium*, genetic variation occurs, as by the parasexual cycle (16). Whatever the mechanisms, the fungus possesses the capacity to vary in pathogenicity and to attack both a variety of hosts and hosts previously resistant to it. In our experience, strains of *Verticillium* pathogenic to tomatoes (varieties like Pearson, Improved Pearson, Bonny Best, and others widely cultivated before development of the VF types) were severely pathogenic to strawberries (*Fragaria ananassa* Duch.); to the stone fruit trees almond (*Prunus communis*), apricot (*P. armeniaca*), peach (*P. persica*), cherry (*P. cerasus*, *P. avium*, and *P. mahaleb*), and plum (*P. domestica*); to trailing blackberry cultivars derivative in part from *Rubus ursinus* Cham. and Schlecht., Boysen (Fig. 2), and Young, but not to Mammoth, Logan, and Olallie blackberries, which are resistant; and to rose rootstocks derivative from *Rosa multiflora* Thumb. and known in the trade as Burr

multiflora, Shafter, Dr. Huey, and Ragged Robin, but not to Manetti (*R. noisettiana* Thory), the common greenhouse rose rootstock. Manetti, however, is susceptible to cotton strains of *Verticillium*.

Verticillium isolates from the weed *Solanum sarachoides* Sendt. are pathogenic to strawberries, including cultivars resistant to tomato isolates, but they are generally nonpathogenic to tomatoes, even to the extremely wilt-susceptible Bonny Best, illustrating the diverse and unpredictable pathogenicity of this variable fungus. Thus, we need what has come to be called the uniform, horizontal, or nonspecific type of resistance that is equal against all strains of the pathogen. This resistance can be developed only by both greenhouse inoculations with isolates of known virulence and by studies in the field where the disease is severe. In the field we also need to be aware of complications that arise from the activities of other soil-borne organisms. Nor can we become complacent after one or two achievements. We must be alert for the first evidence of loss of resistance.

SOME REMARKS ON COTTON.—The Verticillium wilt fungus also infects roots and vascular systems of

Fig. 2. Trailing blackberry cultivar Boysenberry decimated by Verticillium wilt. Land previously had been in tomatoes. Replanted to the resistant Olallie [Black Logan (resistant) × Young (susceptible)] (inset), the replanted field produced abundant crops for 15 years. The nature of the resistance has not been investigated.

many nonsuscepts and of resistant cultivars (14). Thus, capacity to infect roots and even vascular systems is not necessarily linked to virulence. A major reason why progress in the development of wilt-resistant cottons in this country has been slow and the resistance tenuous is that essentially all cultivated cottons and many wild types become systemically infected, either upon inoculation or when grown in infested fields (38), and cotton is susceptible to new infections throughout its entire life. The susceptibility is conditioned by the environment.

Sustained high temperatures during the vegetative period of growth prevent development of wilt symptoms and infection of leaves in cotton and effect recovery of the diseased plant. Sustained high temperatures thus make it possible to cultivate cotton in the presence of the *Verticillium* pathogen, but this phenomenon should not be mistaken for resistance. For example, one of our experimental early-maturing cotton lines, designated B273, develops severe wilt symptoms during much of the growing season, fires its leaves, and collapses prior to harvest. Another line, B149, more indeterminate than the former and later to mature, recovers during the months of highest temperatures and appears to be resistant, but collapses from wilt 4 to 6 weeks before harvest. Recovery growth is associated with lateral vascular cambial activity and the addition of new xylem vessels to the cylinder that remain noninfected until vegetative growth ceases. All of the recovery growth, i.e., the wood added during high temperatures, is susceptible tissue. Thus, both of these cotton lines, although they appear to react so differently to wilt under field conditions, are equally susceptible in our judgment. This may appear strange to the breeder who equates an increase in yield of a cotton line over that of a standard variety with increased wilt resistance. Time and time again we have suffered the adage that 10 percent more yield means 10 percent more wilt resistance. It simply is not so.

Cotton breeders should strive to develop cultivars that are not systemically infected. For us at Berkeley, this involves first a greenhouse inoculation-box test made during a period of favorable environment for wilt pathogenesis, followed by intensive, correlated field studies (Fig. 3). On specific parent lines, and on all F_1, we make several thousand leaf cultures each season because resistance derived from our sources in *Gossypium barbadense* prevents infection of leaves at all growth phases except in the very young plant.

Capacity to prevent leaf infection may be less important in *Gossypium hirsutum* subsp. *mexicanum* var. *nervosum* and its Upland hybrid derivatives than in *G. barbadense* and its derivatives. We cannot speak for the new wilt-resistant varieties Tashkent 1, 2, and 3, which have revolutionized the Soviet Union's industry and put it ahead of ours in production per hectare, but our results with resistant parent material of these cottons and hybrids indicate considerable latent leaf infection (39). Possibly these hybrids obviate injurious effects of ethylene to leaves. Thus, there are two major sources of wilt resistance in cotton: taxa of *G. barbadense* and *G. hirsutum* subsp. *mexicanum* var. *nervosum*.

The concept of genetically variable multiline cultivars.—In general, cotton breeders and breeders of

other crops have worked toward the objective of true-breeding or pure seed lines. California state law requires it for cotton. For the plant pathologist, the true-breeding line often means a short useful life and rapid development of new virulent pathogen strains (17, 31, 32). For instance, within a few years after its release in 1963, the Acala SJ-1 cotton variety had been severely devastated by wilt.

Cotton lends itself ideally to development of genetically variable, wilt-resistant multiline cultivars. With resistance derived from *G. barbadense* dominant over susceptibility, resistant parents yield predominantly resistant offspring. The right selection of parents may yield populations through F_4 or F_5 that without selection combine high yield, earliness, and acceptable fiber

Fig. 3. Portable greenhouse bench supporting deep narrow tapering cotton inoculation boxes. Ten plants of a cotton line are grown in each box and inoculated when approximately 30 cm high. Steel strapping holds boxes together. Tapered side of box is removable by cutting the straps to expose roots for inoculation.

Fig. 4. (Left) Clone of wilt-resistant olive rootstock growing in portion of cotton wilt nursery. (Right) Susceptible clone (dead).

properties and retain sufficient genetic diversity to resist Verticillium wilt. The foundation unit of seed production for the multiline cultivars would be cloned F_1 and selfed, true-breeding lines of the parents. As of our 3rd year of testing the multiline cultivar concept, five of 53 initial populations appear to give near anticipated results (38).

In one part of California, Verticillium wilt of olives is associated with Verticillium wilt of cotton. As cotton encroached upon olive areas, olives succumbed, even trees that had been healthy for 40 or more years. During the past 20 years we have studied approximately 6,000 olive seedlings from world sources and have selected four that are highly resistant and promising as rootstocks (Fig. 4). Many tolerate the infection as "typhoid Marys," displaying no symptoms but carrying infection either generally or in a portion of the tree for a short period each year. Developing rootstock trees is complicated because trees must be able to resist wound infection as a consequence of grafting and transplanting and because Verticillium infection in olives usually cannot be detected by vascular discoloration.

CAPACITY TO ESCAPE.—A side of the *Verticillium* question that has been generally overlooked is that occasional severe outbreaks in new hosts have either disappeared or have not led to anticipated epidemics. For example, the severe outbreaks reported on sugarbeets in Colorado (15), on spinach (29) and on sweet pea seed-producing fields in California (unpublished), on grapes (rootstock variety Müller-Thurgau 5BB) in Germany (35), and on tobacco in Tennessee (41) have either not spread or have disappeared. Verticillium wilt has been destructive to tobacco in New Zealand, where resistant varieties derived from U.S. sources now hold the disease in check (42). Also, Verticillium wilt has been reported recently on Rupestris St. George grapes in New Zealnd (4).

Disappearance of or delimiting *Verticillium* infestations suggests the presence of biological control factors like those effective against *Ophiobolus graminis* (28) and *Streptomyces scabies* (20). Herein lies a pitfall for unwary breeders. There is a capacity among wilt-susceptible hosts, either isolated individuals or groups, be they propagated by seed or clones, to escape infection and remain healthy in the presence of suitable inoculum. Isolated individuals of wilt-susceptible cottons commonly escape the disease in heavily infested lands and occasionally even following inoculation. This escape can be mistaken for resistance. For instance, in diseased cotton fields in California, intensive selection within the Acala 4-42 variety of several hundred of the better looking "resistant" individuals failed in subsequent years of tests to demonstrate any resistance at all (7). These escapes may have fortuitously supported a rhizosphere microflora rich in *Erwinia* and *Agrobacterium* species antagonistic to *Verticillium* (19).

Plants of identical genotype may also show great variation in reaction to Verticillium wilt when exposed in so-called wilt nurseries in the field. Strawberries, for instance, show a strong tendency to escape infection, and this, dramatically illustrated in earlier strawberry breeding work in California, has been confused with resistance. It has also complicated the problem of rating varieties and breeding stocks for resistance (2).

Simplifying the soil biology by preplant fumigation of the wilt nursery with a methyl bromide-chloropicrin mixture adequate to control Verticillium wilt and then inoculating dormant washed plants with clean conidial suspensions of *Verticillium* in water eliminated the escape problem in strawberries. It achieved uniform, rapid disease reactions, made it possible to distinguish resistance from susceptibility, and reduced greatly the number of test plants per clone required for statistical validity (3). Despite these advances and the discovery of high resistance, the ease by which Verticillium wilt of strawberries is controlled by preplant soil application of methyl bromide-chloropicrin mixtures has obviated the need for the resistance and greatly accelerated breeding. Certain wild, glossy-leaved clones of *Fragaria chiloensis* Frez. growing along the California Pacific Coast are highly resistant to Verticillium wilt, whereas others are susceptible. In our tests the pubescent leaf types of *F. chiloensis* indigenous to the Pacific Coast of Peru, Chile, and the Juan Fernandez and Hawaiian Islands are susceptible. These agree in morphology to the one or possibly two plants deposited in the Jardin des Plantes, Paris, by Amédée Frézier in 1714 on his return from Chile. These two plants became the female parent of the present garden strawberry. Whatever wilt resistance is present in the garden strawberry today reflects introgression, we believe, following deliberate hybridization over a century ago and, more recently, hybridization with glossy-leaved forms of *F. chiloensis* collected from the San Francisco Bay area of California (37).

RESISTANCE MECHANISMS.—Depending on the host and time of disease onset, reported mechanisms of resistance may bring about a resistance reaction, i.e., initiate recovery of the infected plant, or may bring about reactions leading to severe symptoms and disease. Vessel occlusion, for instance, may localize the pathogen, contributing to resistance, but it reduces water availability to leaves. Some plants, such as strawberries and trailing blackberries, collapse in the field during periods of stress from high temperatures, as if from lack of water. Plant collapse occurs just prior to and during harvest and coincides with the onset of high early-summer temperatures. The vegetative growth cycle ceases temporarily. Vegetative growth, when resumed after harvest (for strawberries, this is in midsummer in coastal California), consists of the development of new crowns, not of enlargement of the old original crown. Thus, there is a limited vascular system to each crown, and occlusion of any of it could contribute to wilting and then to drying of leaf tissue between the veins, the major symptoms of Verticillium wilt, which, incidentally, are those of a plant lacking water.

There is a mechanism in wilt-resistant cotton that occludes vessels and localizes vascular infection near the base of the main stem. Cotton, in contrast to strawberry, matures late in the season. Leaves of resistant lines such as our Waukena White (*G. barbadense*) are invaded only when the plant is young. Whereas water transport may be reduced by vessel occlusion, an active vascular cambium adds new xylem, and the infected plant recovers. At the same time, the cotton root system penetrates ever deeper, branching into soil containing little inoculum (21), and the plant is exposed to rising soil and air temperatures. Sustained periods of 27°-30° C temperatures prevent leaf infection and cure infected leaves (Burgess and Wilhelm, *unpublished data*). Thus, the onset of high temperatures that causes the strawberry plant to collapse from wilt induces recovery from wilt in resistant cottons. Strawberries have by this time completed a cycle of vegetative growth and are fruiting, whereas cotton is in the phase of its most rapid vegetative growth.

Vascular infection of strawberries, hops, cotton, and other crops also results in an ethylene production greater than that present in noninfected plants (33). When infected in the field, certain cottons develop pronounced epinasty, sometimes followed by abscission. Ethylene may initiate a number of host-physiological responses such as wilting, chlorosis, and abscission, or it may (Alan Smith, *unpublished*) turn off such functions and even cause abnormal growth of fungi. Thus, in the infected host, ethylene may be a key to responses eliciting host injury or host recovery. The environment certainly would override this factor. Ethylene may also induce host production of antifungal polyphenols and phytoalexins (5, 6). In my thinking, however, their role in the resistance of cotton is questionable. For instance, the leaves of one of our most wilt-resistant parent lines, B123, obtained from the Botanical Garden of Bonn, Germany, have a delicate coumarin-like fragrance. *Verticillium* infects the plant only or primarily when it is young, causing yellowing and dehiscence of only one or two of the lowermost leaves. No further symptoms develop in the life of the plant, yet the fungus can be readily isolated in culture from the slightly discolored wood at the base of the main stem of the mature plant at harvest, as if it were simply inhibited from growing by a host-induced chemical or by mechanical localization. In contrast, it is often impossible to isolate the fungus from severely diseased mature wilt-susceptible cottons, including Acala SJ-1. The fungus has been lysed or killed, possibly by host-induced polyphenols; in resistant cottons it survives to plant maturity.

Thus, before much is said about mechanisms of resistance to Verticillium wilt, environmental conditions must be taken into account. Of particular importance are temperature, growth phase, rooting depth, root branching characteristics, whether the host is determinate or indeterminate, and whether it is vegetative or fruiting. There are no simple answers, and I have little faith in attempts to screen for resistance indirectly through rapid detection of host chemicals.

SUMMARY.—Our objectives in developing wilt-resistant crops should be: (a) to identify high resistance in species, taxa, and individuals genetically related by pedigree or ancestry to the wilt-susceptible crops, (b) to strive to achieve resistance that excludes the wilt fungus from the plant, because the infected plant, although it may resist symptom development, challenges the fungus to greater pathogenicity and sustains it, (c) to identify gross anatomical and physiological characters that determine resistance and to investigate how they are interrelated in the fluctuating seasonal environment in which the host grows, and (d) to consider how the information can be applied to the creation of new wilt-resistant crops, and, following the excellent example of J. C. Walker, to go ahead and develop them.

LITERATURE CITED

1. ASHWORTH, L. J., JR., J. E. WATERS, A. G. GEORGE, and O. D. MC CUTCHEON. 1972. Assessment of microsclerotia of Verticillium albo-atrum in field soils. Phytopathology 62:715-719.
2. BAKER, R. E., and G. A. BAKER. 1950. Experimental design for studying resistance of strawberry varieties to Verticillium wilt. Phytopathology 40:477-482.
3. BRINGHURST, R. S., S. WILHELM, and V. VOTH. 1961. Pathogen variability and breeding Verticillium wilt resistant strawberries. Phytopathology 51:786-794.
4. CANTER-VISSCHER, T. W. 1970. Verticillium wilt of grapevine, a new record in New Zealand. N. Z. J. Agr. Res. 13:359-361.
5. CHALUTZ, E., and M. A. STAHMANN. 1969. Induction of pisatin by ethylene. Phytopathology 59:1972-1973.
6. CHALUTZ, E., J. E. DEVAY, and E. C. MAXIE. 1969. Ethylene-induced isocoumarin formation in carrot root tissue. Plant Physiol. (Lancaster) 44:235-241.
7. COOPER, H. B., JR., J. DOBBS, M. LEHMAN, A. C. WILTON, J. H. TURNER, and C. W. SCHALLER. 1968. Breeding and evaluating cottons for higher levels of tolerance to Verticillium wilt, p. 106-153. *In* Cotton Dis. Res. in the San Joaquin Valley, Calif., Rept. for 1966-1967. Univ. Calif. Div. Agr. Sci., Berkeley.
8. CURTIS, K. M. 1944. Annu. Rept., 1943-1944, p. 23. Cawthron Institute, New Zealand.
9. DIMOCK, A. W. 1962. Obtaining pathogen-free stock by cultured cutting techniques. Phytopathology 52:1239-1241.
10. EASTON, G. D., M. E. NAGLE, and D. L. BAILEY. 1969. A method of estimating Verticillium albo-atrum propagules in field soil and irrigation waste water. Phytopathology 59:1171-1172.
11. EASTON, G. D., M. E. NAGLE, and D. L. BAILEY. 1972. Verticillium albo-atrum carried by certified seed potatoes into Washington and control by chemicals. Amer. Potato J. 49:397-402.
12. EVANS, G. 1968. Infection of Xanthium pungens by seedborne Verticillium dahliae. Plant Dis. Reptr. 52:976-978.
13. EVANS, G., S. WILHELM, and W. C. SNYDER. 1966. Dissemination of the Verticillium wilt fungus with cotton seed. Phytopathology 56:460-461.
14. FORDYCE, C., JR., and R. J. GREEN, JR. 1963. Alteration of pathogenicity of Verticillium albo-atrum var. menthae. Phytopathology 53:701-704.
15. GASKILL, J. D., and W. A. KREUTZER. 1940. Verticillium wilt of the sugar beet. Phytopathology 30:769-774.
16. HASTIE, A. C. 1964. The parasexual cycle in Verticillium albo-atrum. Genet. Res. (Cambridge) 5:305-315.
17. JENSEN, NEAL F. 1952. Intra-varietal diversification in oat breeding. Agron. J. 44:30-34.
18. KEYWORTH, W. G. 1939. Verticillium wilt of hops. A summary of observations during 1938, p. 244-249. Rept. East Malling Res. Stn., England.
19. KLINGNER, A. E., D. C. HILDEBRAND, and S. WILHELM. 1971. Occurrence of Erwinia carotovora in the rhizosphere of cotton plants which escape Verticillium wilt. Plant Soil 34:215-218.
20. MENZIES, J. D. 1959. Occurrence and transfer of a biological factor in soil that suppresses potato scab. Phytopathology 49:648-652.

21. PHILLIPS, D. J., and S. WILHELM. 1971. Root distribution as a factor influencing symptom expression of Verticillium wilt of cotton. Phytopathology 61:1312-1313.

22. RAABE, R. D., and S. WILHELM. 1966. Sources of infection in Verticillium wilt of rose. Phytopathology 56:124-127.

23. SACKSTON, W. E., and J. W. MARTENS. 1959. Dissemination of Verticillium albo-atrum on seed of sunflower (Helianthus annuus). Can. J. Bot. 37:759-768.

24. SCHAIBLE, L., O. S. CANNON, and V. WADDOUPE. 1951. Inheritance of resistance to Verticillium wilt in a tomato cross. Phytopathology 41:986-990.

25. SCHNATHORST, W. C., and D. E. MATHRE. 1966. Host range and differentiation of a severe form of Verticillium albo-atrum in cotton. Phytopathology 56:1155-1161.

26. SCHUSTER, M. L., and D. S. NULAND. 1960. Seed transmission of safflower Verticillium wilt fungus. Plant Dis. Reptr. 44:901-903.

27. SHAPOVALOV, M., and B. A. RUDOLPH. 1930. Verticillium hadromycosis (wilt) of cotton in California. Plant Dis. Reptr. 14:9-10.

28. SHIPTON, P. J., R. J. COOK, and J. W. SITTON. 1973. Occurrence and transfer of a biological factor in soil that suppressed take-all of wheat in Eastern Washington. Phytopathology 63:511-517.

29. SNYDER, W. C., and S. WILHELM. 1962. Seed transmission of Verticillium wilt of spinach. Phytopathology 52:365 (Abstr.).

30. SOLOVYEVA, A. I. 1959. Results of tests on world cotton collection for resistance to Verticillium wilt. Vsesoiuznyi Nauchno-Issledovatel'skii Institut Khlopkovodstva, Stantsiia Zashchity Rastenii, Tashkent. Prog. Rep., 1955-1956. 5:142-161. (Trans. title.)

31. STEVENS, N. E. 1942. How plant breeding programs complicate plant disease problems. Science 95:313-316.

32. SUNESON, C. A. 1960. Genetic diversity—a protection against plant diseases and insects. Agron. J. 52:319-321.

33. TALBOYS, P. W. 1972. Resistance to vascular wilt fungi. Proc. B, Roy. Soc. London 181:319-332.

34. THANASSOULOPOULOS, C. C., and W. J. HOOKER. 1970. Leaf and sprout infection by Verticillium albo-atrum. Phytopathology 60:196-203.

35. THATE, R. 1961. Apoplexie der Rebe: eine Verticilliose. Biologische Bundesanstalt für Land-und Forstwirtschaft. Mitteilungen 104:100-103.

36. THOMPSON, R., and A. G. MC LEOD. 1959. Verticillium wilt of tobacco. I. A new disease of tobacco caused by Verticillium dahliae Kleb. N. Z. J. Agr. Res. 2:785-791.

37. WILHELM, S., and J. E. SAGEN. 1974. History of the strawberry—from ancient gardens to modern markets. Agr. Pub., Univ. Calif., Berkeley. 298 p.

38. WILHELM, S., J. E. SAGEN, and H. TIETZ. 1974. Resistance to Verticillium wilt in cotton—sources, techniques of identification, inheritance trends, and the resistance potential of multiline cultivars. Phytopathology (In press.)

39. WILHELM, S., J. E. SAGEN, and H. TIETZ. 1974. Gossypium hirsutum ssp. mexicanum var. nervosum, Leningrad strain—a source of resistance to Verticillium wilt. Phytopathology (In press.)

40. WILHELM, S., and HAROLD E. THOMAS. 1952. Solanum sarachoides, an important weed host to Verticillium albo-atrum. Phytopathology 42:519-520 (Abstr.).

41. WOLF, F. A. 1935. Tobacco diseases and decays. Duke Univ. Press, Durham, N.C. 454 p.

42. WRIGHT, D. S. C. 1968. Verticillium wilt of tobacco. V. Varietal screening for resistance to Verticillium dahliae Kleb., and mode of inheritance of resistance. N. Z. J. Agr. Res. 11:655-664.

The Role of Mutation Breeding in Genetic Control of Plant Diseases

MERRITT J. MURRAY and WINSHIP A. TODD—*A. M. Todd Company, Kalamazoo, Michigan.*

Most biologists think mutations occur so rarely they are unlikely to find a mutation even though they observe millions of individuals. Plant breeders have often labeled the mutants studied by the geneticist as morphological monstrosities or physiological defectives of no practical value. If mutants are rare and almost no mutants are of practical value, then mutation breeding must be considered a foolish venture. Because of this type of thinking, most plant pathologists and plant breeders believe mutation breeding does not justify the effort. They may not express this opinion in public, but their actions attest to it. A retired geneticist and his chemist-businessman sponsor hope this review of our successful mutation breeding research (11, 12) will convince someone that mutation breeding can be worthwhile in other crops.

OBJECTIVES OF MUTATION BREEDING IN PEPPERMINT.—The first step in any mutation breeding project is to decide on a definite, precise, realistic, worthwhile objective. This point cannot be overemphasized. If you do not have a specific objective, you cannot formulate a suitable screening procedure to reach the objective. When we began mutation breeding with Mitcham peppermint (*Mentha piperita* L.), our sole objective was to get a Verticillium wilt-resistant strain without changing the oil quality, yield, or anything else. This specific objective was realistic, since related wild species of *Mentha* were more wilt resistant than was Mitcham peppermint. And this objective was worthwhile because Verticillium wilt, the most important mint disease in the United States, had made production of peppermint uneconomic in Michigan and Indiana. The disease was also rapidly increasing in Oregon and Washington. Extensive hybrid programs were unsuccessful in producing a wilt-resistant strain having acceptable oil quality. Soil fumigation was expensive and its effect could be negated by a dust storm blowing organic soil from one field to another in the Midwest.

SYNOPSIS OF MUTATION BREEDING PROCEDURES.—*Irradiation dosage tests.*—Dormant stolons dug from the field in early April were irradiated at the Brookhaven National Laboratory at Upton, New York, under the direction of Seymour Shapiro. The first stolon material was irradiated in 1955 at 1,000; 2,000; 4,000; 6,000; 8,000; 12,000; and 18,000 R units of X-ray, whereas about a third of the material was irradiated with thermal neutrons in the atomic pile at dosages varying from 2.5×10^{12} to 30×10^{12}. The small-scale dosage tests

conducted in 1955, 1956, and 1957 showed that 5,500 R units of X-ray and 22×10^{12} of neutrons were suitable treatment levels. Ono (16-20) has irradiated seeds, buds, and pollen of several *Mentha* species and shown that the effects of the irradiation differ with the species and the chromosome number of the species, as well as with the part of the plant treated. Developing buds produced more chlorophyll mutations than did dormant seeds.

After observing that *Verticillium* infestation definitely decreased the 3rd-year stand of plants irradiated and field-set in 1955, plans were made to produce 100,000 plants for transplanting in the field during the next 2 years. Assuming 50% survival of irradiated material at the dosages used, this required the production of 100,000 individuals in 1958 and again in 1959. The figures in Table 1 show that this goal was attained and even slightly exceeded.

Methods of growing irradiated material.—The irradiated stolons were cut into 40- to 70-mm pieces and planted in veneer squares in flats during late April or early May at our Kalamazoo greenhouse. Some plants grew quickly and were 10-15 cm tall, whereas others were only beginning to grow. Twenty-five to 50% or more died, depending on the kind of irradiation and the dosage. The disparity in plant size required that the plants be sorted at least four times during the 6- to 10-week growing period. Large plants were transferred to cold frames to retard their growth until they could be planted in the field in mid-June, after all danger of frost had passed.

Stolons subjected to neutron irradiation or to low levels of X-ray irradiation grew as rapidly or even more rapidly than those of the nonirradiated control. In contrast, stolons treated with 5,000-6,000 or more units of X-ray were frequently incapable of immediate growth. The leaves preformed in the bud expanded at once on the shootless and rootless pieces. In the absence of diseases and high temperatures, stolon pieces with radiation-injured meristems often persisted for 3 to 10 weeks and eventually developed adventitious shoots and roots.

Field planting.—Field plants were spaced 0.9×0.45 m ($3' \times 18"$). About 50% of the surviving plants were set in mid-June, 35% on 1 July and 15% on 15 July in the most severely *Verticillium*-infested field found on the 729-hectare (1,800-acre) Mentha Plantation at Mentha, Michigan.

All plants surviving in the flats were transplanted to the farm field. No time was wasted in trying to remove plants with chlorophyll mutations, as natural selection eliminated almost all of these within 3 weeks. Mutation specialists call this diplontic selection, but this term merely means that mutant cells grow slowly, whereas the green normal cells grow very rapidly and produce foliage. Many mutation geneticists would not agree, but our advice is to be malicious in your radiation dosage and stop worrying about smashing chromosomes and creating double mutations or garbage mutants. If natural selection or your screening program does not eliminate steriles and misfits, you can. You can't select mutants unless you've induced them.

TABLE 1. Results of screening procedure I (growing irradiated peppermint on severely *Verticillium*-infested soil with continuous production of this vegetatively propagated perennial species for 6 years on the same plot; stand decreased as disease severity increased)

Year	Number of plants
1st	103,667 planted
2nd	± 6 million[a]
3rd	± 2 million
4th	± 3/4 million
5th	± 1/4 million
6th	58,724[b]
7th	Plot subsequently used for testing individual selections

[a]Overwintering stolons of 1st-year plants produced a 60-fold increase in number of 2nd-year plants, thus forming a solid ground cover of individuals having a tall, largely unbranched, erect plant habit. A stand of this magnitude should persist indefinitely in the absence of weeds, Verticillium wilt, or other diseases.

[b]This figure is a total of all selections made each September in the years 1960-1964.

TABLE 2. Approximate mutation frequency for *Verticillium* resistance in peppermint subject to 5,000-6,000 R units of X-ray with nonwilt selections made at end of 1st year for 1959 material only by screening method II

	Total number	Mutation frequency	Mutants per 10,000
		%	
Plants started in flats	42,588	.031	3.1
Plants set in field	32,981	.039	3.9
Selections made in Sept. 1959	3,085	.421	42.1
Mutants	13		

Spontaneous somatic mutations for hairiness occur rarely in fields of glabrous Mitcham peppermint. Three strains studied by us were as vigorous as the cultivar. In the X-rayed material, there were more hairy mutants in plants that had received 5,000-6,000 R units than in plants that had received 1,000-4,000 R units. No adverse effects on vegetative growth were noted in over 2,000 hairy mutants. Observations in the 1955-1958 period led us to believe that adventitious buds must often come from single cells, since many hairy plants were entirely hairy. Similar conclusions were made by Broertjes et al. (2), who showed from histological preparations that adventitious buds are frequently derived from a single cell. If this is generally true, as Broertjes (1) believes, then any mutation found in a single cell would be present in every cell of the adventitious shoot derived from it. Immediate selection would be possible, for a shoot either would or would not have the desired mutation.

Direct observation indicated that some of the peppermint plants came from adventitious shoots, whereas others came from preformed and uninjured multicellular meristems. Wholly hairy plants must have been derived from single-cell adventitious shoots. Plants having one hairy shoot and one glabrous shoot might have come from two adjacent single-cell adventitious shoots or from one two-cell adventitious shoot. Plants having streaks or irregular patches of a chlorophyll mutant must have come from a mutation in a single cell of a multicellular meristem.

Screening procedure I.—In this first and originally planned screening procedure, a minimum of 100,000 plants were to be planted on very severely wilt-infested soil and allowed to grow for 6 to 10 years in the same plot, with every effort made to increase the severity of the disease year by year.

Increasing the selection pressure gradually seemed necessary to obtaining a plant pure for a *Verticillium*-resistant mutation that had initially occurred in a single cell of a multicellular meristem. A chimeral area on an overwintering stolon theoretically could have a pure branch that produced a 2nd-, 3rd-, or 4th-year plant having the mutation in every cell. Any pure wilt-resistant mutant plant should persist by vegetative propagation despite great increases in selection pressure, whereas normal susceptible plants would be lost from severe wilt damage.

More than 100,000 individuals were planted in rows in

TABLE 3. Approximate mutation frequency for *Verticillium* resistance in peppermint using screening method I (6 years of perennial growth under increasingly severe disease conditions)

	Total number	Mutation frequency	Mutants per 10,000
		%	
Total plants started in flats	138,515	.058	5.8
Total plants set in field	103,667	.077	7.7
Total selections made, Sept. 1960-1964	58,724	.136	13.6
X-rayed plants started in flats	101,596	.079	7.9
X-rayed plants set in the field	72,133	.111	11.1
Mutants	About 80		

a 4.86-hectare (12-acre) field (Table 1). In agricultural practice in southern Michigan, peppermint hay is normally cut and steam-distilled between 15 July and 15 August, and the stolon sod is plowed about 25 cm deep about 10-15 November. The hay was not cut in the experimental plot, but fall plowing without removal of the hay was done to increase the number of *Verticillium* spores in the soil. In this screening procedure, the selection pressure increased each year as the number of *Verticillium* spores increased and the stand of mint decreased. Plants that grew early in the spring were severely damaged by mid- to late-June, with regrowth in July. This cycle repeated itself two or three times during the growing season in all 4th-, 5th-, and 6th-year plants.

Experience showed that the field stand was only 1% in the fall of the 6th year and that the cost of weeding the field during the 7th year would be prohibitive. It was necessary to take a cutting from each nonwilted 6th-year plant and hold it over winter for wilt-testing the next summer as a single plant. Since 6th-year plants were scattered in small clumps, strings were used to delimit long narrow areas to prevent collection of the same plant twice. It was possible to collect the same mutant from two neighboring plants, but the principal selections came from different areas or treatments in the field.

To summarize the time schedule, stolons were irradiated and grown every spring between 1955 and 1959. Sixth-year selections of 1955-irradiated material were made in 1960, whereas those from 1959-irradiated material were made in 1964 in screening procedure I. Selections made each year from 1960 to 1964 were summarized to give the total of 58,724 found in Table 1.

Screening procedure II.—This procedure was not planned, but the presence of completely hairy individuals in 1955-1958 1st-year plants forced us to conclude that adventitious shoots frequently came from single cells whenever X-ray killed the preformed multicellular meristems. Cuttings were made from all nonwilt 1st-year plants of the X-rayed individuals grown in 1959. The cuttings were held over winter in the greenhouse and tested as single plants spaced 1.8 × 1.8 m the following summer. Cuttings were not made from neutron-treated plants, as this irradiation did not kill the multicellular meristems.

Although screening method I can be used only on hardy, vegetatively propagated perennial plants, screening method II can be used on any species that produces adventitious buds from single cells.

Results from screening procedure II.—The 3,085 selections were tested as single plants in 1960 and in replicated 1.8-m plots in 1961 and 1962. Eleven mutants having slight resistance and two having moderate resistance were found, representing a frequency of three mutants per 10,000 individuals (Table 2).

If adventitious shoots in *Mentha* species are almost always derived from a single cell, then it should be possible in a mutation breeding program to inoculate 1st-year *Mentha* shoots with rust as Ralph Green, Jr., is doing at Purdue University. The fact that young shoots should not differ from old plants has led C. E. Horner of Oregon State University to try seedling tests in flats for *Verticillium* resistance in spearmint.

Results from screening procedure I.—A total of 58,724 plants were collected in the fall seasons of 1960-1964, held over winter in the greenhouse, and then wilt-tested the following summer as single plants spaced 1.8 × 1.8 m. On the average, 80-85% of the single plants were wilted in this test. Plants that had no wilt symptoms or only slight symptoms were vegetatively propagated and retested the next year in two 1.8-m plots. The most effective method was to have a plot in one field having a moderate amount of wilt and one plot in a second field having such severe wilt that it would kill all susceptible plants by 1 August and then rot the stolons.

The final 100 strains were tested in 20-50 replications of 1.8-m plots in Indiana, with a similar test conducted by C. E. Horner in Oregon.

In Table 3, the approximate mutation frequency for wilt resistance using screening method I was about one-third that obtained using screening method II in Table 2 with the mutation frequency based on the number of selections that had to be individually tested. Screening method I had a higher frequency than screening method II with the frequency based on the number of plants started in flats or the number of field plants. Neutron irradiation as performed gave very few mutants. Assuming that all mutations came from X-ray, the frequency increased from 7.7 to 11.1 mutants per 10,000 field-grown individuals. Where usable, screening method II is less expensive, requires less time, and can be performed on seedlings or young plants.

Three types of wilt-resistant plants were recognized. Sixty-eight strains had a slight degree of wilt resistance, and these differences were consistent over a 5-year test period. Ten of these strains were sufficiently resistant to be worth growing if nothing better could be obtained. A few of these strains were recognizable from the cultivar due to changes in leaf teeth, leaf surface, plant habit, or plant color, but most were indistinguishable.

Five coarse-leaved vigorous strains with moderate wilt resistance were obtained, and these strains were distinguishable in appearance from the cultivar.

The seven high-resistance strains had erect, less-branched stems and darker green foliage and sometimes produced flowers 5-7 days earlier than the cultivar. The six best strains were included in regional tests subject to normal agricultural methods and farm distillation. C. E. Horner of Oregon State University had a .405 ha test near Corvallis, Oregon, and C. B. Skotland of Washington State University had a similar test at Prosser, Washington.

Ralph Green, Jr., will attest to the authenticity of the Midwest test on the William Gehring and Sons Farms at Medaryville, Indiana, since he made detailed observations on the development of *Verticillium* symptoms. The center plot was Verticillium wilt-susceptible Mitcham. This plot was not worth harvesting, whereas the wilt-resistant strains on each side had normal yields and oil quality for the area. One resistant strain was released in January 1972 as Todd's Mitcham, and this variety is being distributed by the Indiana, Michigan, Oregon, and Washington Mint Certification programs (14). The agricultural use of these strains should help limit the increase of spores in the soil. Martinson and Horner (10) and Green (3) have clearly shown that the severity of the wilt disease is directly related to the number of spores in the soil. None of our mutant strains was immune, but most fields should have less than 5% wilted plants if the growing conditions are reasonable. Nelson (15) believed that mint *Verticillium* strains differed in virulence, but no problems due to different pathogen strains have been encountered up to this time.

Word of caution regarding mutation breeding.—Do not assume that a selected mutant strain has only one mutation or that the mutation does not have multiple effects. To illustrate, we found that certain wilt-resistant mutants were more susceptible to frost or to insect damage than was the nonirradiated cultivar. Certain strains may have heterosis for herbage growth and produce 50% more oil, but the quality of the oil may be substandard.

Care had to be taken in evaluating the oil quality of all mutant strains, since peppermint oil has over 80 constituents (9) and since single genes can interfere with normal development in a series of compounds (13). Oil from each of the mutant strains was evaluated in products by the major U.S. peppermint oil users. The amounts of 12 major oil constituents were studied for each strain and cultivar throughout the complete growing season in Indiana. Similar data were secured at Prosser.

CONCLUSIONS.—A) Spontaneous mutations are rare, but X-ray irradiation increases the mutation rate, and this increase is a billionfold greater than the spontaneous rate. Anyone who doubts this unequivocal statement would be ill-advised to stand in front of an X-ray machine without first learning the medical uses of irradiation in cancer therapy.

B) A high X-ray dosage will produce morphological monstrosities, physiological defectives, chromosomal abnormalities, and double mutations, but natural selection, a suitable screening method, or the experimenter can eliminate the garbage mutations.

C) Resistance to a specific disease or disease strain is a definite objective that lends itself to a screening program based on inoculation of seedlings, cuttings, or mature plants.

D) The fact that an adventitious shoot usually comes from a single cell means that a shoot either does or does not have the desired mutant. One may screen and select shoots derived from the adventitious buds formed on *Mentha* stolons or other plant stems if the multicellular meristems (buds) have been inactivated by X-ray. One may also test all shoots derived from adventitious buds formed on leaves in a large series of monocotyledonous and dicotyledonous species that produce them (2).

E) Purification and stabilization of mutants arising in the multicellular meristems of vegetatively propagated perennial species probably can be accomplished in a screening program that gradually increases selection pressure. This method is not greatly different from a program based on irradiated seed where bulk lots of M-1, M-2, and M-3 self-pollinated seed are planted and increasing selection pressure is used to screen for disease resistance.

F) If seed irradiation does not produce the desired mutants in monocotyledonous plants that do not develop adventitious shoots, then a geneticist suggests that genic male-sterile strains should be pollinated with irradiated pollen.

G) Mutation breeding is the only method that can be used on apomictic species or on highly sterile species that produce neither seed nor pollen.

H) All varieties arising from mutation breeding must be carefully evaluated before release, since many mutations have multiple effects. *Verticillium* resistance was associated with frost susceptibility, insect damage, and a change in oil chemistry in specific mutant strains of mint.

I) Don't blame mutation breeding if you use too little irradiation, do not have a definite objective, have an unrealistic objective, use a poor screening method, or study too few individuals.

J) Read the information booklets of the International Atomic Energy Agency (4-8) and of the Food and Agriculture Organization and make use of irradiation to modify otherwise valuable cultivars for a specific trait like disease resistance, insect resistance, or chemical composition.

K) In screening adventitious shoots in *Mentha*, one has the apparent possibility of finding three to seven *Verticillium*-resistant individuals per 10,000. As enthusiastic supporters of mutation breeding, we hope this report will stimulate others to use it properly.

LITERATURE CITED

1. BROERTJES, C. 1968. Mutation breeding of vegetatively propagated crops, p. 139-165. Proc. 5th Eucarpia Cong., Milan, Italy, 1968.

2. BROERTJES, C., B. HACCIUS, and S. WEIDLICH. 1968. Adventitious bud formation on isolated leaves and its significance for mutation breeding. Euphytica 12:321-344.

3. GREEN, R. J., JR. 1969. Survival and inoculum potential of conidia and microsclerotia of Verticillium albo-atrum in soil. Phytopathology 59:874-876.

4. INTERNATIONAL ATOMIC ENERGY AGENCY. 1966. Mutations in plant breeding. Proc. Panel Vienna 1966. Vienna. 271 p.

5. INTERNATIONAL ATOMIC ENERGY AGENCY. 1968. Mutations in plant breeding II. Proc. Panel Vienna 1967. Vienna. 315 p.

6. INTERNATIONAL ATOMIC ENERGY AGENCY. 1969. Induced mutations in plants. Proc. Symp. Pullman 1969. Vienna. 748 p.

7. INTERNATIONAL ATOMIC ENERGY AGENCY. 1970. Manual on mutation breeding. Tech. Rept. Ser. 119. IAEA, Vienna. 237 p.

8. INTERNATIONAL ATOMIC ENERGY AGENCY. 1971. Mutation breeding for disease resistance. Proc. Panel Vienna 1970. Vienna. 249 p.

9. LAWRENCE, B. M., J. W. HOGG, and S. J. TERHUNE. 1972. Essential oils and their constituents. X. Some new trace constituents in the oil of Mentha piperita L. Flavour Ind. 3:467-472.

10. MARTINSON, C., and C. E. HORNER. 1964. Colonization of plant debris in soil by Verticillium dahliae. Phytopathology 54:900 (Abstr.).

11. MURRAY, M. J. 1969. Successful use of irradiation breeding to obtain Verticillium-resistant strains of peppermint, Mentha piperita L., p. 345-371. *In* Induced mutations in plants. IAEA, Vienna.

12. MURRAY, M. J. 1971. Additional observations on mutation breeding to obtain Verticillium-resistant strains of peppermint, p. 171-195. *In* Mutation breeding for disease resistance. IAEA, Vienna.

13. MURRAY, M. J., and D. E. LINCOLN. 1970. Genetic basis of acyclic oil constituents in Mentha citrata Ehrh. Genetics 65:457-471.

14. MURRAY, M. J., and W. A. TODD. 1972. Registration of Todd's Mitcham peppermint. Crop Sci. 12:128.

15. NELSON, R. 1950. Verticillium wilt of peppermint. Mich. Agr. Exp. Stn. Tech. Bull. 221. 259 p.

16. ONO, S. 1970. Studies on the radiation breeding in the genus Mentha. VI. Dose-response curve for root growth and interspecific difference in radiosensitivity. Sci. Rept. Fac. Agr. Okayama Univ. 35:1-6.

17. ONO, S. 1971. Studies on the radiation breeding in the genus Mentha. IX. Effective irradiation techniques to induce mutation in mint. Sci. Rept. Fac. Agr. Okayama Univ. 37:9-15.

18. ONO, S. 1971. Studies on the radiation breeding in the genus Mentha. XII. Variation in quantitative characters of X_2 plant after hybridization and irradiation in mint. Sci. Rept. Fac. Agr. Okayama Univ. 38:1-8.

19. ONO, S. 1972. Studies on the radiation breeding in the genus Mentha. XIII. Effects of irradiation on pollen. Sci. Rept. Fac. Agr. Okayama Univ. 39:1-8.

20. ONO, S. 1972. Studies on the radiation breeding in the genus Mentha. XV. A comparison of biological effects between γ-rays and X-rays on mint seeds. Sci. Rept. Fac. Agr. Okayama Univ. 40:1-8.

Breeding for Resistance in Dicotyledonous Plants to Root Rot Fungi

D. H. WALLACE and R. E. WILKINSON—*Departments of Vegetable Crops and Plant Breeding, and Department of Plant Pathology, Cornell University, Ithaca, New York.*

Most crops are attacked by one or more cortical root rot fungi. Soils on which a crop has not been previously grown may be nearly free from pathogens affecting that crop, but inoculum often increases rapidly and root rot becomes a major problem after one or a few plantings. Continuous cropping will almost certainly result in root rot. Without resistant varieties, a crop rotation of 5-10 years is often the only feasible control. Although less obvious than direct loss from root rot, refraining from growing a crop on otherwise adapted soil is also uneconomical. Chemicals can be effective, but field use is usually too costly. Breeding resistant varieties is the best way to control root rot when resistance is available. This paper reviews breeding efforts with six dicotyledonous crops, identifies other species with recognized resistance, and generalizes as to how root rot resistance breeding can be most efficiently conducted.

ALFALFA (*Medicago sativa* L.).—Alfalfa root rot caused by *Phytophthora megasperma* was identified in 1954 (20, 21). Frosheiser and Barnes (24) reviewed techniques used in breeding a resistant variety released in 1973. This resistance permits alfalfa to be grown on wet, poorly drained soils. Lehman et al. (41) had previously released populations with plants varying from susceptible to high-level resistance. Testing was mostly in a leveled field, with minimum water at emergence and seedling stages but much water during later plant growth.

Alfalfa is a 4n species. Populations were derived from selfs, intercrosses, and test crosses (pollinations with susceptible lines) of each of 11 lines having varied levels of resistance (48). All populations segregated, the data indicating genetic control by one tetrasomic gene with incomplete dominance. Highly resistant plants had the nulliplex genotype (*pm pm pm pm*), and highly susceptible plants were quadruplex (*Pm Pm Pm Pm*). Simplex, duplex, and triplex genotypes were increasingly more susceptible. Certain lines had modifying genes, but resistance from diverse sources was controlled primarily by the *pm* gene (48).

The released variety and other populations with improved resistance were developed by intercrossing resistant selections. Frosheiser and Barnes (24) and Lu et al. (48) recommend recurrent selection as the most appropriate breeding method, because a small to moderate proportion of many varieties of plants have some resistance.

SAFFLOWER (*Carthamus tinctorius* L.).—Safflower root rot was observed in Nebraska about 1947, and the causal organism was identified as *Phytophthora drechsleri* (19). Low-level resistance in cultivar Western Oilseeds 14 was shown in crosses with Nebraska-10 to be controlled by a dominant gene (80). From the first backcross to Nebraska-10, a selected line proved to carry resistance and exceed Nebraska-10 in yield and other characters. It was named Gila and released in 1958, 6 years after breeding began (79). Cultivar US 10 was derived from the sixth backcross to Nebraska-10 and released in 1959 (77). Greenhouse and field procedures were both used in testing and selection (79). In the greenhouse, inoculum was smeared into the hypocotyl and positioned just below the soil surface. Artificial and natural infestation were both employed in the field.

Pathogenic races of *P. drechsleri* have been identified (78), but Gila and US 10 were, in 1963, as resistant to all races as any cultivar (77). About this time, selections from P.I. 250724 and 253538 at Biggs, California, were found to have a resistance to all known races that was controlled by a recessive gene (78, 80, 81). This recessive gene and an earlier identified dominant gene were both identified in crosses to susceptible Nebraska-10. The composite of resistant lines from P.I. 250724 and 253538 is commonly referred to as Biggs safflower and was released as a breeding line under the name USB. Late maturity and low seed oil content make it commercially unacceptable (81). The recessive resistance, particularly under extensive or prolonged irrigation, is superior to the dominant resistance.

SOYBEAN (*Glycine max* [L.] Merr.).—Breeding of root rot-resistant soybeans has been briefly reviewed by Schmitthenner (65). This root rot, caused by *Phytophthora megasperma* var. *sojae* and identified in 1951, occurs in all soybean-producing areas. The near immunity found in variety Mukden in 1954 (6) is controlled by a dominant gene, *Rps* (27, 34, 39). Since 1963, many resistant varieties have been released. They include Clark 63, Chippewa 64, Amsoy 71, etc., names identifying both the year of release and the variety into which resistance was incorporated by backcrossing. Lam-Sanchez et al. (39) crossed Mukden with eight other resistant lines from diverse locations. All F_2 progenies failed to segregate, indicating that all resistant lines had the *Rps* gene. These workers concluded that resistance was widespread and involved only gene *Rps*.

The genetics, however, is more complex, as was shown when Hartwig et al. (27) reported both a second race of *P. megasperma* and a third allele of *Rps*. Genotype *Rps Rps* was resistant to races 1 and 2, whereas rps^2 rps^2 was resistant to 1 but susceptible to 2, and *rps rps* was susceptible to both. Order of dominance was $Rps > rps^2 > rps$. Morgan and Hartwig (58) identified 20 varieties with resistance to races 1 and 2, this being more common than resistance to race 1 only. Schmitthenner (65) recently reported a race 3 that is pathogenic to some *Rps Rps* varieties and nonpathogenic to others. Some but not all $rps^2 \ rps^2$ lines were resistant to race 3. Other agronomically unsatisfactory lines of unknown genotype were resistant to all three races. These results suggest that a second locus controls resistance to race 3, but verification is required. Root rot on *Rps Rps* varieties is increasingly frequent, and protection against race 3 requires determining the genetics of resistance and incorporation into acceptable varieties. There is further evidence that other genes besides *Rps* influence resistance to Phytophthora root rot. Smith and Schmitthenner (71) reported that Blackhawk × Capital always gave more susceptible F₃ and F₄ plants than other *Rps Rps × rps rps* crosses; they concluded that one or more modifier genes prevented full expression of *Rps*. From all possible crosses among Harosoy (*rps rps*), Dunfield (*rps rps*), and Mukden (*Rps Rps*), Averre (4) concluded that Dunfield carries the dominant allele of a gene *Pl* that reduces the expressed resistance of *Rps*. With field plantings of soybean lines that were near-isogenic (except for *Rps* and *rps* and some seed differentiating characters), Buzzell and Haas (13) found seed yield of susceptible *rps rps* relative to resistant *Rps Rps* to be 75% for 2 years and 35% for a 3rd year. Also with near-isogenic lines, Caviness and Walters (14) found yields did not differ in the absence of Phytophthora root rot, but yields of *rps rps* varieties were reduced as much as 32% where root rot was present. Meyer and Sinclair (53) showed that, even in the absence of usual stem symptoms, *P. megasperma* reduced root growth and seed yield of *rps rps* varieties, but had little effect on near-isogenic *Rps Rps* resistant counterparts.

In searching for resistance, identifying pathogenic races, and backcrossing resistance into susceptible soybean cultivars, testing has generally been in the greenhouse. In one study, small amounts of fungus were inserted into a slit in the hypocotyl several days after emergence (66). About 4 days later, resistant varieties showed slight wound discoloration, whereas susceptible varieties showed a brown, soft decay with subsequent seedling death.

As the use of Phytophthora root rot-resistant soybeans has increased, other pathogens have also been shown to cause root rot. In descending order of frequency, Schmitthenner (64) isolated *Pythium ultimum, Rhizoctonia, Fusarium,* and *Thielaviopsis* from rotted roots of *Phytophthora* resistant lines. Thomson et al. (82) showed that *Pythium aphanidermatum, P. debaryanum,* and *P. ultimum* were all pathogenic on soybeans. Laviolette and Athow (40) found some resistance to *P. ultimum* but were unable to determine the genetics because variation in susceptibility caused by plant age was not accounted for. Saharan and Gupta (63) reported a collar (hypocotyl) rot caused by *Fusarium semitectum.*

TOBACCO (*Nicotiana tabacum* Linn.).—Clayton (17) reviewed the breeding of tobacco resistant to black root rot, caused by *Thielaviopsis basicola.* His review and work with black shank caused by *Phytophthora parasitica* var. *nicotianae* are summarized here because the tobacco work illustrates the many problems, frustrations, contradictions, and possible achievements from breeding for root rot resistance.

Breeding for black root rot resistance began about 1920. No resistant *N. tabacum* was found in Mexico or Central or South America, where tobacco originated. However, some North American varieties had moderate resistance, and a few from the Near East had high resistance. These resistances were not successfully bred into acceptable varieties. The moderate resistance was controlled by numerous recessive genes and was difficult to recover from segregating progenies. It was easy to recover the high-level resistance, which was apparently controlled by the same recessive plus additional dominant genes, but incorporation into acceptable varieties was not achieved because genetic linkage with small leaf size was never broken.

Testing *Nicotiana* species for resistance to black root rot began in 1934. Surprisingly, of 31 species, 14 were immune and others were highly resistant. In 1941, when the F₁ (a 4n alloploid) of *N. tabacum × N. debneyi* was found to be immune, transfer of resistance from *debneyi* to tobacco was initiated. This immunity was dominant and monogenic. In backcrosses 8 and 9 in 1951, the first 100% immune lines were obtained. The first black root rot-resistant burley-type tobacco variety was released in 1965. Homozygous resistant lines were never obtained from comparable efforts with broadleafed and flue-cured types. Subsequently, a root rot-immune burley was crossed with a broadleafed type. The F₂, F₃, and F₄ progenies failed to segregate with expected monogenic ratios, but in a first and second backcross, regular monogenic, dominant segregation occurred.

Testing was all in the greenhouse during fall, winter, and spring, when low temperatures could be maintained. Infested field soil was first cropped with susceptible tobacco. Testing was then done by growing plants in a mixture of this stock soil with sterilized greenhouse compost.

The many races of *Thielaviopsis basicola* vary widely in virulence on different host genotypes. Some varieties are resistant in one locality and susceptible in another. As of now, the *debneyi* monogenic immunity is effective against all races.

Black shank of tobacco, caused by *Phytophthora parasitica* var. *nicotianae,* appeared in Florida about 1915 but was not serious until a very susceptible cultivar, Round Tip, was planted extensively in 1924. It remains the major threat to tobacco production throughout the southeastern United States (15, 22). Tisdale (83) selected within and crossed local cigar wrapper types and in 1931 released some black shank-resistant cultivars. One of these, Florida 301, is the resistance source for other tobacco types. Transfer into other types is difficult because genetic control is multigenic, and there is undesirable linkage. Full expression is only moderate, and only about 60% of full expression can be transferred to other tobacco types without excessive sacrifice of quality or yield (16).

The related species *N. longiflora* and *N. plumbaginifolia* are highly resistant. A monogenic dominant resistance has been successfully transferred into tobacco from both species (3, 18). Unfortunately, some *P. parasitica* isolates, now termed race 1, were fully virulent on both species supplying this resistance and also on tobacco lines carrying it (3, 84). Rapid multiplication of race 1 in fields heavily infested with race 0 has limited use of this high-level monogenic resistance, so black shank is controlled with cultivars carrying the moderate-level, multigenic resistance derived from within *N. tabacum* (22, 74).

PEAS (*Pisum sativum* L.).—Sundheim and Wiggen (76) have identified *Aphanomyces* root rot as the most important disease of pea, occurring wherever peas are grown. They indicated that yield reductions in infected fields are obvious and that growing peas only once every 10 years, since long-time crop rotation is the only control so far available, causes extensive economic losses. Field soil taken to the greenhouse is commonly used in conjunction with a susceptible pea variety to determine a root rot index (75) that is highly correlated with field infection. Sundheim and Wiggen (76) report that more than one-fourth of the sampled fields in Norway give root rot indexes too high for growing acceptable pea crops.

In 1927, both Haenseler (25) and Lindford (43) reported a root rot caused by *Aphanomyces euteiches*. Pea root rot is commonly called the root rot complex to imply involvement of many causal organisms. *Aphanomyces euteiches* is the more common pathogen, but Schroeder (68) also listed *Fusarium solani* f. *pisi*, *Pythium ultimum* and other *Pythium* species, and the sterile or *Rhizoctonia* stage of *Pellicularia filamentosa* as causal organisms.

The earliest reports (1960) of resistance to *Aphanomyces* root rot were from Michigan (44, 45, 47), with a 1972 report from New York (49). These investigators used a greenhouse test in which zoospores were placed around recently emerged seedlings. Lockwood and Ballard (47) tested 805 accessions and found a few with low-level resistance. They had little success at determining the genetics of resistance, because the small differential between susceptibility and the low-level resistance made it difficult to evaluate segregating pea progenies. Lockwood (45) reported that some F_6-F_9 selections carried the resistance, and that three or four backcrosses had been made to commercial varieties.

Marx et al. (49) indirectly studied the genetics. They used selection G213, P.I. 175227, which consistently had more resistance than any other line, although it also was only low-level resistance which the authors preferred to call tolerance. Backcross, F_1, and F_2 data indicated that resistance was dominant and genetically linked with three horticulturally undesirable plant and seed characters, each controlled by a dominant allele of three unlinked loci. Substitution of the recessive allele, i.e., desirable plant or seed character, for the dominant allele giving horticultural unacceptability, decreased intensity of the already low-level resistance. A small proportion of segregating populations did have at least one horticulturally acceptable character and some tolerance, but none possessed both all three recessive horticultural characters and the full complement of low-level resistance. These data indicate that resistance to

Aphanomyces root rot is controlled by several to many genes and that genetic linkage to undesirable characters limits breeding progress.

In addition to finding low-level *Aphanomyces* root rot resistance in peas, Lockwood (46) also tested numerous lines for resistance to *Fusarium solani* f. *pisi*. Several had resistance, that of P.I. 140165 being slightly higher and more stable under varied environments (46). King et al. (33) and Hagedorn (26) also found resistant lines. Knavel (35) studied segregation for resistance in the cross P.I. 140165 × Early Perfection. His data suggested partial maternal inheritance with nuclear control by several genes, the latter conclusion agreeing with data from crosses between other resistant and susceptible lines (33). Kraft and Roberts (38) later showed that P.I. 140165 and four other P.I. lines simultaneously carried resistance to both *F. solani* f. sp. *pisi* and *Pythium ultimum*. Other lines were resistant to *Fusarium* or *Pythium*, but not to both. Resistance was associated with unacceptable characters, i.e., with tall plants, purple flowers, and speckled or pigmented seed coats (38, 52).

Muehlbauer and Kraft (59) crossed five root rot-resistant and five susceptible varieties in all combinations. The 90 crosses, including reciprocals, were screened in the greenhouse in naturally infested soil and in soils artificially infested with *F. solani* or *P. ultimum*. Reciprocals behaved alike. There was correlation between root rot scores in soil infested with *F. solani* and *P. ultimum*, a correlation the authors interpreted as an indication that resistance against the two organisms might be controlled by the same genes. Highly significant general combining ability indicated much additive, i.e., fixable, genetic variation. Broad sense heritabilities were 44, 32, and 23%, respectively, for naturally infested field soil and soils artificially infested with *F. solani* or *P. ultimum*. These results support feasibility of breeding root rot-resistant varieties, but other factors suggest this will not be easy. For one thing, there was lack of agreement between results from naturally infested and soils artificially infested with *F. solani* or *P. ultimum*. In these tests, P.I. 140165, previously shown to have about the best available resistance, showed the most root rot of all varieties in the naturally infested soil. This suggests that either additional strains of *F. solani* or *P. ultimum* were present in naturally infested soil or that expression of root rot resistance under field conditions requires resistance not only to *F. solani* and *P. ultimum*, but also to other pathogens like *Rhizoctonia solani*, *Thielaviopsis basicola*, or *Fusarium oxysporum*.

Little is known about races of pea root rot pathogens. Beute and Lockwood (7) identified at least two races of *A. euteiches*, as differentiated by a 6.5-fold difference in disease index. Bolton and Donaldson (8) recently concluded that *F. solani* f. *pisi* and *F. oxysporum* f. *pisi*, the causal organisms of root rot and wilt in peas, respectively, are morphological-physiological variants of the same fungal species. At least five races of *F. oxysporum* have been identified.

BEANS (*Phaseolus vulgaris* L.).—Resistance to dry root rot of beans, caused by *Fusarium solani* f. *phaseoli*, was reported by Burkholder (12) in 1919, shortly after he identified the disease and its causal organism. Varieties

Flat Marrow and Scarlet Runner, the latter of species *P. coccineus*, had low-level resistance. Flat Marrow was crossed with commercial beans, but resistance was not successfully transferred into acceptable varieties.

About 1950, lines were screened in a field that had been infested since about 1920 with *F. solani* f. *phaseoli* and annually planted to beans during subsequent years. The field was therefore highly infested with several root rot fungi. T. L. York identified line N203 as having moderate- to high-level resistance (reported in 87, 88). At a high inoculum level in the greenhouse, N203 can be badly rotted, but at a low level and in the field it will be almost free, whereas commercial varieties will be badly rotted. Azzam (5) compared N203 with P.I. 165435, which had better resistance in the greenhouse using pure cultures of *F. solani*; N203 had better resistance in the field. N203 has resistance not only to *F. solani*, but also to *Pythium* spp., *Rhizoctonia solani* (69), and *Thielaviopsis basicola* (29, 90). The only resistance reported superior to N203 was line 2114-12 (10), from a *Phaseolus vulgaris* × *P. coccineus* (variety Scarlet Runner) cross made about 1950 (86). In the highly infested field described above, it had slightly less root rot than N203 (10).

McCrostie (51) concluded that the resistance of Flat Marrow was controlled by two recessive genes. On the other hand, Yerkes and Freytag (91) concluded from a *P. vulgaris* × *P. coccineus* cross that resistance was dominant. Azzam (5) supported the hypothesis of two or three recessive genes, and Smith and Houston (70) hypothesized that there was one recessive and one dominant gene. Slow breeding progress and other studies (10, 23, 87) have failed to support these simple conclusions of single genetics and instead indicate multigenic control.

During early field growth and short-duration greenhouse tests, F_1 plants had intermediate root rot compared with the resistant and susceptible parents, suggesting additive gene action (28, 29). But, particularly in the field, the susceptible parent continued with time to accrue symptoms more rapidly than did the resistant parent or the F_1. At maturity, therefore, the resistant parent and F_1 were similar, with far less root rot than the susceptible parent, suggesting dominant gene action.

The proportion of the variance for *Fusarium* root rot scores among 9- to 13-week-old field-grown F_2 plants that was caused by genetic segregation (broad sense heritability) was about 75%; that of 5-week-old greenhouse plants was 60% (28). The remaining proportion was of environmental origin. Earlier, after but 5 weeks in the field, the genetic effect had been only 10-30%, whereas the environmental effect had been 70-90%, suggesting that the shorter exposure to pathogens was insufficient to facilitate infection of all susceptible segregants and full expression of resistance of all resistant segregants. The regressions of F_3 progeny means on parental F_2 plant means (selection effectiveness, i.e., narrow sense heritabilities) were much lower (26-44%). Similar broad and narrow sense heritabilities were obtained for *Thielaviopsis* resistance, which is also under multigenic control (29). N203 and 2114-12 are both resistant to both *Thielaviopsis* and *Fusarium*, but these resistances are not controlled by the same genes (30).

As with peas and tobacco, breeding progress has been limited by genetic linkage between resistance and unacceptable characters. N203 is viney and late with black seed, and, in segregating populations, high-level resistance has obvious correlation with these characters. Earlier maturing lines with high-level resistance were ultimately developed, one of them being the indeterminate line 2114-12, which derives its resistance from *P. coccineus* and has brown rather than black seed. A determinate line with high-level resistance has not been found. Red kidney-type segregates commonly have low resistance. Data from many progenies clearly show that resistance decreases with increasing seed size.

After nearly 50 years of effort, the first red kidney types, with only intermediate resistance compared with the parents, are currently (1973) being tested in yield trials. Burke and Nelson (11) have reported progress toward development of root rot-resistant Red Mexican- type beans.

OTHER CROPS.—New plant species-pathogen species associations are constantly being identified or found in new areas (36, 54, 56, 57, 61, 73). Besides these six, recent references have been found to root rot of about 20 dicotyledonous species caused by an approximately equal number of pathogens. Resistances have been reported for the following plant-pathogen associations: seedling rots of lima bean (*Phaseolus lunatus*) caused by *Rhizoctonia solani* (89), root and stem rot of soybean caused by *Phytophthora parasitica* (67), crown rot of strawberry caused by *Colletotrichum fragariae* (32), foot rot of black pepper (*Piper nigrum*) caused by *Phytophthora palmivora* (2), root rot of pepper (*Capsicum annuum*) caused by *Phytophthora capsici* (62, 72), root rot of cotton (*Gossypium hirsutum*) caused by *Thielaviopsis basicola* (42, 50), damping-off of cotton seedlings caused by *Rhizoctonia solani* (60), damping-off of watermelon (*Citrullus vulgaris*) caused by *Fusarium semitectum* (55), root and crown rot of sugar beet (*Beta vulgaris*) caused by *Rhizoctonia* spp. (31), and stem and root rot of cacao caused by *Phytophthora palmivora* (92). Some of these resistances were in related rather than in the named species.

DISCUSSION AND CONCLUSIONS.—A major conclusion is that a unique approach, research investigation, and ultimate solution are required for each specific association of plant species and pathogen (or pathogens). Breeders and pathologists must cooperatively exercise imagination and initiative to find all available resistance and to develop testing, breeding, and selection methods because each host and pathogen have unique developmental, physiological, and genetical expressions, and therefore unique host-pathogen interactions. Some pathogens have many races and others have few or no races, or at least no identified races.

Several factors emphasize the requirement for unique programs: (a) some resistances are controlled by single recessive genes, some by single dominant genes, (b) some genes have multiple alleles, some have only a resistant and a susceptible allele, and (c) some resistances are controlled by two or three recessive or dominant genes, whereas some are under multigenic control with the indistinguishable effects of single genes. In other plant species, major and minor (modifier) gene effects can be distinguished. Two or more genetic systems for resistance are available in some species, one of which may or may not give resistance against all known pathogen races. In some crops, intraspecies resistance genes are effective. In others, intraspecies genes exist, but interspecific transfer has proven more effective. In still other species, or for a different pathogen within the same species, interspecific gene transfer has not been successful. For each plant species and pathogen, then, transfer of interspecific resistance genes presents unique problems.

Some intraspecific resistances, either monogenic or multigenic, are tightly linked with unacceptable characters, whereas others of either type, although linkage is perhaps more frequent and difficult to break with multigenic control, present no such difficulty. When undesirable linkages may be present, the breeder should probably try to incorporate resistance into several plant types to increase the probability of success. Transfer into one type may provide the bridge needed for transfer into other types. Such linkage may also, and frequently will, impede progress with interspecific transfer of resistance.

Some resistances approach immunity, with segregation into only susceptible and resistant classes, whereas others express only low-level tolerance that is difficult or impossible to select for in segregating progenies.

In some species, resistance genes are found in only one or a few lines. In others, they are widely distributed in many varieties and plant introductions. Some resistances are found in lines from geographical locations where pathogens are endemic, whereas in the same or other species, resistance comes from geographical locations with no evolutionary expectation for its existence.

It has been hypothesized that some genes or multigene groups may give resistance against two (or more) pathogens, whereas others are effective against only one pathogen. More research is needed in this area.

The many variabilities of both host and pathogen preclude recommendation of a preferred or most efficient breeding method. Pedigree selection will sometimes be effective. Backcrossing will continue to be useful, especially with mono- or digenic control. Recurrent selection and intercrossing of selected plants will be most useful with multigenic control or with polyploid species. Given the current state of knowledge, however, there are no compelling arguments for selecting only mono- or digenic (vertical) or only multigenic (horizontal) resistance when a choice is available (1, 85).

Breeding of root rot-resistant alfalfa, safflower, and soybean varieties each required about 10 years and used intraspecific resistance, whereas black root rot resistance in tobacco required about 45 years and used interspecific gene transfer. Approximately 50 years of effort with peas and beans and related species have not yet yielded an acceptable resistant variety. In tobacco, early work with intraspecies resistance using both moderate- and high-level resistance, both with multigenic control, failed because the moderate-level resistance could not be followed through segregating generations and the high-level resistance was tightly linked with small leaf size. No conceivable foresight could predict that, after overcoming difficulties of interspecific gene transfer, resistance with simple monogenic control, in contrast to

intraspecific multigenic control, would be derived from the related species *N. debneyi* and would be successfully used in breeding black root rot-resistant tobacco varieties with resistance against all known pathogen races. Similarly, it was not predictable that the high-level monogenic interspecific resistance against tobacco black shank would be overcome by a new pathogen race, making multigenic moderate-level intraspecific resistance the more useful resistance.

After reviewing the literature of many plant species on breeding for root rot resistance, no generalizations that would expedite root rot breeding in all species could be derived. Breeding progress depends largely on trial and error research for each species.

Even though greenhouse and laboratory testing and selection are often more effective and rapid in the short run, in the long run, reliable methods for field testing and selection should be developed and used. The resistance desired is field expression, so greenhouse and laboratory results must correlate with field results. Such correlation can be easily evaluated if a reliable field testing procedure exists, but it may be difficult if infection is erratic among locations within the field or from year to year. Requirement for field testing also arises because field root rot is commonly caused by several races or several pathogens by the so-called root rot complex. Greenhouse and laboratory results may not correlate with field results because not all races or species that infect in the field are used or are not applied in appropriate relative quantities.

Reliable field testing procedures are especially essential with multigenic, i.e., quantitative genetic, control, where only progeny testing of many individual plant selections permits identification of homozygous resistant genotypes (10, 28, 29, 30, 87). The large populations required are much easier grown and evaluated in the field. Field testing methods must include data collection and analysis techniques appropriate to quantitative characters, which usually means scoring each individual plant. It may be necessary to do this at imminent maturity if seed is to be collected from the best plants. There must be adequate replication of each line, including comparison with sister and related progeny lines. Two or more years (generations) of data are usually required to make final decisions. Obviously, much added work and confusion arise if many plants escape infection.

Repeatable field testing procedures will greatly improve the effectiveness of root rot breeding programs (10, 24, 28, 37, 87). Uniformly and highly infested field plots used with necessary control of soil moisture or other environmental factors for expression of root rot with or without minimal escapes is essential to rapid and effective breeding. *Such field plots are rare and should be considered valuable assets of the states or nations in which they exist.*

A thorough intra- and interspecies search for resistance should be made early in any breeding program. It is desirable to compare all possible resistances for intensity, ease of transfer, and undesirable linkage. Transgressive segregates superior to all known resistances may arise from crosses between and among the different resistances.

Serving as a parent is the major use of resistant lines. An alternative use is as root rot-resistant stocks onto which susceptible commercial varieties can be grafted.

This has been effective with corky root of tomato caused by *Mycelia sterilia* (9), with foot rot of black pepper caused by *Phytophthora palmivora* (2), and with walnut trees and other wilt and nematode diseases (9). Resistant genotypes might be so used whenever field or greenhouse plant spacings and populations and grafting compatibilities permit. Their use may give more immediate and effective root rot control than breeding for resistance.

In many plant species, type and/or rate of phytoalexin synthesis have been correlated with root rot resistance (see Paxton, this volume). Many of us had hoped that expediting of root rot breeding would be derived from this knowledge. Unfortunately, the role of phytoalexins is complex. There are many different phytoalexins and, as of now, they are too difficult to analyze to make a contribution to breeding and selection procedures.

LITERATURE CITED

1. ABDALLAH, M. M. F., and J. G. T. HERMSEN. 1971. The concept of breeding for uniform and differential resistance and their integration. Euphytica 21:351-361.
2. ALCONERO, R., F. ALBUQUERQUE, N. ALMEYDA, and A. G. SANTIAGO. 1972. Phytophthora foot rot of black pepper in Brazil and Puerto Rico. Phytopathology 62:144-148.
3. APPLE, J. L. 1962. Physiological specialization within Phytophthora parasitica var. nicotiana. Phytopathology 52:351-354.
4. AVERRE, C. W. 1963. Differential host-parasite relation between Glycine max and Phytophthora megasperma var. sojae. Ph.D. thesis, Purdue Univ., Lafayette, Ind. 101 p.
5. AZZAM, H. A. 1957. Inheritance of resistance to Fusarium root rot in Phaseolus vulgaris L. and Phaseolus coccineus L. Ph.D. thesis, Oreg. State Coll., Corvallis. 80 p.
6. BERNARD, R. L., P. E. SMITH, M. J. KAUFMANN, and A. F. SCHMITTHENNER. 1957. Inheritance of resistance to Phytophthora root and stem rot in the soybean. Agron. J. 49:391.
7. BEUTE, M. K., and J. L. LOCKWOOD. 1967. Pathogenic variability in Aphanomyces euteiches. Phytopathology 57:57-60.
8. BOLTON, A. T., and A. G. DONALDSON. 1972. Variability in Fusarium solani f. pisi and F. oxysporum f. pisi. Can. J. Plant Sci. 52:189-196.
9. BRAVENBOER, L., and G. PET. 1963. Control of soil borne diseases in tomatoes by grafting on resistant rootstocks. p. 317-324. Proc. 16th Int. Hort. Congr., 1962.
10. BRAVO, A., D. H. WALLACE, and R. E. WILKINSON. 1969. Inheritance of resistance to Fusarium root rot of beans. Phytopathology 59:1930-1933.
11. BURKE, D. W., and C. E. NELSON. 1967. Response of field beans to nitrogen fertilization on Fusarium infested and noninfested land, p. 1-5. Wash. Agr. Exp. Stn. Bull. 687.
12. BURKHOLDER, W. H. 1919. The dry root rot of bean. N. Y. Agr. Exp. Stn. (Cornell) Mem. 26:999-1033.
13. BUZZELL, R. I., and J. H. HAAS. 1972. Natural and mass selection estimates of relative fitness for the soybean rps gene. Crop Sci. 12:75-76.
14. CAVINESS, C. E., and H. J. WALTERS. 1971. Effect of Phytophthora rot on yield and chemical composition of soybean seed. Crop Sci. 11:83-84.
15. CLAYTON, E. E. 1953. Control of tobacco diseases through resistance. Phytopathology 43:239-244.
16. CLAYTON, E. E. 1953. Developments in growing tobacco, p. 540-548. USDA Yearbook of Agr.
17. CLAYTON, E. E. 1969. The study of resistance to the black root rot disease of tobacco. Tobacco Sci. 168:30-37.
18. COLLINS, G. B., P. D. LEGG, C. C. LITTON, and M. J. KASPERBAUER. 1971. Inheritance of resistance to black shank in Nicotiana tabacum L. Can. J. Genet. Cytol. 13:422-428.
19. ERWIN, D. C. 1952. Phytophthora root rot of safflower. Phytopathology 42:32-35.
20. ERWIN, D. C. 1954. Root rot of alfalfa caused by Phytophthora cryptogea. Phytopathology 44:700-704.
21. ERWIN, D. C. 1965. Reclassification of the causal agent of root rot of alfalfa from Phytophthora cryptogea to P. megasperma. Phytopathology 55:1139-1143.
22. FLOWERS, R. A., and J. W. HENDRIX. 1972. Population density of Phytophthora parasitica var. nicotiana in relation to pathogenesis and season. Phytopathology 62:474-477.
23. FRAZIER, W. A. 1966. Proc. 17th Int. Hort. Congr., Vol. I. (Abstr. 451).
24. FROSHEISER, F. I., and D. K. BARNES. 1973. Field and greenhouse selection for Phytophthora root rot resistance in alfalfa. Crop Sci. 13:735-738.
25. HAENSELER, C. M. 1927. Pea root rot studies. N.J. Agr. Exp. Stn. Annu. Rept. 39:334-339.
26. HAGEDORN, D. J. 1960. Testing commercial pea varieties for reaction to Fusarium root rot, Fusarium solani f. pisi. Phytopathology 50:637 (Abstr.).
27. HARTWIG, E. E., B. L. KEELING, and C. J. EDWARDS, JR. 1968. Inheritance of reaction to Phytophthora rot in the soybean. Crop Sci. 8:634-635.
28. HASSAN, A. A., D. H. WALLACE, and R. E. WILKINSON. 1971. Genetics and heritability of resistance to Fusarium solani f. phaseoli in beans. J. Amer. Soc. Hort. Sci. 96:623-627.
29. HASSAN, A. A., R. E. WILKINSON, and D. H. WALLACE. 1971. Genetics and heritability of resistance to Thielaviopsis basicola in beans. J. Amer. Soc. Hort. Sci. 96:628-630.
30. HASSAN, A. A., R. E. WILKINSON, and D. H. WALLACE. 1971. Relationship between genes controlling resistance to Fusarium and Thielaviopsis root rots in beans. J. Amer. Soc. Hort. Sci. 96:631-632.
31. HECKER, R. J., E. G. RAPPEL, and J. O. GASKILL. 1972. Development of sugarbeets with resistance to Rhizoctonia root and crown rot. Agron. Abstr. 1972:10.
32. HORN, H. L., K. R. BURNSIDE, and R. B. CARVER. 1972. Control of the crown rot phase of strawberry anthracnose through sanitation, breeding for resistance, and benomyl. Plant Dis. Reptr. 56:515-519.
33. KING, T. H., H. G. JOHNSON, H. BISSONNETTE, and W. A. HAGLUND. 1960. Development of lines of Pisum sativum resistant to Fusarium root rot and wilt. Proc. Amer. Soc. Hort. Sci. 75:510-516.
34. KLARMAN, W. L., and J. W. GERDEMANN. 1963. Induced susceptibility in soybean plants genetically resistant to Phytophthora sojae. Phytopathology 53:863-864.
35. KNAVEL, D. E. 1967. Studies on resistance to Fusarium root rot Fusarium solani f. sp. pisi (F. R. Jones) in Pisum sativum L. Proc. Amer. Soc. Hort. Sci. 90:260-267.
36. KRAFT, J. M. 1972. Chickpea, a new host of Fusarium solani f. sp. pisi. Plant Dis. Reptr. 56:110-111.
37. KRAFT, J. M., and J. W. BERRY. 1972. Artificial infestation of large field plots with Fusarium solani f. sp. pisi. Plant Dis. Reptr. 56:398-400.
38. KRAFT, J. M., and D. D. ROBERTS. 1970. Resistance in peas to Fusarium and Pythium root rot. Phytopathology 60:1814-1817.
39. LAM-SANCHEZ, A., A. H. PROBST, F. A.

LAVIOLETTE, J. F. SCHAFER, and K. L. ATHOW. 1968. Sources and inheritance of resistance to Phytophthora megasperma var. sojae in soybeans. Crop Sci. 8:329-330.

40. LAVIOLETTE, F. A., and K. L. ATHOW. 1971. Relationship of age of soybean seedlings and inoculum to infection by Pythium ultimum. Phytopathology 61:439-440.

41. LEHMAN, W. F., D. C. ERWIN, and E. H. STANFORD. 1967. Root rot tolerance of new alfalfa strains available to plant breeders. Calif. Agr. 21:6.

42. LINDERMANN, R. G., and T. A. TOUSSOUN. 1968. Breakdown in Thielaviopsis basicola root rot resistance in cotton by hydrocinnamic (3-phenylpropionic) acid. Phytopathology 58:1431-1432.

43. LINDFORD, M. B. 1927. Additional hosts of Aphanomyces euteiches, the pea root rot fungus. Phytopathology 17:133-134.

44. LOCKWOOD, J. L. 1960. Pea introductions with partial resistance to Aphanomyces root rot. Phytopathology 50:621-624.

45. LOCKWOOD, J. L. 1960. Progress and problems in breeding peas resistant to root rots, p. 358-366. Mich. Agr. Exp. Stn. Quart. Bull. 43.

46. LOCKWOOD, J. L. 1962. A seedling test for evaluating resistance of peas to Fusarium root rot. Phytopathology 52:557-559.

47. LOCKWOOD, J. L., and J. C. BALLARD. 1960. Evaluation of pea introductions for resistance to Aphanomyces and Fusarium root rots, p. 704-713. Mich. Agr. Exp. Stn. Quart. Bull. 42.

48. LU, NANCY SHYH-JANE, D. K. BARNES, and F. I. FROSHEISER. 1973. Inheritance of Phytophthora root rot resistance in alfalfa. Crop Sci. 13:714-717.

49. MARX, G. A., W. T. SCHROEDER, R. PROVVIDENTI, and W. MISHANEC. 1972. A genetic study of tolerance in pea (Pisum sativum L.) to Aphanomyces root rot. J. Amer. Soc. Hort. Sci. 97:619-621.

50. MATHRE, D. E., and J. D. OTTA. 1967. Sources of resistance in the genus Gossypium to several soilborne pathogens. Plant Dis. Reptr. 51:864-866.

51. MC CROSTIE, G. P. 1921. Inheritance of disease resistance in the common bean. Amer. Soc. Agr. J. 13:15-32.

52. MC DONALD, W. C., and H. H. MARSHALL. 1961. Resistance to pre-emergence damping-off in garden peas. Can. Plant Dis. Surv. 41:275-279.

53. MEYER, W. A., and J. B. SINCLAIR. 1972. Root reduction and stem lesion development on soybeans by Phytophthora megasperma var. sojae. Phytopathology 62:1414-1416.

54. MICHAIL, S. H., M. A. ABD-EL-REHIM, H. ELAROSI, and E. A. KHAIRY. 1971. Dry rot of garlic cloves in U.A.R. (Egypt). Phytopathol. Medit. 10:202-205.

55. MICHAIL, S. H., I. A. IBRAHIM, M. A. ABD-EL-REHIM, and F. M. FADEL. 1971. Damping-off of Cucurbitaceous plants in U.A.R. III. Fusarium semitectum Berk. et Rav., a damping-off causal organism of watermelon. Phytopathol. Medit. 10:46-49.

56. MILDENHALL, J. P., R. G. PRATT, P. H. WILLIAMS, and J. E. MITCHELL. 1971. Pythium brown root and forking of muck-grown carrots. Plant Dis. Reptr. 55:536-540.

57. MIRCETICH, S. M., and H. L. KEIL. 1970. Phytophthora cinnamomi root rot and stem canker of peach trees. Phytopathology 60:1376-1382.

58. MORGAN, F. L., and E. E. HARTWIG. 1965. Physiologic specialization in Phytophthora megasperma var. sojae. Phytopathology 55:1277-1279.

59. MUEHLBAUER, F. J., and J. M. KRAFT. 1973. Evidence of heritable resistance to Fusarium solani f. sp. pisi and Pythium ultimum in peas. Crop Sci. 13:34-36.

60. NAIM, M. S. 1964. Pathogenicity of Rhizoctonia solani Kühn associated with the damping-off of Egyptian cotton varieties. Phytopathol. Medit. 3:129-134.

61. ORMROD, D. J., and W. D. CHRISTIE. 1972. Phomopsis root rot of greenhouse cucumbers in British Columbia. Plant Dis. Reptr. 56:53-55.

62. POLACH, F. J., and R. K. WEBSTER. 1972. Identification of strains and inheritance of pathogenicity in Phytophthora capsici. Phytopathology 62:2-26.

63. SAHARAN, G. S., and V. K. GUPTA. 1972. Pod rot and collar rot of soybean caused by Fusarium semitectum. Plant Dis. Reptr. 56:693-694.

64. SCHMITTHENNER, A. F. 1964. Fungi associated with root necrosis of Phytophthora-resistant soybeans. Phytopathology 54:906 (Abstr.).

65. SCHMITTHENNER, A. F. 1972. Evidence for a new race of Phytophthora megasperma var. sojae pathogenic to soybean. Plant Dis. Reptr. 56:536-539.

66. SCHOEN, J. F. 1967. A modified pathological test to distinguish Phytophthora root rot resistance among soybean varieties. Proc. Assoc. Off. Seed Anal. 57:130-131.

67. SCHOEN, J. F. 1971. Reaction of six soybean varieties to a pathogenic isolate of Phytophthora parasitica from white clover. Plant Dis. Reptr. 55:130-131.

68. SCHROEDER, W. T. 1953. Root rots, wilts and blights of peas, p. 401-408. USDA Yearbook of Agr.

69. SCHROTH, M. N., and R. J. COOK. 1964. Seed exudation and its influence on pre-emergence damping-off of bean. Phytopathology 54:670-673.

70. SMITH, F. L., and B. R. HOUSTON. 1960. Root rot resistance in common beans sought in plant breeding program. Calif. Agr. 14(9):8.

71. SMITH, P. E., and A. F. SCHMITTHENNER. 1959. Further investigations of the inheritance of resistance to Phytophthora root rot in the soybean. Agron. J. 51:321-323.

72. SMITH, P. G., K. A. KIMBLE, R. G. GROGAN, and A. H. MILLETT. 1967. Inheritance of resistance in peppers to Phytophthora root rot. Phytopathology 57:377-379.

73. STANGHELLINI, M. E., and E. L. NIGH. 1972. Occurrence and survival of Pythium aphanidermatum under arid soil conditions in Arizona. Plant Dis. Reptr. 56:507-510.

74. STOKES, G. W., and C. C. LITTON. 1966. Source of black shank resistance in tobacco and host reaction to races 0 and 1 of Phytophthora parasitica var. nicotiana. Phytopathology 56:678-680.

75. SUNDHEIM, L. 1972. Physiological specialization in Aphanomyces euteiches. Physiol. Plant Pathol. 2:301-306.

76. SUNDHEIM, L., and K. WIGGEN. 1972. Aphanomyces euteiches on peas in Norway. Sci. Reports, The Agr. Univ. Norway. Vol. 51. 17 p.

77. THOMAS, C. A. 1964. Registration of US 10 safflower. Crop Sci. 4:446-447.

78. THOMAS, C. A., and J. M. KLISIEWICZ. 1963. Selective pathogenesis within Phytophthora drechsleri. Phytopathology 53:368.

79. THOMAS, C. A., D. D. RUBIS, and P. S. BLACK. 1960. Development of safflower varieties resistant to Phytophthora root rot. Phytopathology 50:129-130.

80. THOMAS, C. A., and D. E. ZIMMER. 1970. Resistance of Biggs safflower to Phytophthora root rot and its inheritance. Phytopathology 60:63-64.

81. THOMAS, C. A., and D. E. ZIMMER. 1971. Registration of USB safflower germplasm. Crop Sci. 11:606.

82. THOMSON, T. B., K. L. ATHOW, and F. A. LAVIOLETTE. 1971. The effect of temperature on the

pathogenicity of Pythium aphanidermatum, P. debaryanum, and P. ultimum on soybean. Phytopathology 61:933-935.

83. TISDALE, W. B. 1931. Development of strains of cigar-wrapper tobacco resistant to black shank (Phytophthora nicotiana Breda de Haan). Fla. Agr. Exp. Stn. Bull. 226. 45 p.

84. VALLEAU, W. D., E. M. JOHNSON, and G. W. STOKES. 1959. Black shank resistance from wild species of Nicotiana, p. 10-11. Ky. Agr. Exp. Stn. 72nd Annu. Rept.

85. VAN DER PLANK, J. E. 1968. Disease resistance in plants. Academic Press, New York and London. 349 p.

86. WALL, J. R., and T. L. YORK. 1960. Genetic diversity as an aid in interspecific hybridization in Phaseolus and in Cucurbita. Proc. Amer. Soc. Hort. 75:419-428.

87. WALLACE, D. H., and R. E. WILKINSON. 1965. Breeding for Fusarium root rot resistance in beans.

Phytopathology 55:1227-1231.

88. WALLACE, D. H., and R. E. WILKINSON. 1966. Origin of N203 (P.I. 203958). Annu. Rept. Bean Improvement Coop. 9:38-39.

89. WARREN, H. L., R. M. HELFRICH, and V. L. BLOUNT. 1972. Evaluation of lima bean varieties and selections for resistance to Rhizoctonia solani. Plant Dis. Reptr. 56:268-270.

90. WILKINSON, R. E., J. KRIKUN, and D. H. WALLACE. 1963. Resistance to Thielaviopsis basicola in beans. Annu. Rept. Bean Improvement Coop. 6:29.

91. YERKES, W. D. J., and G. P. FREYTAG. 1956. Phaseolus coccineus as a source of root-rot resistance for the common bean. Phytopathology 46:32 (Abstr.).

92. ZENTMYER, G. A. 1968. Resistance of four cacao varieties to Phytophthora palmivora. Phytopathology 58:554 (Abstr.).

Phytoalexins, Phenolics, and Other Antibiotics in Roots Resistant to Soil-Borne Fungi

JACK D. PAXTON—*Department of Plant Pathology, University of Illinois, Urbana.*

A multitude of compounds have been identified in plants. Some, especially phenolics, have the potential of restricting the growth of microorganisms in plants. Most have been identified from the aerial portions, probably because these portions can be more readily observed and because a hatchet rather than a shovel probably was used to sample the plant. There is little reason to believe that most compounds so far identified only in the aerial portions of plants are restricted from the subterranean portions; rather, their observation in stems and leaves reflects the hatchet phenomenon.

Many phenolic compounds isolated and identified from roots and other subterranean portions of various plants inhibit bacteria, fungi, nematodes, and even other plants. However, there are unique problems in critically testing the role of these compounds in disease resistance. It is virtually impossible to duplicate under controlled conditions the complex biological and physical environment around even a small segment of root growing in the field. Because of this problem and the problems of determining the concentration of a compound in a plant cell and because of the problem in determining the interaction of this compound with all the other compounds in its environment, little other than circumstantial evidence for the role of these compounds in plant disease resistance has been reported.

Nevertheless, there has been some fine research into the role of specific compounds in the disease resistance of subterranean portions of plants. Much of this work has been reviewed recently by Deverall (12), Ingham (20), Kuć (26), Rhode (38), Stoessl (43), and Woods (51), to mention only a few. I will review some of the more recent examples as well as a few of the classic examples in which specific compounds play a role in root disease resistance. For brevity, some of these examples are not cited directly, but reference to them can be found in the reviews and articles, especially in those of Ingham (20), Kuć (26), and Stoessl (43).

Current feeling is that much of plant disease resistance is determined by phytoalexins. These are compounds that inhibit microorganisms but are found in the plant in exceedingly small concentrations until after plant tissue has been subjected to stress, such as attempted invasion. The critical event in disease resistance actually may be the initial stimulation of phytoalexin production.

ORCHID.—The first investigations into the chemical nature of the disease resistance of plants was the work of Bernard in 1909 and 1911 on orchids. He noticed that some orchids require infection by mycorrhizal fungi and, although the roots are invaded, the bulbs seldom become infected. Bernard and later Nobecourt found that an inhibitory substance was formed in the infected tissues (20). This substance, hircinol (13) (Fig. 1), inhibited the growth of *Rhizoctonia repens*, a mycorrhizal fungus of this orchid. This lead to the concept of controlled parasitism, in which a delicate balance is maintained between the mycorrhizal fungus and its host. Hircinol was the first phytoalexin studied because this antifungal compound is not found in tissues that have been heat-killed or treated with chloroform.

Gaümann and Jaag (20) confirmed these results and studied the chemical defense reaction in *Orchis militaris* and *O. morio*. In these orchid species, two more inhibitory compounds were found (15). Orchinol (Fig. 1) is fungistatic and of primary importance in the resistance of *O. militaris* bulbs to *R. repens*. Para-hydroxybenzylalcohol is only slightly inhibitory, and loroglossol (Fig. 1), which is closely related to orchinol and hircinol, is inactive.

Orchinol persists in *Orchis* bulbs for several months but occurs in parts of the bulb distant from mycorrhizal infection in concentrations that are usually inadequate to prevent fungal growth. Not all soil fungi stimulate orchinol production, and *Rhizoctonia solani* can degrade orchinol, so inoculation of bulbs of *O. militaris* by *R. solani* leads to their rapid infection and destruction.

Orchinol has been isolated from the orchid genera *Orchis*, *Serapias*, *Gymnadenia*, *Loroglossum*, *Coeloglossum*, *Anacamptis*, and *Nigritella*. Orchinol is inhibitory to a number of soil fungi and is probably important in the disease resistance of orchids.

ONION.—Link et al. (28, 29) found that catechol and protocatechuic acid (Fig. 1) are present in the dead outer scales of red and yellow onions. Although these compounds are not themselves colored, they diffuse out of the scales of colored onions and prevent spore germination and invasion by *Colletotrichum circinans*. The scale pigments (flavones and anthocyanins) are not inhibitory, even though resistance is associated only with the colored onions. If the colored scales are removed, the bulbs are attacked as readily as are those of white varieties. On the other hand, *Aspergillus niger*, which is not affected by catechol or protocatechuic acid, attacks both colored and white varieties. Protocatechuic acid also may be responsible for the resistance of onions to *Diplodia natalensis*. Thus, preformed compounds can play an important role in plant disease resistance.

PEA.—The first phytoalexin to be completely chemically characterized was pisatin (Fig. 2). Pisatin has been extensively studied by Hadwiger (8), Cruickshank and Perrin (10), and Bailey (20). They found that pisatin occurs in roots as well as in most other parts of the pea (*Pisum sativum*) plant, that it has a wide spectrum of antibiotic activity, and that it occurs in the plant in concentrations high enough to account for resistance to many microorganisms. Pathogens of peas are usually less sensitive to pisatin in vivo, possibly because some

PLANT		COMPOUNDS	
Common Name	Scientific Name	Compound Name	Structure
ORCHID	*Orchis militaris*	orchinol	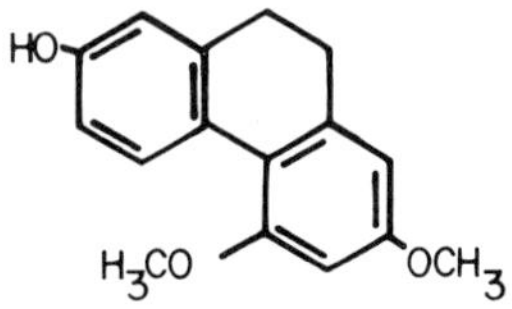
	Loroglossum hircinun	hircinol	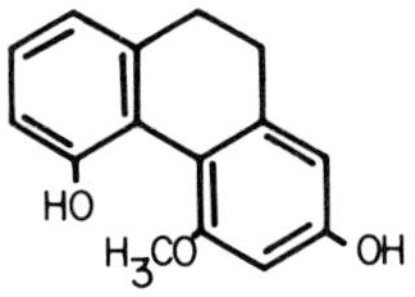
		loroglossol	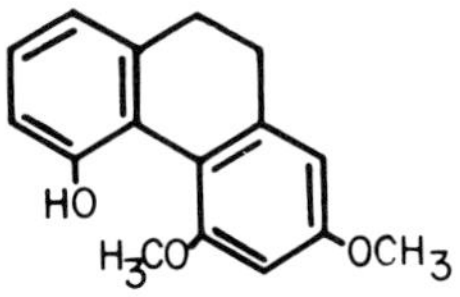
ONION	*Allium cepa*	catechol	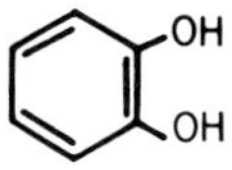
		protocatechuic acid	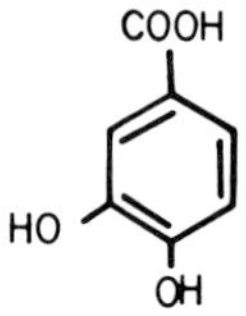
AVOCADO	*Persea borbonia*	borbonyl acetate	$H_3C-C(=O)-O-C(=CH_2)-C(=O)-CH=CH-(CH_2)_{11}-CH_3$
MAHONIA	*Mahonia* spp.	berberine	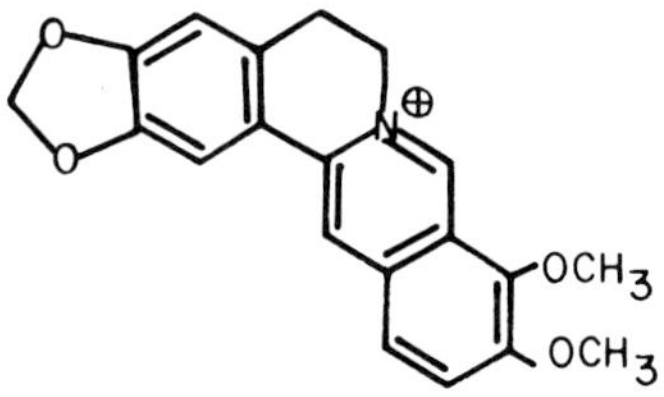
TULIP	*Tulipa* spp.	tulipalin A	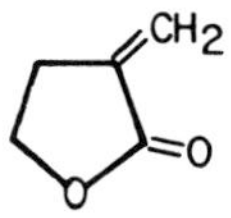
WALNUT	*Juglans* spp.	juglone	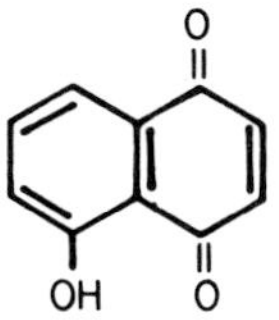
		hydrojuglone glucoside	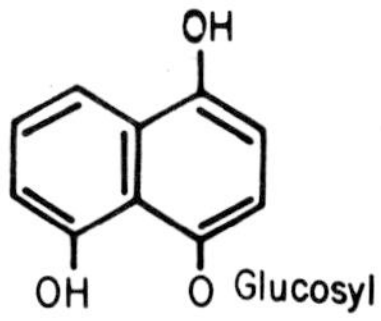

Fig. 1.

pathogens of peas degrade pisatin (50).

The situation with bacteria is less clear, as phytopathogenic bacteria like *Xanthomonas phaseoli* and *Pseudomonas pisi* are not inhibited by pisatin. The effect of pisatin on other bacteria is quite variable, as is the ability of bacteria to stimulate pisatin formation.

PLANT		COMPOUNDS	
Common Name	Scientific Name	Compound Name	Structure
PEA	Pisum sativum	pisatin	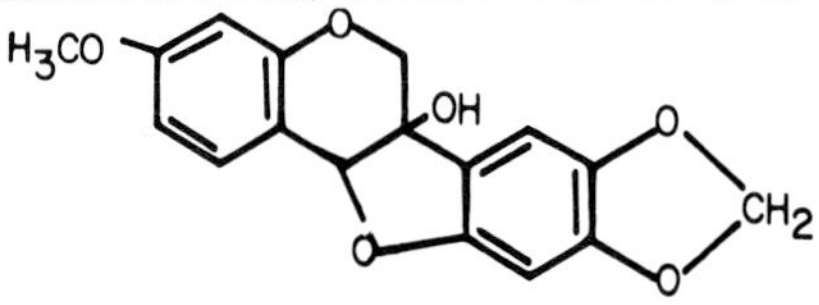
BEAN	Phaseolus vulgaris	phaseollin	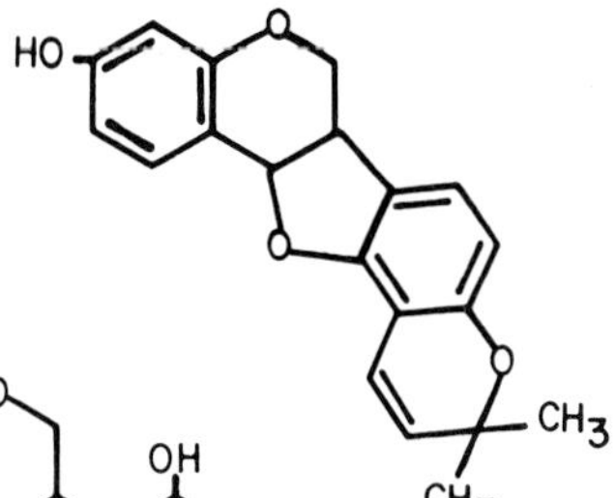
		kievitone (substance II)	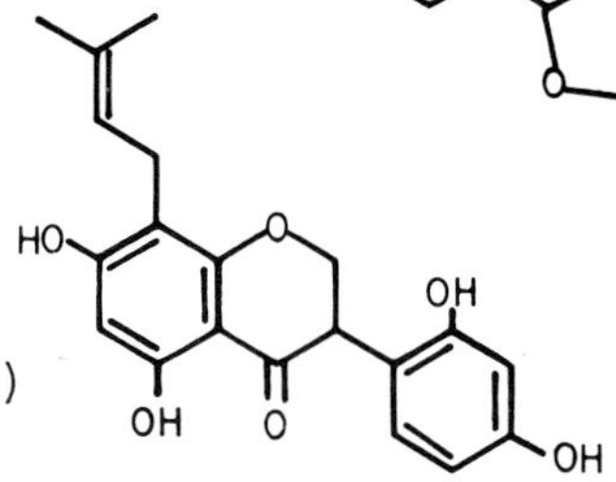
		phaseollinisoflavan	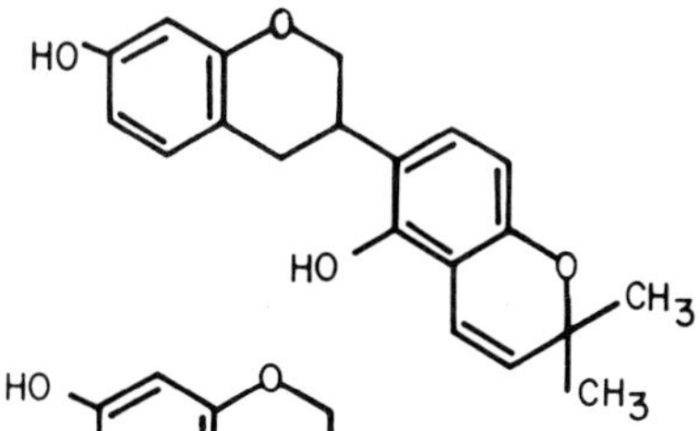
		2'-methoxyphaseollinisoflavan	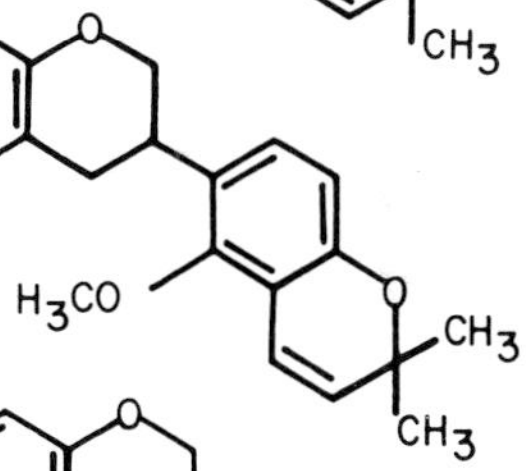
SOYBEAN	Glycine max	hydroxyphaseollin	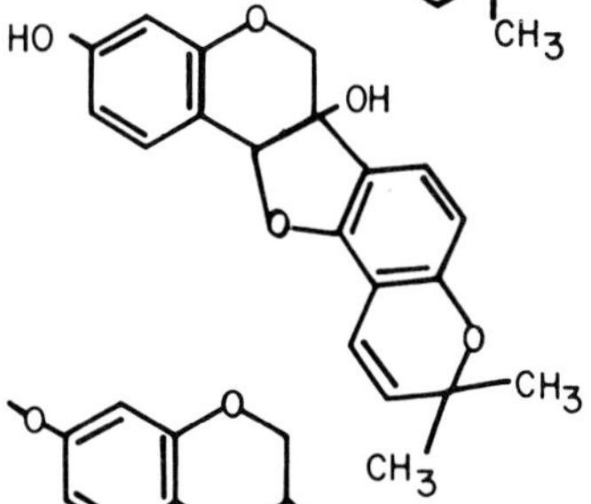
RED CLOVER	Trifolium pratense	trifolirhizin	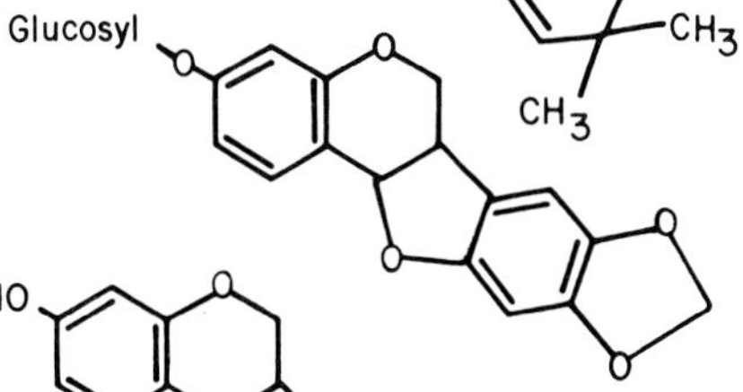
		maackiain	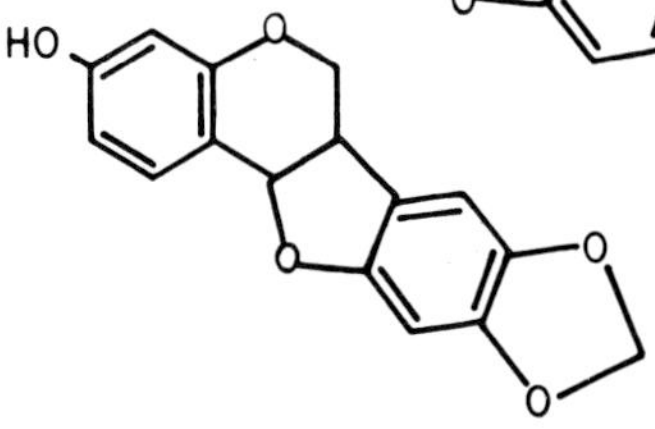

Fig. 2.

Hadwiger (8) and others (20) have studied the biosynthesis of pisatin and developed an interesting hypothesis on the initiation of pisatin production. More recently, Christenson and Hadwiger (8) questioned the role of pisatin in Fusarium foot rot of pea and suggested that pisatin production in pea seedlings is coupled with

PLANT		COMPOUNDS	
Common Name	Scientific Name	Compound Name	Structure
CARROT	Daucus carota	6-methoxymellein (isocoumarin)	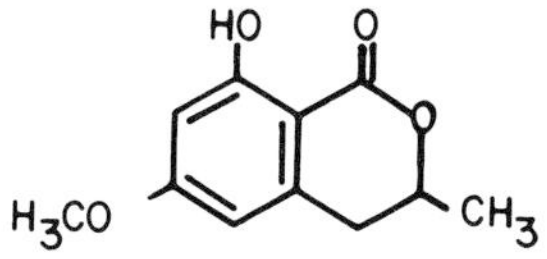
		chlorogenic acid	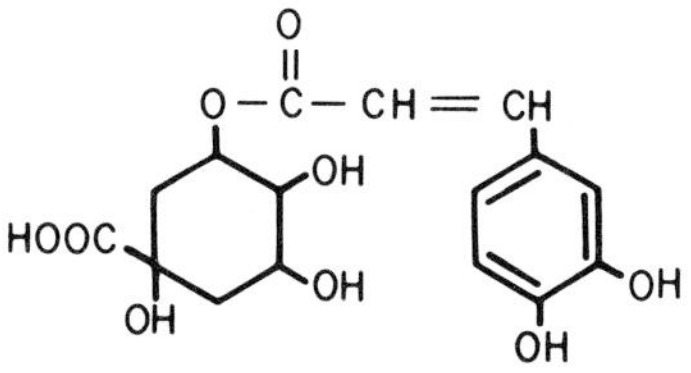
POTATO	Solanum tuberosum	rishitin	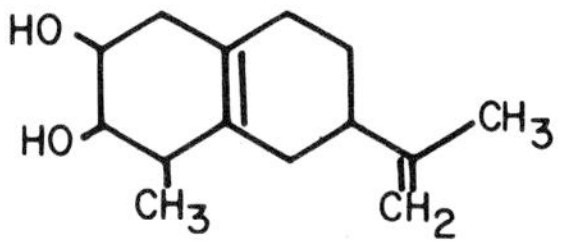
		lyubimin	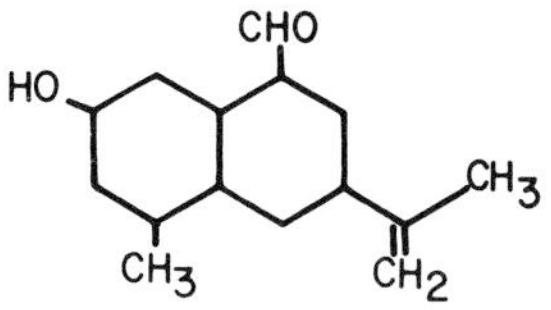
SWEET POTATO	Ipomoea batatas	ipomeamarone	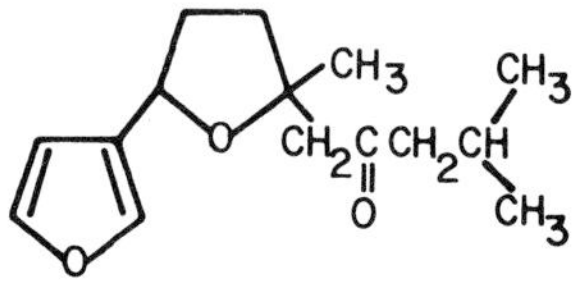
COTTON	Gossypium spp.	gossypol	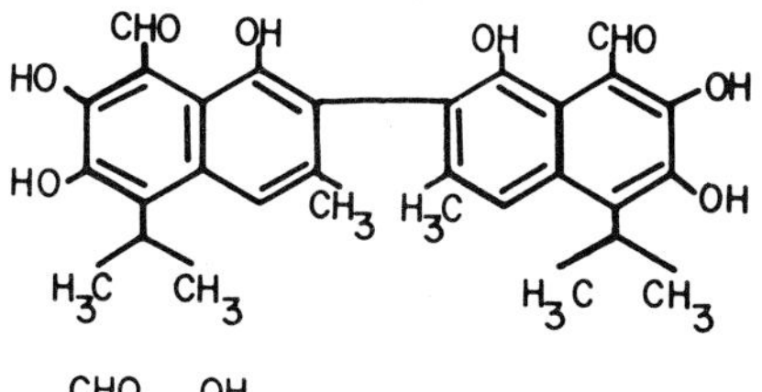
		vergosin	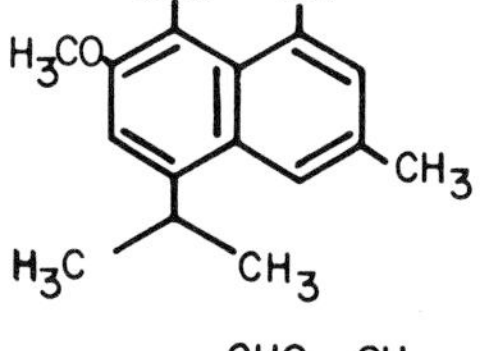
		hemigossypol	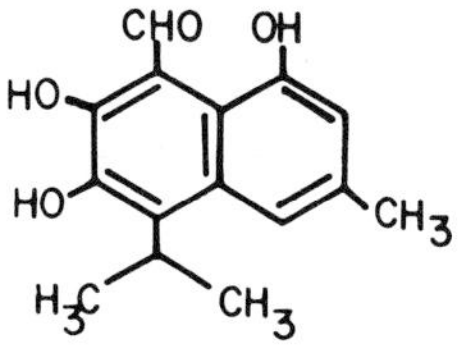
SAFFLOWER	Carthamus tinctorius	safynol	$H_2C(OH)-CH(OH)-CH=CH-[C\equiv C]_3-CH=CH-CH_3$

Fig. 3.

additional host responses that cumulatively render the pea plant resistant to *Fusarium solani*.

It is unfortunate that most of this work has been done with detached pea pods and fungi or compounds that seldom come into contact with endocarp tissue.

BEAN.—Phytoalexins also have been studied extensively in the garden bean (*Phaseolus vulgaris*). In fact, the bean was the second plant from which Müller obtained evidence for his phytoalexin theory (33). Phaseollin (Fig. 2) is produced by a wide range of *P. vulgaris* cultivars as well as by *P. lunatus*, *P. radiatus*, and *P. leucanthus*, and it has a wide spectrum of inhibitory activity against fungi (11). Phaseollin may not inhibit bacteria, but some bacteria induce phaseollin production.

Kuć (26) and others have worked on induced plant immunity, in which phaseollin plays an important role. Cruickshank and Perrin (20) isolated a small protein from *Monilinia fructicola* which they called monilicolin A. This interesting protein is capable of stimulating phaseollin production in bean pods but not pisatin production in pea.

Van Etten and Bateman (46) have done the only studies on the mechanism of action of a phytoalexin (phaseollin). Phaseollin damages membranes in the bean plant as well as in fungi.

More recent work has implicated several more compounds in the disease resistance of bean. Several workers (5, 42) have isolated at least three additional biologically active compounds from bean hypocotyls. These compounds are kievitone, phaseollinisoflavan, and 2'-methoxyphaseollinisoflavan (Fig. 2). It will be interesting to see whether they are also produced and are active in the disease resistance of roots.

Partridge and Keen (*personal communication*) found that kievitone is produced by cowpea hypocotyls after inoculation with fungi.

SOYBEAN.—Phytoalexin production in soybeans was first reported by Uehara, who used soybean seed pod cavities inoculated with spore suspensions of *Fusarium* spp. Klarman and Gerdemann (24) showed that a water-soluble phytoalexin could be leached out of the hypocotyls of intact plants. Additional studies were made by Klarman and others, but it wasn't until later that Keen et al. (23) identified the phytoalexin from soybeans as 6-α-hydroxyphaseollin (Fig. 2). Biehn (26) showed that several phenolic compounds are formed in soybean-fungus interactions.

Phytoalexin production is reversibly blocked in soybeans by heat treatment, and a synchronous loss and return of disease resistance accompanies this effect on phytoalexin production. Disease resistance can also be lost by leaching the hypocotyls with string wicks or by growing plants in the dark (24). If the compounds leached out of the hypocotyls are returned to the hypocotyl, disease resistance is partially restored (7). Frank and Paxton found another, yellow-colored phytoalexin that is produced very early in the Phytophthora-soybean interaction (20). Several fungi produced compounds in culture media that stimulate production of this phytoalexin in soybean (14), much as monilicolin A stimulates phaseollin production in bean. This

phytoalexin may play a role in the cross-protection of soybeans against *Phytophthora* spp. Meyer et al. (20) reported that soybean roots can also produce phytoalexin. Keen and Horsch (22) found no difference between the rates of production of hydroxyphaseollin in resistant and susceptible soybean root systems inoculated with *Phytophthora megasperma* var. *sojae*. They accordingly warned against the use of "unnatural" host parasite systems, even though the field expression of disease resistance of soybeans to this fungus occurs primarily in the root system. Their point is well taken, insofar as most of the systems studied in disease resistance and phytoalexin production are far from "natural."

RED CLOVER AND ALFALFA.—Several antifungal compounds have been isolated from red clover, including formononetin, biochanin A, and trifolirhizin (Fig. 2). Maackiain (Fig. 2) occurs in the roots of red clover as the glucoside trifolirhizin and is found in root extracts only after these extracts have been allowed to incubate with their natural glucosidases. It is my opinion that many biologically inhibitory compounds are stored in tissues as their relatively inactive glycosides, to be released upon cell injury and thus to act in protecting tissue from invasion.

Another pterocarpan has been isolated from alfalfa. This phytoalexin, medicarpin (20), has not been studied in roots.

The structural similarity of many of the phytoalexins isolated from the leguminosae is striking. It may imply that legumes had a common ancestor (as taxonomists believe) and evolved similar biochemical mechanisms of disease resistance.

CARROT.—Carrot roots accumulate 6-methoxymellein (Fig. 3), earlier known as carrot isocoumarin, when inoculated with several different fungi or even when stored (18). Various forms of stress cause the production of a range of phenolic compounds in carrot (9).

The most effective elicitor of 6-methoxymellein production in carrot is *Ceratocystis fimbriata*, a fungus that can produce closely related compounds. *Stemphylium radicinum*, a storage pathogen of carrot, does not induce 6-methoxymellein production. Other compounds must play a role in the resistance of carrot roots to *C. fagacearum*, *C. pilifera*, and *C. coerulescens* (19). Several other antifungal compounds, including benzoic acid and chlorogenic acid (Fig. 3), have been isolated from carrots. Chalutz et al. (6) doubt that 6-methoxymellein is important in the disease resistance of carrots.

POTATO.—The phytoalexin concept was first proposed by Müller and Börger (33) when they were studying the nature of resistance in potato to *Phytophthora infestans*. They could not explain the differential resistance of potato varieties to races of this fungus in terms of preformed compounds. They postulated that postinfectional antifungal metabolites of potato are responsible for the observed resistance. Nevertheless, a multitude of preformed compounds that are inhibitory to microorganisms in potato have been studied. Probably one of the most common antifungal compounds found in plants, including potato, is chlorogenic acid (Fig. 3) (27). However, its role in disease resistance is still being debated (21, 36). Chlorogenic acid occurs in potato skins in concentrations high enough to be inhibitory to some fungi (2). Differences in concentration between resistant and susceptible varieties have been small enough to send researchers looking for other compounds, including quinone oxidation products from chlorogenic acid and polyphenols, to explain resistance.

Other inhibitory phenols in potato tissue are caffeic acid, scopolin, α-solanine and α-chaconine, ρ-hydroxybenzoic acid, o-hydroxybenzoic acid, and 4-hydroxy-3-methoxybenzoic acid. The role of these compounds in disease resistance has been overshadowed by the discovery of the phytoalexin rishitin (Fig. 3) by Tomiyama and others (20, 40). Rishitin may have been the compound upon which Müller and Börger (33) built the phytoalexin theory, but they did not isolate it. Rishitin has been studied by several workers and is now generally considered to be important (40). Injury induces α-solanine and α-chaconine accumulation, but their synthesis is suppressed when rishitin accumulates in inoculated tissue. Rishitin synthesis is not induced by heavy metal ions, other compounds, or injury. More study is necessary to determine the role of rishitin in the disease resistance of aerial parts of the potato plant. Most of the work has been done on tuber slices, and none has been done on roots.

Two more phytoalexins, lyubimin (Fig. 3) (32) and phytuberin, have been partially characterized from potato. Metlitskii et al. (32) reported that production of lyubimin can be induced by some proteins isolated from *P. infestans*. Recent work by Zalewski and Sequeira (53) indicates that yet additional antimicrobial compounds are active in resistance to *Pseudomonas solanacearum*.

Another member of the Solanaceae, tomato, produces rishitin (41) as well as α-tomatine. The steroid glycoalkaloid α-tomatine may play a role in the resistance of tomato to *Fusarium oxysporum* f. sp. *lycopersici*. Brueske and Dropkin (4) suggest that free phenols may be involved in the resistance of tomato roots to the root knot nematode.

SWEET POTATO.—Ipomeamarone (Fig. 3) has been isolated from sweet potato roots infected by *Ceratocystis fimbriata*, and its biochemistry has been studied extensively (1, 25). Healthy uninoculated roots do not contain ipomeamarone, but inoculated resistant roots produce as much as 40 mg/gram fresh tissue weight; susceptible roots produce less than half as much. Weber and Stahmann (49) were able to induce immunity to *C. fimbriata* in sweet potato roots, but they were unable to detect significant amounts of ipomeamarone in this immune tissue.

Many other compounds, including chlorogenic acid and scopolin, have been isolated from sweet potato roots, but little is known about their role in disease resistance.

COTTON.—Cotton roots contain and secrete gossypol (Fig. 3). Bell (20) ascribed resistance against *Verticillium albo-atrum* to the rapid increase in gossypol

concentration in the xylem sap of resistant plants. Gossypol is also toxic to insects and animals, as might be expected for a biologically active compound.

Two more active antifungal compounds, vergosin (Fig. 3) and hemigossypol (Fig. 3), have been isolated from inoculated cotton stems by Zaki et al. (52). They state that de novo production of gossypol-related compounds may constitute a defense mechanism of cotton to *Verticillium*, but more work is needed to prove this (52). Schnathorst and Mathre (20) were able to cross-protect cotton with strains of *V. albo-atrum*.

CEREALS.—Virtanen et al. (48) reported the occurrence of 2(3)-benzoxazolinone in rye seedling roots. This compound and another closely related compound, 6-methoxy-2(3)-benzoxazolinone, obtained from maize and wheat, inhibited the growth of *Fusarium nivale*. However, it is more likely that the glucoside and aglycone precursors of these compounds function in disease and insect resistance (47).

Barley produces hordatines that are antifungal compounds (30). Stoessl (43) also isolated ρ-coumaroylagmatine from barley, but it was only weakly antifungal.

SAFFLOWER.—Safynol (Fig. 3) is the only phytoalexin that has been isolated from the Compositae (44). Its occurrence in hypocotyls and its role in disease resistance have been studied by several workers (20). The occurrence of safynol in roots and related compounds in other composites is an area that needs study.

AVOCADO.—A. I. Zaki (*personal communication*) has recently isolated borbonyl acetate (Fig. 1), an antifungal compound, from the roots of *Persea borbonia*. It is active against *Phytophthora cinnamomi*, a pathogen of the edible avocado, *P. americana*, which does not have this compound in its roots. A concentration of 10 ppm is approximately the ED_{50} against mycelial growth of *P. cinnamomi*.

MAHONIA.—Greathouse and Watkins (16) found that berberine (Fig. 1) is important in the resistance of *Mahonia trifoliolata* and *M. swaseyi* to *Phymatotrichum omnivorum*.

TULIP.—Tulipalin A (Fig. 1) inhibits *Fusarium oxysporum* f. sp. *tulipae*, preventing invasion of the white skin of tulip bulbs (3).

WALNUT.—Most tissues of the walnut tree contain a relatively high concentration of hydrojuglone glucoside (Fig. 1). The roots of walnut trees inhibit many other plants (17), and isolating soil microorganisms like *Phytophthora* spp. from its roots can be difficult, probably due to the release of juglone (Fig. 1) from hydrojuglone glucoside when walnut cells are ruptured (17, 37). Phytotoxic materials affecting tomatoes can diffuse into the soil from walnut leaves. This is another example of a toxic compound that is stored in the plant as a relatively harmless glycoside but is released when the cells are injured.

MARIGOLDS.—Two nematicides, thiophene I and II, have been isolated from the roots of marigold (45), but whether these are the only compounds responsible for the suppression of nematode populations around marigolds remains undetermined.

OTHER PLANTS.—Several other plants contain less studied compounds in their roots that may play important roles in disease resistance. Strawberry roots apparently produce phytoalexins in response to *Phytophthora fragariae* (35). Peach roots contain amygdalin, which may yield hydrolysis products toxic to many organisms (38). Asparagus that is resistant to *Trichodorus christiei* contains a toxic glycoside in its roots (39). Scheffer contends that *Eragrostis curvula* is resistant to *Meloidogyne* spp. because of high pyrocatechol concentrations in the roots (38). Hydroxymatairesinol is formed during the interaction of *Fomes annosus* and Norway spruce and may play an important role in disease resistance. Volatile compounds like terpenes affect other plants and soil bacteria and therefore play a role in plant-plant and plant-microorganism interactions (34). And a multitude of compounds that have no known roles in the plant have been isolated from roots.

CONCLUSION.—The physiology and biochemistry of disease resistance in plant roots are ripe areas for research. Plant roots in natural soil are surrounded by a multitude of microorganisms in a relatively moist environment that is seemingly ideal for plant infection. Yet relatively few organisms invade a given root system. Evolution has, of course, fostered this situation. In proportion to what we know about disease resistance in the aerial portions of plants, however, we know little about the below-ground portions. It certainly is possible that the same compounds formed by the plant above ground can be formed below ground, yet there are examples where this is not true (nicotine in *Nicotiana* spp., for example). The effect of mycorrhizal infections on plant roots and their diseases, first studied in orchids, is an area that should be studied more in the future (31). In the long run, humankind will be the beneficiary of an increased understanding of "nature's fungicides."

LITERATURE CITED

1. AKAZAWA, T., I. URITANI, and H. KUBOTA. 1960. Isolation of ipomeamarone and two coumarin derivatives from sweet potato roots injured by the weevil Cylas formicarius elegantulus. Arch. Biochem. Biophys. 88:150-156.
2. ALLEN, E. H. 1970. The nature of antifungal substances in the peel of Irish potato tubers. Phytopathol. Z. 69:151-159.
3. BERGMAN, B. H. H., J. C. M. BEIJERSBERGEN, J. C. OVEREEM, and A. KAARS SIJPESTEIJN. 1967. Isolation and identification of α-methylenebutyrolactone, a fungitoxic substance from tulips. Rec. Trav. Chim. Pays-Bas 86:709-714.
4. BRUESKE, C. H., and V. H. DROPKIN. 1973. Free phenols and root necrosis in Nematex tomato infected with the root knot nematode. Phytopathology 63:329-334.
5. BURDEN, R. S., J. A. BAILEY, and G. W. DAWSON. 1972. Structures of three new isoflavanoids from Phaseolus vulgaris infected with tobacco necrosis virus. Tetrahedron Letters 41:4175-4178.
6. CHALUTZ, E., J. E. DEVAY, and E. C. MAXIE. 1969. Ethylene-induced isocoumarin formation in carrot root tissue. Plant Physiol. (Lancaster) 44:235-241.
7. CHAMBERLAIN, D. W., and J. D. PAXTON. 1968. Protection of soybean plants by phytoalexin. Phytopathology 58:1349-1350.
8. CHRISTENSON, J. A., and L. A. HADWIGER. 1973. Induction of pisatin formation in the pea foot region by pathogenic and nonpathogenic clones of Fusarium solani. Phytopathology 63:784-790.
9. COXON, D. T., R. F. CURTIS, K. R. PRICE, and G. LEVETT. 1973. Abnormal metabolites produced by Daucus carota roots stored under conditions of stress. Phytochemistry 12:1881-1885.
10. CRUICKSHANK, I. A. M., and D. R. PERRIN. 1965. Studies on phytoalexins. IX. Pisatin formation by cultivars of Pisum sativum L. and several other Pisum species. Austral. J. Biol. Sci. 18:829-835.
11. CRUICKSHANK, I. A. M., and D. R. PERRIN. 1971. Studies on phytoalexins. XI. The induction, antimicrobial spectrum and chemical assay of phaseollin. Phytopathol. Z. 70:209-229.
12. DEVERALL, B. J. 1972. Phytoalexins. Annu. Proc. Phytochem. Soc. 8:217-233.
13. FISCH, M. H., B. H. FLICK, and J. ARDITTI. 1973. Structure and antifungal activity of hircinol, loroglossol and orchinol. Phytochemistry 12:437-441.
14. FRANK, J. A., and J. D. PAXTON. 1971. An inducer of soybean phytoalexin and its role in the resistance of soybeans to Phytophthora rot. Phytopathology 61:954-958.
15. GÄUMANN, E., and H. KERN. 1959. Über die Isolierung und den chemischen Nachweis des Orchinols. Phytopathol. Z. 35:347-356.
16. GREATHOUSE, G. A., and G. M. WATKINS. 1938. Berberine as a factor in the resistance of Mahonia trifoliolata and Mahonia swaseyi to Phymatotrichum root rot. Amer. J. Bot. 25:743-748.
17. GRIES, G. A. 1934. Juglone—the active agent in walnut toxicity. No. Nut Growers Assn. Annu. Rept. 34:52-55.
18. HAMPTON, R. 1962. Changes in phenolic compounds in carrot root tissue infected with Thielaviopsis basicola. Phytopathology 52:413-415.
19. HERNDON, B. A., J. KUC, and E. B. WILLIAMS. 1966. The role of 3-methyl-6-methoxy-8-hydroxy-3,4-dihydroisocoumarin in the resistance of carrot root to Ceratocystis fimbriata and Thielaviopsis basicola. Phytopathology 56:187-196.
20. INGHAM, J. L. 1972. Phytoalexins and other natural products as factors in plant disease resistance. Bot. Rev. 38:343-424.
21. JOHNSON, G., and L. A. SCHAAL. 1952. Relation of chlorogenic acid to scab resistance in potatoes. Science 115:627-629.
22. KEEN, N. T., and R. HORSCH. 1972. Hydroxyphaseollin production by various soybean tissues: a warning against use of "unnatural" host-parasite systems. Phytopathology 62:439-442.
23. KEEN, N. T., J. J. SIMS, D. C. ERWIN, E. RICE, and J. W. PARTRIDGE. 1971. 6-α-Hydroxyphaseollin: an antifungal chemical induced in soybean hypocotyls by Phytophthora megasperma var. sojae. Phytopathology 61:1084-1089.
24. KLARMAN, W. L., and J. W. GERDEMANN. 1963. Resistance of soybeans to three Phytophthora species due

to the production of a phytoalexin. Phytopathology 53:1317-1320.

25. KUBOTA, T., and T. MATSUURA. 1953. Chemical studies on the black rot disease of sweet potato. VI. On the chemical structure of ipomeamarone. J. Chem. Soc. Japan 74:248-251.

26. KUĆ, J. 1972. Phytoalexins. Annu. Rev. Phytopathol. 10:207-232.

27. KUĆ, J., R. E. HENZE, A. J. ULLSTRUP, and F. W. QUACKENBUSH. 1956. Chlorogenic and caffeic acids as fungistatic agents produced by potatoes in response to inoculation with Helminthosporium carbonum. J. Amer. Chem. Soc. 78:3123-3125.

28. LINK, K. P., H. R. ANGELL, and J. C. WALKER. 1929. The isolation of protocatechuic acid from pigmented onion scales and its significance in relation to disease resistance in onions. J. Biol. Chem. 81:369-375.

29. LINK, K. P., and J. C. WALKER. 1933. The isolation of catechol from pigmented onion scales and its significance in relation to disease resistance in onions. J. Biol. Chem. 100:379-383.

30. LUDWIG, R. A., E. Y. SPENCER, and C. H. UNWIN. 1960. An antifungal factor from barley of possible significance in disease resistance. Can. J. Bot. 38:21-29.

31. MARX, D. H. 1972. Ectomycorrhizae as biological deterrents to pathogenic root infections. Annu. Rev. Phytopathol. 10:429-454.

32. METLITSKII, L. V., O. V. OZERTSKOVSKAYA, N. S. VULFSON, and L. I. CHALOVA. 1971. The role of lyubimin in potato resistance to Phytophthora infestans and its chemical identification. Mikol. Fitopatol. 5:439-443.

33. MÜLLER, K. O., and H. BÖRGER. 1940. Experimentelle Untersuchungen über die Phytophthora—Resistenz der Kartoffel. Arb. Biol. Reichsanstalt. Landu. Forstw. (Berlin) 23:189-231.

34. MÜLLER, W. H. 1965. Volatile materials produced by Salvia leucophylla: effects on seedling growth and soil bacteria. Bot. Gaz. 126:195-200.

35. MUSSELL, H. W., and R. C. STAPLES. 1971. Phytoalexin-like compounds apparently involved in strawberry resistance to Phytophthora fragariae. Phytopathology 61:515-517.

36. PATIL, S. S., R. L. POWELSON, and R. A. YOUNG. 1964. Relation of chlorogenic acid and free phenols in potato roots to infection by Verticillium albo-atrum. Phytopathology 54:531-535.

37. PAXTON, J. D., and E. E. WILSON. 1965. Anatomical and physiological aspects of branch wilt disease of Persian walnut. Phytopathology 55:21-26.

38. RHODE, R. A. 1972. Expression of resistance in plants to nematodes. Annu. Rev. Phytopathol. 10:233-252.

39. RHODE, R. A., and W. R. JENKINS. 1958. Basis for resistance of asparagus officinalis var. altilis L. to the stubby root nematode Trichodorus christiei Allen. Md. Agr. Exp. Stn. Bull. A-97. 19 p.

40. SATO, N., K. KITAZAWA, and K. TOMIYAMA. 1971. The role of rishitin in localizing the invading hyphae of Phytophthora infestans in infection sites at the cut surface of potato tubers. Physiol. Plant Pathol. 1:289-295.

41. SATO, N., K. TOMIYAMA, N. KATSUI, and T. MASAMUNE. 1968. Isolation of rishitin from tomato plants. Ann. Phytopathol. Soc. Japan 34:344-345.

42. SMITH, D. A., H. D. VAN ETTEN, and D. F. BATEMAN. 1973. Kievitone: the principal antifungal component of "substance II" isolated from Rhizoctonia-infected bean tissues. Physiol. Plant Pathol. 3:179-186.

43. STOESSL, A. 1970. Antifungal compounds produced by higher plants. Rec. Advan. Phytochem. 3:143-180.

44. THOMAS, C. A., and E. H. ALLEN. 1970. An antifungal polyacetylene compound from Phytophthora-infected safflower. Phytopathology 60:261-263.

45. UHLENBROEK, J. H., and J. D. BIJLOO. 1959. Investigations on nematocides. II. Structure of a second nematocidal principle isolated from Tagetes roots. Rec. Trav. Chim. Pays-Bas 78:382.

46. VAN ETTEN, H. D., and D. F. BATEMAN. 1971. Studies on the mode of action of the phytoalexin phaseollin. Phytopathology 61:1363-1372.

47. VIRTANEN, A. I. 1961. Some aspects of factors in the maize plant with toxic effects on insect larvae. Acta Chem. Fenn. 34B:29-31.

48. VIRTANEN, A. I., P. K. HIETALA, and O. WAHLROOS. 1957. Antimicrobial substances in cereals and fodder plants. Arch. Biochem. Biophys. 69:486-500.

49. WEBER, D., and M. STAHMANN. 1966. Induced immunity to Ceratocystis infection in sweet potato root tissue. Phytopathology 56:1066-1070.

50. WIT-ELSHOVE, A. DE, and A. FUCHS. 1971. The influence of the carbohydrate source on pisatin breakdown by fungi pathogenic to pea (Pisum sativum). Physiol. Plant Pathol. 1:17-24.

51. WOODS, F. W. 1960. Biological antagonisms due to phytotoxic root exudates. Bot. Rev. 26:546-569.

52. ZAKI, A. I., N. T. KEEN, and D. C. ERWIN. 1972. Implication of vergosin and hemigossypol in the resistance of cotton to Verticillium albo-atrum. Phytopathology 62:1402-1406.

53. ZALEWSKI, J. C., and L. SEQUEIRA. 1973. Inhibition of bacterial growth by extracts from potato tissues. Phytopathology 63:942-944.

COLLOQUIUM I

SOIL FUNGISTASIS AND LYSIS

COLLOQUIUM II

INTEGRATION OF PESTICIDE-INDUCED AND
BIOLOGICAL DESTRUCTION OF SOIL-BORNE PATHOGENS

Colloquium on Soil Fungistasis and Lysis: Summary and Synthesis

JOHN L. LOCKWOOD—*Department of Botany and Plant Pathology, Michigan State University, East Lansing.*

This colloquium marked the 20th anniversary of the first report of the general and widespread fungistatic effect of soil (4), and provided an opportunity for us to measure our progress in understanding this phenomenon. The large number of participants reflected a healthy resurgence of interest in fungistasis. Following the initial publication by Dobbs and Hinson in 1953 (4), attention focused on attempts to demonstrate the presence of inhibitory factors in soil. However, these proved elusive, and interest declined. In the late 1960's, the proposal of nutrient relations as an alternative hypothesis (8) led to a revival of interest, and this has now been followed by a new quickening of interest due to the discovery of volatile inhibitory compounds in some soils (6, 7). Three general hypotheses to explain fungistasis were put forward in this colloquium: (a) preformed nonvolatile inhibitors, (b) preformed volatile inhibitors, and (c) inhibition resulting from microbial utilization of propagule exudates.

FUNGISTASIS.—*Preformed nonvolatile inhibitors.*—Vaartaja and Basith, and Vaartaja, presented evidence for the presence of fungal inhibitors in aqueous soil extracts. When such extracts were fractionated using molecular sieve chromatography, inhibitory fractions with molecular weights ranging from $< 2,000$ to $> 80,000$ were found from a number of different soils. Thus far, relatively few propagules other than mycelia of *Pythium ultimum* have been used in bioassays, but if the inhibitors can be shown to be of widespread occurrence and to affect propagules of a wide range of fungi, such results may represent a major advance.

Preformed volatile inhibitors.—Perhaps the most provocative recent new line of work in the fungistasis field is that dealing with volatile inhibitors. Five abstracts provided evidence for the involvement of these materials in fungistasis, among which were allyl alcohol (Balis, Kouyeas), an inorganic compound (Ko and Hora), an unidentified substance released by elevating soil pH (Hora and Baker), and ethylene (Smith). Allyl alcohol was identified in emanations from an Irish soil. In a model system, the compound inhibited germination of spores, and its inhibitive effect was nullified by nutrients. Kouyeas showed that germination of spores of several fungi was improved when $AgNO_3$, which complexes with unsaturated hydrocarbons, was interposed between soil and spores on agar placed above the soil. Based on the differential effect of $AgNO_3$ solutions of different pH in nullifying the volatile inhibitors from various soils, Kouyeas postulated the existence of several inhibitors, but considered allyl alcohol to be the most important.

Ethylene was reported widespread in Australian soils of pH 7.0. Passing an air stream over soil resulted in mold growth on the soil; this was prevented by introducing ethylene in the air stream at concentrations occurring naturally in the soil. The inorganic volatile compound in alkaline soils of Hawaii (Ko and Hora) and the inhibitor released on raising the pH of soil (Hora and Baker) may possibly be ammonia released from ammonium salts. Ko, in remarks from the floor, reported preliminary evidence that the inorganic volatile is derived from NH_4Cl, and B. Schippers verbally reported that chitin-amended soil evolved ammonia that was inhibitory to *Aspergillus flavus*.

Inhibition resulting from microbial utilization of propagule exudates.—Research by Bristow and Lockwood, Blakeman, Jackson and Knight, and Green and Ayanru showed that model systems imposing a diffusion stress reduce germination of propagules that do not require an exogenous source of nutrients. Such model systems appear to operate through removal of spore nutrients that are capable of initiating germination of such propagules. They are designed to imitate a similar stress imposed through microbial utilization of spore nutrients. That exudates from spores of this kind are readily utilized by microorganisms was shown by the rapid evolution of $^{14}CO_2$ from soil following amendment with labeled exudate (Bristow and Lockwood), by the development of a halo of bacteria surrounding conidia of *Botrytis cinerea* on leaves (Blakeman), and by electron micrographs showing spores of *Helminthosporium sativum* rapidly colonized by bacteria and actinomycetes in soil (Old and Wong).

The exudation of germination stimulatory nutrients by nutrient-independent propagules may be of survival value by providing the propagule a signal of the competitive condition of the soil; under severe competitive stress resulting in rapid microbial utilization of spore nutrients, germination would tend to be withheld.

The operation of this system of inhibition may only be temporary. Conidia of *Helminthosporium* or *Botrytis* incubated in conditions of diffusion stress that resulted in a depletion of endogenous nutrients became dependent on a source of exogenous nutrients after but a few days (Bristow and Lockwood, Blakeman). Similarly, Dix showed that spores of *Drechslera rostrata* lost electrolytes as they aged in vitro, and that such loss was correlated with an increased sensitivity to fungistasis. The fact that inhibition tends to be greater on soil than in the leaching model systems (Bristow and Lockwood, Jackson and Knight) suggests that inhibitory products of microbial metabolism of spore exudates may participate in the inhibition, although direct evidence for such products is so far lacking.

With the onset of nutrient dependence, the simplest extension of the mechanism proposed above to account for sustained inactivity would be the maintenance of an energy-deprived environment through microbial activity. Such a mechanism could, of course, also apply to propagules initially dependent upon exogenous nutrients (8).

Residual fungistasis.—This type, in which an inhibition persists after soil sterilization, was discussed by Hora and Ko and by Dobbs. It apparently occurs in relatively fewer soils and is causally distinct from the generally occurring microbially induced fungistasis. From Dobbs's earlier work (3) and from that of Hora and Ko, it appears that the primary cause is inorganic ions of different kinds. Both groups have shown that vegetation can remove the inhibitory effect; in Hawaii (Hora and Ko), this appears to be accomplished by uptake of an inhibitory Al ion by vegetation.

Lysis.—Two types of lysis occurring in soil were discussed: (a) the perforation of melanized spores after long incubation and (b) the rapid and very common maceration of nonmelanized fungal structures. Two most interesting papers by Old and Wong and by Clough and Patrick reported the colonization and apparent penetration of melanized culture- and soil-produced spores by soil microorganisms. Although it has not yet been possible to reproduce this effect with microorganisms isolated from the spores or soil, a causal relationship is strongly suspected. The problem is of great interest because of the possibilities of an effect on survival of the spores and of the involvement of unusual kinds of microorganisms. Clough and Patrick found that spores in soils in which perforations occurred lost viability more quickly than those in soils where it did not. Moreover, perforated spores were always dead. For unknown reasons, previously killed spores apparently are not subject to perforation.

The general, rapidly destructive lysis has many features in common with fungistasis in that its occurrence in soils is widespread, it is of microbial origin, and its onset is delayed by nutrients. The latter point was confirmed by the report of Bumbieris. Honour and Tsao showed that oogonia of *Phytophthora parasitica*, in contrast to antheridia and hyphae, are selectively melanized in soil, and this appears to protect them against lysis (1).

COMMENTS.—*Fungistasis.*—Future work must be directed toward assessment of the relative contribution of the various factors proposed in the fungistasis phenomenon, including the extent to which different mechanisms may be involved in different situations, and the possibility that there predominates a single fundamental causal factor upon which secondary factors may be superimposed. Some directions toward this end that were suggested by this colloquium are outlined below.

A) The question of nutrient levels and distribution in soil is critical to any distinction between the nutrient deprivation hypothesis and one based on preformed inhibitors. Considerations of known energy input into the soil, microbial population levels, and energy requirements for their maintenance indicate that the soil must be chronically deficient in energy (5). Accordingly, such nutrients as are available are likely to be concentrated largely around such comparatively nutrient-rich foci as root surfaces and undecomposed organic matter. This view conforms with a nutritional interpretation of fungistasis in which germination is primarily restricted to such comparatively nutrient-rich substrates. By contrast, the inhibitor hypothesis requires that both nutrients and inhibitors be distributed widely throughout the soil. These alternatives could be tested.

B) It may be useful to differentiate between pre-existing inhibitors and those formed in situ by microbial metabolism of spore exudates. Results of Bristow and Lockwood and of Jackson and Knight indicate indirectly that inhibitory substances might be formed from exudates of nutrient-independent spores. This seems almost a necessity if one is to propose an inhibitor hypothesis for fungistasis on leaves. Leaves, incidently, represent a much simpler system than soil and thus may have advantages for studying fungistasis. A point made verbally by S. H. F. Chinn—that it is difficult to account for inhibition persisting in soil diluted 200-fold with silica on the basis of preformed inhibitors—is also pertinent in this context.

C) Knowledge of whether propagules require an exogenous energy source for germination may help clarify the fungistatic mechanism. For example, failure of nutrient-dependent propagules to germinate in natural or sterilized subsoils might be attributed to the presence of an inhibitor (in sterilized subsoil a "residual" one), when it could as well be due simply to a deficiency of exogenous energy (14). Conclusions drawn from the use of nutrient-dependent propagules as assay organisms on agar disks removed after incubation on soil are subject to the same constraint. The inclusion of one or more kinds of nutrient-independent propagules for comparison would help provide evidence one way or the other in such situations.

D) The results obtained with inhibitory soil extracts and emanations need to be related ecologically to the soil microsites in which the propagules exist, with respect to inhibitor concentrations and the involvement of other factors such as nutrients and pH. Much was said in the colloquium about germination or its absence being the result of a balance between nutrients that stimulate germination and inhibitors resulting from microbial metabolism of the nutrients (15). Since the proportion of inhibitors to nutrients rather than absolute amounts is assumed to be critical, fungistasis in high or low energy-containing soil is explained. Such systems were postulated by Smith for ethylene-induced fungistasis and by Bumbieris for lysis of *Phytophthora cinnamomi* hyphae. Possibly, the fungal inhibition by yeasts on plant leaves in the presence of pollen may also represent such a system (Fokkema). Attractive as the hypothesis may be, there are as yet too few data by which to judge its applicability.

The use of $AgNO_3$ as a presumptive diagnostic reagent for the presence of unsaturated volatiles (Balis, Kouyeas) may be complicated by the possible partial sterilization of soil by the vapor phase. Moreover, $AgNO_3$ was not effective with ethylene (Smith). One of the major points of Smith's argument for ethylene involvement in fungistasis was that it is "on scale" with respect to the concentration necessary for inhibition of fungal growth when passed over soil. However, one needs to know whether soil might accumulate a concentration higher than that in the air stream.

E) Expanded use of experimental models would permit quantitative evaluation of isolated components of the soil system, singly or in combinations, without confusion by other factors. Examples are the aqueous leaching models to impose nutrient stress, and systems using specific inhibitors and nutrients. Comparison of fungistatic spectra of such models with those of soil should provide information useful in sorting out the significant factors. In the final analysis, findings derived from model systems and from the use of "unnatural" materials and conditions must eventually be related to field situations. This was brought out verbally by W. C. Snyder with respect to the use of culture-grown spores and of soil in petri dishes, and by Griffin and Pass, who warned that loading up the soil with too many spores can result in crowding inhibition, as occurs in vitro.

F) Questions were rightly raised with respect to whether we have used the natural stimulants of germination in annulment studies. Many have reported that complex organic amendments are more effective than simple materials like glucose (10). In recent work, fatty acids were shown to be components of bean rhizospheres and to stimulate germination of *Thielaviopsis basicola* (11). Blakeman, in this colloquium, reported that gibberellic acid is present in leaf exudates and is capable of stimulating germination of *Botrytis* spores. Another unanswered question is whether germination stimulants always act as energy sources or through some other mechanism. Answers to such questions are clearly important to understanding mechanisms of fungistasis and in relating our findings to natural situations. Chinn pointed out from the floor that annulment might be effected not only through direct stimulation of propagules, but also by "neutralizing" an inhibitory agent or through suppression of the activity of soil microorganisms.

LYSIS.—There was little discussion of the mechanism of lysis in soil, i.e., whether it occurs by heterolysis or by autolysis, and if by autolysis, the nature of its induction. Ko and I previously reported (9) on evidence that led us to view it as autolytic and induced by carbon starvation. There is also the possibility that toxic microbial metabolites of mycelial exudates might induce autolysis, but, as in the case of fungistatic products of spore exudates, this has not yet been demonstrated. Alternatively, a school of thought represented mainly by Alexander and his colleagues (13) views it basically as heterolytic, although very few instances of lysis of living mycelium typical of that occurring in soil by enzyme preparations are known.

There is a great need for further work on these basic questions. If lysis in soil is an autolytic phenomenon, one may ask why fungi should have evolved such a self-destructive system. Instead, lysis may represent a survival mechanism that is not always successful. It often accompanies formation of resistant structures and may provide for a recycling of materials from which resistant structures can be synthesized during periods of energy shortage. Christias and I (2) induced sclerotial formation where the only possible source of synthetic material was mycelium that underwent concurrent massive lysis. Further work might be directed toward discovering whether this is a general phenomenon or whether carbon starvation is but one of several stimuli for resting structure formation.

CONCLUSIONS.—Our continued progress in understanding fungal development and survival in soil depends on our talent in framing definitive questions and in devising and adapting methods that will provide answers. Perhaps noteworthy in the work reported in this colloquium are the use of (a) ultraviolet spectrophotometry and gas chromatography to detect volatile inhibitors, (b) molecular sieve chromatography to separate different nonvolatile inhibitors and germination-stimulating substances in soil extracts, (c) scanning and transmission electron microscopy to examine intimately the fungal cell and its neighbors in soil, (d) model systems that allow the isolation and quantitative manipulation of factors in the soil system, and (e) radioactively labeled propagules to study the significance of propagule exudates.

I myself feel that the primary causes of fungistasis will turn out to be few, though in so stating I am aware that this is not a unanimously held opinion. The report of Basith and Vaartaja of multiple inhibitors may suggest the opposite, and Dobbs, with his soil particle test, is telling us that different soils may have selective and specific effects on fungi that may be ignored by searching for generalized explanations for fungistasis. However, soil inhibitory spectra, though admittedly sparse (12), and the regularity of occurrence of fungistasis seem to me more compatible with few and simple factors than with many and complex ones. Further, I have been impressed by the similarities in the soil conditions in which fungistasis, lysis, and resistant structure formation are induced. This suggests to me that the same crucial environmental factors may be causally involved in all three phenomena. At least this is an hypothesis worth testing.

Finally, whereas much progress has been made in the past 20 years, final answers obviously are not yet in. But the stage now appears set for solid future progress in our field in terms of a sufficient accumulation of background information and a corps of able researchers. Moreover, there prevailed in this colloquium a certain restraint, even on the part of those who usually are the most partisan contributors. Comments were neither strident nor petty, and conclusions were tentative, suggesting we have reached a stage of maturity in our field that should enhance the prospects for cooperation and rapid advancement.

LITERATURE CITED

1. BLOOMFIELD, B. J., and M. ALEXANDER. 1967. Melanins and resistance of fungi to lysis. J. Bacteriol. 93:1276-1280.

2. CHRISTIAS, C., and J. L. LOCKWOOD. 1973. Conservation of mycelial constituents in four sclerotium-forming fungi in nutrient-deprived conditions. Phytopathology 63:602-605.

3. DOBBS, C. G., and M. J. GASH. 1965. Microbial and residual mycostasis in soils. Nature (London) 207:1354-1356.

4. DOBBS, C. G., and W. H. HINSON. 1953. A widespread fungistasis in soils. Nature (London) 172:197-199.

5. GRAY, T. R. G., and S. T. WILLIAMS. 1971. Microbial productivity in soils, p. 255-286. In D. E. Hughes and A. H. Rose (ed.), Microbes and biological productivity. Cambridge Univ. Press, Cambridge.

6. HORA, T. S., and R. BAKER. 1972. Extraction of a volatile inhibitor from soil inducing fungistasis. Phytopathology 62:1475-1476.

7. KO, W. H., and F. K. HORA. 1972. The nature of a volatile inhibitor from certain alkaline soils. Phytopathology 62:573-575.

8. KO, W. H., and J. L. LOCKWOOD. 1967. Soil fungistasis: relation to fungal spore nutrition. Phytopathology 57:894-901.

9. KO, W. H., and J. L. LOCKWOOD. 1970. Mechanism of lysis of fungal mycelia in soil. Phytopathology 60:148-154.

10. LOCKWOOD, J. L. 1964. Soil fungistasis. Annu. Rev. Phytopathol. 2:341-362.

11. PAPAVIZAS, G. C., and M. F. KOVACS, JR. 1972. Stimulation of spore germination of Thielaviopsis basicola by fatty acids from rhizosphere soil. Phytopathology 62:688-694.

12. SCHUEPP, H., and E. FREI. 1969. Soil fungistasis with respect to pH and profile. Can. J. Microbiol. 15:1273-1279.

13. SKUJINS, J. J., H. J. POTGIETER, and M. ALEXANDER. 1965. Dissolution of fungal cell walls by a streptomycete chitinase and β-(1→3) glucanase. Arch. Biochem. Biophys. 111:358 363.

14. STEINER, G. W., and J. L. LOCKWOOD. 1970. Soil fungistasis: mechanism in sterilized, reinoculated soil. Phytopathology 60:89-91.

15. WATSON, A. G., and E. J. FORD. 1972. Soil fungistasis—a reappraisal. Annu. Rev. Phytopathol. 10:327-348.

ABSTRACTS

The Nature of Volatile Inhibitors Involved in Soil Fungistasis

C. BALIS—Department of Industrial Microbiology, University College, Dublin, Ireland.

Agar disks pre-exposed to soil volatile emanations restricted spore germination of Penicillium chrysogenum. Inhibition was decreased when $AgNO_3$ solution was present, provided it was not in contact with either the soil or the disks. Ultraviolet spectroscopy of water condensates collected from soil headspace showed absorption at 200 nm. After treatment with $AgNO_3$ or bromine, absorption bands appeared at 225 nm and 265 nm, respectively. These results and those reported by Balis and Kouyeas (1968, Ann. Inst. Phytopathol. Benaki. 8:145-149) indicate that a water-soluble, volatile substance with unsaturated carbon bonds is involved in soil fungistasis. Allyl alcohol showed absorption bands similar to the water condensates and at a concentration of 10 ppm inhibited spore germination; its inhibitory effect was nullified in the presence of nutrients.

Inhibition of Pythium ultimum Extracts From Soils

M. BASITH and O. VAARTAJA—Forest Ecology Research Institute, Environment Canada, Ottawa, Ontario.

Pythium ultimum was grown in agar amended (1:1) with either distilled water or soil extracts. Sterilized or unsterilized extracts from soils reduced mycelial growth. Extracts from autoclaved, but not yet recolonized, soils had a lower degree of inhibition. Extracts from dry or abundantly leached soils were not highly inhibitory; some were stimulatory. After a brief lag, during which the inhibition was low, the inhibition became very high for soils amended with sucrose (0.15 mg/cm³ or more) and $(NH_4)_2HPO_4$ (0.03 mg/cm³ or more). With smaller amounts and different ratios of these, the inhibition was less and differed among three soils tested. The inhibition was reduced after (a) addition of nutrients to the extracts, (b) heating the extracts, and (c) filtration, particularly with several layers of paper filters. Pythium ultimum was inhibited by extracts that stimulated Fusarium oxysporum. The data support the hypothesis that inhibitors produced biologically are one cause of soil mycostasis.

Evidence for Fungistasis on Leaf Surfaces

J. P. BLAKEMAN—Department of Botany, University of Aberdeen, Scotland.

The inhibition of germination of Botrytis cinerea spores observed on leaves of chrysanthemum and red beetroot is due, in part, to the presence of bacteria surrounding the spores. Although spores of B. cinerea germinate well in deionized water in vitro, evidence indicated that failure to germinate on leaves was due to removal of nutrients from the spores by bacteria. To simulate the possible action of bacteria, spores were leached on membranes with water. Spores failed to germinate while being leached, and only a small proportion of leached spores could subsequently germinate under static conditions in water. Leached spores could, however, germinate in dextrose or Pfeffer-dextrose solutions. Spores germinated well on membranes during leaching with very dilute dextrose or Pfeffer-dextrose solutions. In tests on agar, most isolates of bacteria obtained from leaves showed a marked ability to absorb and concentrate aniline blue dye, which was associated with development of the polysaccharide sheath. Uptake of nutrients into the polysaccharide sheath of bacteria situated near spores may lead to an increase in concentration gradient from within to outside the spore. Failure of spores to germinate may be due to exhaustion of endogenous nutrients or to uptake by bacteria (simulated by leaching) competing for soluble metabolites and thus preventing incorporation into proteins and other substances normally synthesized during germination. These observations suggest similarities with fungal spores inhibited in soil by fungistasis.

Role of Soil Microbial Energy Sink in Fungistasis of Nutrient-Independent Spores

P. R. BRISTOW and J. L. LOCKWOOD—*Department of Botany and Plant Pathology, Michigan State University, East Lansing.*

Conidia of *Helminthosporium victoriae* and *H. sativum* germinated in distilled water, but lost this capability after 7 days of incubation on soil or on leached sand in a model system designed to imitate the microbial energy sink of soil. Exudates from these conidia stimulated germination when used as the leaching solution in the model system. Exudates contained glucose and ninhydrin-positive materials that were metabolized within minutes by soil microorganisms, preventing their utilization by the spores in germination. To determine the contribution of the energy sink to fungistasis, agar disks containing 0.1% glucose were incubated on soil and on the model system. Germination of *Thielaviopsis basicola* endoconidia was determined on disks removed at intervals of up to 24 hours. Germination per unit of glucose remaining was lower for disks incubated on soil than on the model system. However, no reduction in germination occurred on nutritionally inert discs of polyacrylamide gel similarly incubated and assayed with *H. victoriae* conidia. This suggested that the reduced germination using glucose-agar disks may have resulted from the production of microbial metabolites. Microbial utilization of spore exudates may prevent germination of nutrient-independent spores initially through removal of germination stimulants and the possible production of fungistatic metabolites, and in the long term by rendering the spores nutrient-dependent.

Some Factors Affecting Soil Fungistasis and Lysis

M. BUMBIERIS—*Waite Agricultural Research Institute, University of Adelaide, South Australia.*

Previous work (Bumbieris and Lloyd, 1967, Austral. J. Biol. Sci. 20:103-112; 1169-1172) has shown that the rate at which fungal hyphae lyse in soil is affected by soil moisture, fertility, and certain nutrients. Further work with *Phytophthora cinnamomi* in three different soils showed that the distance hyphae were able to grow before becoming lysed was shortest in sand with the least microbial activity and longest in a pine forest soil where microbial activity was highest. This suggested that the balance between inhibitory factors and available suitable nutrients may affect the behavior of fungal structures in soil. This view has also been expressed by Watson and Ford (1972, Annu. Rev. Phytopathol. 10:327-348). The above assumption is supported to some extent by the following. When the inhibitory factor was diluted by mixing soil with acid-washed kaolin, numbers of germinated zoospores increased in proportion to the quantity of kaolin added. Addition of glucose at various concentrations also increased the number of germinated zoospores, and the distance hyphae were able to grow before becoming lysed increased in proportion to the concentration of the added nutrient.

Microorganisms Associated With Perforated Chlamydospores of Thielaviopsis basicola *in Soil*

KATHERINE S. CLOUGH and Z. A. PATRICK—*Department of Botany, University of Toronto, Ontario, Canada.*

Chlamydospores of the fungus *Thielaviopsis basicola*, a pathogen of tobacco, which remain in field or greenhouse soil for periods of 6 weeks or more, show perforations in the cell wall. The perforations range in size from 0.1 μm to 0.5 μm and extend through the cell wall into the lumen of the cell. The appearance of perforations is accompanied by loss in spore viability. Electron microscope studies describe the structure of the chlamydospore wall as it appears in field soil and the size and morphology of bacteria associated with the cell wall and perforations. We believe that microorganisms cause these perforations; attempts to isolate these organisms are described.

Relationship of Soil Microflora Activities to Mycostasis and Mycolysis

HOANG G. DIEM—*Laboratoire de Botanique et de Microbiologie, Université de Nancy, France.*

To explain the failure of nutritionally independent spores to germinate, Ko and Lockwood (1967) introduced the notion of nutrient sink set up by soil microorganisms. Diem (1974) observed that spores of *Cladosporium cladosporioides* suspended in water could normally germinate despite the presence of dense populations of *Bacillus pumilus* or other bacteria. These bacteria did not aggregate around spores prior to germination. Similar results were obtained by Old (1965) with *Helminthosporium sativum*. Under experimental conditions, these spores did not act as nutrient substrates to be depleted by bacteria. In the hypothesis of Ko and Lockwood, soil microorganisms resemble ectoparasites because their development is ensured to the detriment of sporic substances and hinders the normal functioning of spores. Numerous studies have emphasized the role of volatile substances in mycostasis (Romine and Baker, 1972; Hora and Baker, 1972). Diverse information and the characteristics of studied soils lead us to think that ammonification might be responsible in some cases for soil mycostasis and mycolysis. Watson and Ford (1972), emphasizing the idea of the balance of stimulators and inhibitors, differentiated three dynamic stages of mycostasis: induction, maintenance, and release. But if we consider the three states of spores in soil (degeneration, resting, and germination), the resting state might be regarded as corresponding to the equilibrium position of a beam balance. Spores germinate or die according to whether the effect of stimulators overcomes or does not overcome that of unfavorable factors. Therefore, unfavorable factors are not only inhibitory to germination, but they might also become or include lethal factors to spores.

Change in Response to Soil Fungistasis as Spores Age

N. J. DIX—*Biology Department, The University of Stirling, Scotland.*

The colonization of substrates in soil by fungi depends initially upon the ability of spores to overcome the fungistatic property of soil. The long-term survival of spores as viable propagules in soil depends upon little change in this ability as time passes. However, the response to fungistasis by *Drechslera rostrata* spores changes with age. Initially, during a maturation phase lasting about 3 weeks, spores aged by being stored dry in the laboratory declined in sensitivity from a high level in newly harvested spores to a low level. Over a further period of aging, sensitivity increased again, reaching a high level in 10-week-old spores. Indirect evidence was obtained to support the theory that changes in sensitivity with age are due to the diffusion of electrolytes from spores, resulting in a change in nutrient status. Conductivity of electrolytes diffused from spores in deionized water declined as spore age increased during the maturation period. Sensitivity to soil fungistasis measured by percentage reduction in performance by three parameters of germination was significantly and positively correlated with conductivity over this period.

Residual Soil Mycostasis and the Soil Particle Test

C. G. DOBBS—*School of Plant Biology, University College of North Wales, Bangor, Caernarvonshire, United Kingdom.*

A brief survey in South Australia in 1969 by the agar diffusion method, added to earlier work at Bangor and by D. A. Griffiths in Malaya, plus more recent reports by Ko and Hora in Hawaii, all indicate that residual (i.e., inorganic) soil mycostasis is a widespread, probably worldwide, heterogeneous phenomenon in a small minority of soils. Different test fungi show a wide range of susceptibility, in some cases related to their origin and ecology. A soil particle test (1971, Biol. du Sol 13:31-33) showed that "live" soils tested inhibit the growth of their own quick-growing fungi from particles suspended in agar and may also stimulate the growth of bacteria and actinomycetes from the particles. In controls over glass, some soils gave a significant interaction between microbial and fungal growth from the particles; other did not. Certain autoclaved soils also stimulated such microbial growth. Where this was accompanied (not necessarily always causatively) by a marked suppression of fungal growth, the condition has been referred to as residual (microbial) mycostasis. This, although an artifact due to autoclaving, may have some bearing on what occurs in desert soils or in other soils after fires. The soil particle test, both as a self-soil and reciprocal soil particle test, is now being used at Bangor to study soil fungal and microbial ecology. The selective influence of a soil on the organisms that survive, grow, and interact within it may have been somewhat obscured because of the preoccupation with generalized theories of causation of a necessarily complex phenomenon. A minor point to consider might be the influence of fungal and microbial lysis and nutrient loss on mycostasis, especially in spore masses and mycelial colonies.

Fungistatic Factors in Relation to Two Fungal Pathogens

R. S. DWIVEDI, D. SINGH, and R. P. RAI—*Botany Department, Banaras Hindu University, India.*

Survival of *Fusarium oxysporum* f. sp. *lycopersici* causing tomato wilt and lysis of hyphae and sporangia of *Phytophthora colocasiae* causing leaf blight of *Colocasia antiquorum* were studied in vitro in relation to fungistatic factors of soils. Supplementation of staling growth products of some dominant fungi from tomato fields with nutrient agar reduced the radial growth of *F. oxysporum* f. sp. *lycopersici*. Small segments of tomato roots previously colonized by *F. oxysporum* f. sp. *lycopersici* and buried in sterile soil with inocula of various fungi from tomato fields allowed colonization of only a few fungi. Survival of this pathogen in soil and its colonization of roots of tomato may be attributed to its tolerance to fungistatic factors of soil and to fungistatic products of various fungi growing in its association. Sterile soil samples preinoculated with some dominant soil fungi and actinomycetes caused varying amounts of lysis of hyphae and sporangia of *P. colocasiae*. This was confirmed by (a) inoculating the same microbes in sand mixed with suitable nutrients and placing upon them hyphal inocula of the test fungus, (b) placing hyphae of the pathogen directly on agar substrate supplemented with staling products of fungi, and (c) suspending washed hyphae in liquid cultures of different genera of actinomycetes.

Saprophytic Fungi and Yeasts as Antagonists Against Helminthosporium sativum *on Agar Plates and on Rye Leaves With Pollen*

N. J. FOKKEMA—*Phytopathologisch Laboratorium "Willie Commelin Scholten," Baarn, The Netherlands.*

About 50 isolates of each of the main saprophytic fungal colonizers of rye leaves were tested for growth inhibition of *Helminthosporium sativum* on agar plates. The inhibition at a distance in vitro was most pronounced with *Aureobasidium pullulans*, followed by *Sporobolomyces* spp. Only a few isolates of *Cryptococcus* spp. inhibited at a distance, and all *Cladosporium* isolates grew until contact was made. In contrast to this diversity in antagonistic action in vitro, representative isolates of all four groups behaved similarly in vivo. In the presence of pollen, addition of saprophytes to the inoculum reduced the superficial mycelial development of *H. sativum* and the resulting necrotic leaf area to about 50%. The discrepancy between the antagonistic effects in vitro and in vivo was explained by the finding that the inhibition in dual culture on agar was not derived from the presumed antagonist but from the *Helminthosporium* colony itself.

Effect of Soil Fungistasis on Germination Patterns of Sclerotia of Macrophomina phaseolina

R. J. GREEN, Jr., and D. K. G. AYANRU—*Department of Botany and Plant Pathology, Purdue University, Lafayette, Indiana, and Department of Biological Science, University of Benin, Nigeria.*

Sclerotia of *Macrophomina phaseolina* exhibited an abnormal germination (AG) on the surface of nonsterile soil. Sclerotia on nonsterile soil amended with various sugar, amino acids, and organic acids exhibited mainly normal germination (NG), and hyphae invaded the soil surface. An intermediate germination (IG) pattern occurred with sclerotia undergoing AG followed by addition of appropriate soil amendments. Autoclaved soil induced NG (98%), but there was no mycelial soil invasion. Continuous leaching of sclerotia with sterile deionized water was fungistatic. The following sugars, amino acids, and organic acids stimulated NG (80-98%): maltose, D-ribose, DL-aspartic acid, glutamic acid, L-leucine, DL-phenylalanine, L-serine, and citric, malonic, and tartaric acids. Lactose, L-cystine, urea, and all inorganic mineral salts tested were ineffective, whereas DL-methionine was inhibitory to germination. Alterations in the germination pattern of sclerotia under different soil conditions suggest an internal mechanism for controlling germ tube initiation.

Soil Fungistasis: Comparative Exogenous Carbon and Nitrogen Requirements for Spore Germination in Soil and Axenic Culture

G. J. GRIFFIN and T. PASS—*Virginia Polytechnic Institute and State University, Blacksburg, Virginia, and Morehead State University, Morehead, Kentucky.*

Much greater amounts of exogenous organic C and inorganic N were required per spore for both low ($< 3\%$) and complete ($> 90\%$) chlamydospore germination by *Fusarium solani* in contact with nonsterile soil (indigenous inorganic N = 13.3 $\mu g/g$ soil) than in axenic culture. Increasing the spore density decreased greatly the degree of spore germination in rewetted nonsterile soil (for *F. solani* and *F. solani* f. sp. *phaseoli* macroconidia) and in certain axenic media (for *F. solani* macroconidia and chlamydospores and *Aspergillus flavus* conidia). In soil and axenic culture, *A. flavus* conidia required greater amounts and more complex sources of exogenous C and N for germination than did chlamydospores of fusaria. Large differences in the level of spore germination by these fungi were observed in the rhizosphere and other natural soil habitats. We suggest (a) that the exogenous C and N requirements for spore germination are much greater in the rhizosphere than in axenic culture, (b) that spore germination studies in nonsterile soil as currently performed, at high spore densities (10^5-10^6 spores/g soil), may significantly underestimate the percentage of spore germination that occurs in soil habitats at natural population levels, and (c) that the differential qualitative and quantitative exogenous C and N requirements for spore germination in soil may be related to the occurrence of spore germination in natural soil habitats by *A. flavus* and *F. solani*.

Lysis of Phytophthora parasitica Oospores in Soil

RICHARD C. HONOUR and PETER H. TSAO—*Department of Plant Pathology, University of California, Riverside.*

Oogonia of *Phytophthora parasitica* with associated antheridia and containing oospores were colonized by bacteria and actinomycetes in natural soils. Isolation and enumeration of soil microorganisms lytic to *Phytophthora* hyphal walls revealed that almost all were *Streptomyces* spp. Also, *Streptomyces* spp. were favored in enrichment cultures when a liquid medium, in which oogonia containing oospores were used as the sole carbon source, was seeded with the remains of partially degraded oogonia recovered from natural soil. The walls of oogonia, which were initially exposed to sterilized soils, extracts of sterilized soils, or plant root extracts, became melanized and thickened and persisted in soil for long periods. Antheridia and hyphal remnants were not melanized and were lysed from nonmelanized or melanized oogonia in natural soils or in sterilized soils reinfested with a known lytic *Streptomyces* sp. As observed by light microscopy and scanning electron microscopy, nonmelanized oogonia were colonized by *Streptomyces* spp. within 5 days after exposure to natural soils, antheridia and hyphal remnants were lysed within 10 days, and most oogonia were devoid of oospores within 40 days. The remaining oogonia that contained oospores were slightly melanized. By 120 days, only empty, collapsed, or partially degraded oogonia remained. On the other hand, both the oospores and their melanized and thickened oogonia remained intact after 120 days in reinfested sterilized soil. Oospores with nonmelanized oogonia that were not exposed to any soil treatment remained unchanged. Melanization and thickening of the oogonium wall increased the resistance of oospores to lysis by microorganisms in soil.

Effect of Forestation on Fungitoxicity in Certain Acid Soils

FRANCES K. HORA and W. H. KO—*Department of Plant Pathology, University of Hawaii, Beaumont Agricultural Research Center, Hilo.*

Germination of *Neurospora tetrasperma* ascospores, which are insensitive to the widespread soil fungistasis, was inhibited on topsoils of agricultural Hydrol Humic Latosol but was not inhibited on similar soils collected from a nearby planted forest. All agricultural and some forest subsoils were inhibitory to ascospore germination. There was no correlation between soil pH and intensity of soil fungitoxicity, which is caused by the presence of a free Al ion in soil. In general, forest soils contained a higher percentage of organic matter than agricultural soils. However, there was no correlation between amount of soil organic matter and inhibition of spore germination on agricultural and forest soils collected at different sites or from different depths. All major plant species in the forest contained higher Al concentrations than the sugar cane in the agriculture area. Therefore, it was postulated that a decrease in fungitoxicity in the forest soil results from continuous accumulation of a free Al ion from soil.

Nonbiological Generation of a Volatile Fungistatic Factor in Soil

T. S. HORA and RALPH BAKER—*Department of Botany and Plant Pathology, Colorado State University, Fort Collins.*

Lime, as $Ca(OH)_2:CaCO_3(1:1)$, was applied to raw soil (pH 6.3) and to soil either before or after autoclaving. Soil pH was adjusted to 7.0, 7.5, 8.0, and 8.5 by the addition of lime. Water agar disks were suspended over moistened soil samples in closed petri dishes at 25° C for 24 hours, and germination of four soil fungi was subsequently followed on these disks. Volatile fungistatic activity was observed in 9 of the 10 treatments tested. Application of $MgO:MgCO_3(1:1)$ to soil was equally effective in releasing the volatile inhibitor. A sterile extract was obtained from an aqueous suspension (1:1) of a slightly acid soil. Volatile fungistatic activity was again detectable when 0.2 M Tris [tris (hydroxymethyl)-amino methane] solution or $Ca(OH)_2:CaCO_3(1:1)$ suspension was added to the sterile soil extracts (final pH 9.0 and 8.8, respectively). These observations indicate that at least one step in the series of reactions leading to the formation of volatile fungistatic factor can be induced in soil in the absence of biological activity.

Conidial Germination of Cochliobolus sativus on Soil and Under Artificial Nutrient Stress

R. M. JACKSON and R. A. KNIGHT—*Department of Biology, University of Surrey, Guildford, Surrey, England.*

The theory that soil fungistasis may be due to depletion of nutrients resulting from a sink action of the soil was examined using *Cochliobolus sativus*. Conidia were collected from cultures by a suction apparatus; they had no contact with an aqueous medium until the start of each experiment. Germination was inhibited over soil and also when conidia were exposed to an artificial nutrient depletion gradient by placing them on the surface of dialysis tubing through which distilled water was flowing. Germination occurred slowly under artificial nutrient stress, but within 192 hours the percentage of germinated conidia nearly equaled that of control conidia incubated for 24 hours. When incubated on moist filter paper following exposure to nutrient depletion, conidia germinated better than controls did. Conidia incubated over soil, in contrast to those on dialysis tubes, did not continue to germinate after 24 hours. When those ^{14}C-labeled conidia were washed on a membrane filter in a continuous flow of water, 11.4% of the total label (primarily as glucose) was lost within 31 hours, and nearly half of this amount was lost during the first 10 minutes. Labeled conidia were incubated on dialysis tubes containing static water, flowing water, and an unsterile soil suspension. Inhibition of germination was greatest over the soil suspension, but loss of label to this suspension was least. Germination was greatest over static water, but label recovered from this water was similar to that from the flowing water. These results make it difficult to explain inhibition of germination over soil solely on the basis of nutrient depletion.

Characteristics of a Volatile Inhibitor Isolated From an Alkaline Soil

W. H. KO and FRANCES K. HORA—*Department of Plant Pathology, University of Hawaii, Beaumont Agricultural Research Center, Hilo.*

A volatile inhibitor was extracted with water from an alkaline soil. The inhibitor in the aqueous extract volatilized at pH 8 but not at pH 2. It was heat stable and dialyzable and was exchangeable by cation but not by anion exchange resins. The volatile inhibitor was purified by steam distillation. The acidified aqueous extract was evaporated at 100°C to dryness. The resulting dry powder released the volatile inhibitor when suspended in water at pH 8. The volatile inhibitor in the powder remained active after ignition at 380°C for 12 hours. This suggests the possibility that the volatile inhibitor in the alkaline soil may be an inorganic compound.

The Volatile Fungistatic Factors in Soil

V. KOUYEAS—*Department of Microbiology, Agricultural College of Athens, Greece.*

By using the agar disk technique of Romine and Baker (1972, Phytopathology 62:602-605), it was shown for four soils and eight soil fungi that the presence of $AgNO_3$ solution in the free space between the soil and the assay disk always improved spore germination. The constant enhancement of spore germination by the presence of $AgNO_3$ is considered as indicating the involvement in soil fungistasis of a single or a group of closely related volatile inhibitors.

Lysis of Conidia of Cochliobolus sativus by Soil Microorganisms

K. M. OLD and J. N. F. WONG—*Department of Biological Sciences, Dundee University, Scotland.*

An investigation was made into the lysis of conidia of *Cochliobolus sativus* in natural soil and their colonization by soil microorganisms. Scanning and transmission electron microscopical methods were used. The surfaces of conidia were rapidly colonized by soil bacteria and actinomycetes; conidia were penetrated, and pits and perforations varying in diameter from 0.5 μ to 3.0 μ were observed. Having gained entry to conidia, bacteria colonized cell lumina and walls. The pigmented outer layer of the walls was most resistant to lysis. After 100 days of incubation on natural soil at 25°C, most conidia were merely empty shells. A variety of microorganisms took part in spore breakdown, some being of unusual morphology. One group of bacteria having a helically lobed appearance was isolated from lysing conidia and cultured. Another group of microorganisms commonly observed in spore cell walls had a characteristic fine structure, with two concentric unit membranes being present. They appeared to produce hundreds of membrane-bound vesicles when growing within the cell wall matrix. So far, these microorganisms have not been isolated. Attempts to reproduce perforation and lysis of conidia by incubation with selected bacteria have not been successful.

Ethylene as a Cause of Soil Fungistasis

A. M. SMITH—*Biological and Chemical Research Institute, Rydalmere, New South Wales, Australia.*

Volatile inhibitors of fungistasis were identified by gas chromatography from 50 soils with different origins and histories that were collected throughout Australia. Ethylene was the only material identified in all soils in concentrations likely to be biologically active. The ethylene is formed by microbial activity. Ethylene inhibited the germination of propagules of several soil fungi, including sclerotia of *Sclerotium rolfsii* and spores of *Helminthosporium sativum*. If an air stream was swept over the soil surface to remove the volatile factor, fungistasis was removed, but if ethylene at 1 ppm was added to the air stream, the inhibition remained. More ethylene is produced by nutrient-rich (high organic matter) soils, but such soils require higher levels of ethylene to inhibit germination of fungal propagules. Thus, nutrient is the main stimulator of propagule germination, and ethylene is the main inhibitor. The balance between the two determines whether a soil is fungistatic at any given time. Latest data indicate that ethylene is produced by spore-forming bacteria in anaerobic microsites in soil that develop as a consequence of the proliferation of aerobic microorganisms. The implications of ethylene in the occurrence and control of soil-borne plant diseases are clear, but ethylene also affects bacteria, actinomycetes, and nematodes in soil. Therefore, it will have a controlling influence on most biological processes in soil, including energy turnover from organic matter and mineralization of nitrogen. Ethylene appears to be a major regulating factor in soil microbiology that has previously been overlooked.

Mycelial Growth and Lysis on Cellulose Films

M. J. SWIFT—*Botany Department, Birkbeck College, University of London, England.*

Transparent cellulose film (cellophane) provides a useful, though selective, substrate for the study of fungal activity in soil, as has been demonstrated in the succession studies of Tribe (1957, Symp. Soc. Gen. Microbiol. 7:287-298) and others. Quantitative estimation of fungal growth can be made by microscopic or chemical analytic methods. By suitable marking of the film and systematic recovery from the growth environment and re-examination, the rate of mycelial lysis and decay can also be determined.

Inhibition of Pythium ultimum in Different Molecular Fractions From Gel Filtration of Soil Extracts

O. VAARTAJA—*Forest Ecology Research Institute, Environment Canada, Ottawa, Ontario.*

Filtered aqueous extracts from clay and sandy Ontario soils were fractionated using molecular sieve chromatography (Sephadex, Bio-Gels). Growth responses to the fractions were bioassayed with mycelia of *Pythium ultimum*. Of 66 extracts, 48% contained one or several fractions that fully or partly inhibited growth, 5% contained stimulatory fractions, and 9% contained fractions with both kinds of action. In 38%, every fraction was without significant ($P < 0.01$) effects. Of 12 extracts that in toto were not inhibitory, four nevertheless contained inhibitory fractions. Discrepancies between in toto and fractional effects were frequent, resulting from (a) special affinity and hence only gradual, delayed elution in unbuffered gels of molecules with ionic charges, and (b) counteraction by either stimulatory or nonstimulatory substances. Such counteraction was demonstrated with 36 fractions bioassayed both in purified agar (all 36 inhibited) and in agar amended with 5% sucrose and 0.5% yeast extract (only one inhibited)). The chromatographic data indicated different molecules (molecular weight estimates for 80 inhibitory fractions: 15%, 100-2,000; 21%, 3,000-9,000; 35%, 10,000-20,000; 18%, 25,000-55,000; 11% > 80,000). The stimulations occurred in fractions of molecular weight 2,000-50,000. These results are considered the first strong and direct evidence of nonvolatile inhibitors contributing to mycostatic situations in different soils.

Colloquium on Integration of Pesticide-Induced and Biological Destruction of Soil-Borne Pathogens: Summary and Synthesis

COLIN D. MC KEEN—*Research Branch, Canada Agriculture, Central Experimental Farm, Ottawa.*

During the many years that the important soil-borne fungal pathogens attacking our agricultural and forestry crops have been studied, much research has gone into biological control; somewhat less has been devoted to chemical control. To date, practitioners (and this would include those experimenters who have tested agricultural techniques) have apparently been more successful in using biological control than have soil ecologists in explaining what biological changes take place and the mechanisms that bring them about. Regardless of the disease control obtained, it has long been thought that until these soil microbiological phenomena were more clearly elucidated, biological control would continue to be unpredictable, erratic, and founded on a narrow empirical base that did little to enhance the prestige of plant and forest pathologists who recommended this means of crop protection.

Reduction in root rot severity has often been attempted by the introduction of microbial antagonists or through the incorporation into infested soils of organic amendments favoring specific antagonists. Inconsistent results have usually been obtained. Presumably, the microbial balance established by the prevailing environmental conditions resists the increase of introduced organisms.

Varying degrees of success against preemergence and early postemergence damping-off and in protecting the upper part of the root system and the lower stem regions have been obtained with chemical seed dressings and by applying chemicals in bands alongside the seeds at planting. With the advent of systemic fungicides of the carbathiin and benomyl groups, which showed activity against pathogens previously unaffected by the older protectant fungicides, new hopes for improved control of certain soil-borne pathogenic fungi arose.

Where high cash value crops are being grown, soil treatment with broad spectrum fumigants that destroy virtually all the microflora and fauna have been used. Although costly, fumigants like methyl bromide, chloropicrin, Vorlex, Vapam, allyl alcohol, and a few others are used economically and effectively as cropping practices in glasshouses and nurseries. However, satisfactory disease control is seldom effective for more than one or two cropping seasons, so the practice is usually carried out annually or biennially.

The purpose of this colloquium was to look at the biologic control of a few soil-borne pathogens, not by way of the old, often-tried approaches, but rather by following the use of specific chemicals to see where and how effectively the two methods might be integrated.

EFFECT OF SPECIFIC FUNGICIDES ON SOIL MICROFLORA AND DISEASE CONTROL.—Following up some previous research that determined the fate of thiram in soil and its effect on the incidence of damping-off, Richardson showed that in treated soil, bacteria increased and fungi decreased, but the proportion of saprophytic fungi, particularly *Trichoderma* and *Penicillium*, increased. Also, the thiram treatment rendered the soil more difficult to infest artificially with *Pythium* and also deterred the increase of pathogens during repeated cropping. Since the protection of seedlings persisted long after the concentration of thiram was reduced to ineffective levels, the beneficial influence of certain microflora was suggested. Richardson presented evidence to show that certain combinations of insecticides and fungicides applied as seed dressings to control damping-off in peas gave rise to additive protective effects. A synergistic response was revealed when the systemic fungicide chloroneb was applied to pea seeds in combination with either thiram or captan. Being taken up by the seedling roots and accumulated in the lower stem, chloroneb protects the parts of the seedling that extend beyond the zone of protection provided by the standard fungicide.

Bollen presented evidence to show the differential effect of benomyl on soil-borne pathogens infecting the lower stem of rye plants sprayed 2 months previously. Whereas infection of stem bases by *Fusaria* was markedly decreased, infection by *Rhizoctonia solani* was considerably increased. In vitro tests showed that benomyl inhibited *Fusaria* and *Cercosporella*, and this inhibitive phenomenon was considered important in increasing the yields of sprayed plots.

In soils treated with Panogen PX at 5 ppm, Chinn found that treatment reduced the ratings of disease caused by *Cochliobolus sativus* in wheat seedlings from nearly 30 to 3%. This lead to increased populations of *Penicillium* spp.

DELETERIOUS EFFECT OF CERTAIN HERBICIDES ON DISEASE CONTROL.—That soils treated with chemicals occasionally predispose plants to more disease than untreated soils do has long been recognized. Altman presented evidence to show that the use of preplant herbicides [Pyramin (5-amino-4-chloro-2-phenyl-3-(2H)pyridazinone) and Ro-Neet (S-ethyl-N-ethyl-thiocyclohexanecarbamate)] applied to soils at 3 ppm increased damping-off of sugarbeets by 50 to 100%. These herbicides caused increased permeability of cells of the hypocotyl, increased their exudate of glucose, and simultaneously reduced the calcium levels of these cells. Lower than normal calcium levels have been associated with increased susceptibility to certain damping-off diseases.

EFFECT OF BROAD SPECTRUM FUMIGANTS ON FUNGAL PATHOGENS.—Varghese and Chew reported that the root disease of tea caused by *Ganoderma* sp. can be controlled by fumigating the patches of infested soil with methyl bromide. In the fumigated patches, the infected tea stumps showed increased colonization with *Trichoderma, Aspergillus, Penicillium,* and *Spicaria; Ganoderma* was not isolated from stumps heavily colonized by these fungi.

In studying the reaction of *Armillaria mellea* after exposure to sublethal doses of methyl bromide, Ohr and Munnecke observed that *A. mellea* underwent a lag period before growth resumed, whereas *Trichoderma,* an associated antagonist, grew immediately. *Trichoderma* was approximately twice as tolerant of methyl bromide as was *Armillaria.* Fumigation caused leakage of ionic substances from the cells of *A. mellea,* and these were measured by conductivity. Ohr and Munnecke postulated that sublethal doses of fumigation stressed *Armillaria,* reducing its antibiotic production, and permitted *Trichoderma* to exert its antagonistic mortal action. Smith found that exposure of sclerotia of *Sclerotium rolfsii* to sublethal doses of carbon disulfide caused them to leak small quantities of sugars and amino acids.

Exposure to propylene oxide caused less leakage. A postfumigation drying period caused sclerotia exposed to either fumigant to leak massive quantities of nutrients. Fumigation followed by drying rendered sclerotia very susceptible to colonization and to destruction by antagonistic microorganisms.

CONCLUSIONS.—Of the findings brought forth in this colloquium, it is noteworthy that improved biochemical techniques are now available for assessing the effect of systemic and protectant fungicides applied to plants or plant parts. These effects can be studied where the chemicals have been applied as seed dressings or sprays or to infested soils by these or other means. The techniques can be used in conjunction with bioassays and other monitoring procedures. Also, recognition of the importance of the phenomenon of stress resulting from the application of sublethal dosages of broad-spectrum fumigants and the further alteration of microbial antagonism induced by short postfumigation drying treatments resulting in enhanced biological control of certain pathogens is a notable step forward in the field of soil fumigation. With this injection of new developments, plant pathologists have found new hope for a firm belief in the integration of biological and chemical control of soil-borne pathogens. For plant pathologists who have the perseverance and patience to do research in this difficult field, the prospect of significant progress looms brighter.

ABSTRACTS

Interactions of Pesticides in Predisposing Plants to Disease

JACK ALTMAN—*Department of Botany and Plant Pathology, Colorado State University, Fort Collins.*

Pyramin [5-amino-4-chloro-2-phenyl-3-(2H)pyridazine] and Ro-Neet (S-ethyl-N-ethyl-thiocyclohexanecarbamate) used as preplant herbicides at 3 ppm resulted in increased glucose exudates at the hypocotyl-soil interface. This increase in glucose was detectable at 2, 4, and 10 weeks after treatment. Hypocotyl extracts were analyzed for glucose spectrophotometrically using the Kornberg-Horecker method for true glucose. Increases in electrical conductivity were also observed from hypocotyls of plants grown in herbicide-treated soil. This increase reflected an increase in cell membrane permeability. Sugarbeets grown in herbicide-treated soils had lower calcium levels in the hypocotyl area 1 week after germination than plants grown in nontreated soil. The effect persisted for 3-10 weeks after emergence, but calcium levels in hypocotyls of plants grown in treated soil gradually increased to the levels of plants grown in control soil. This lower calcium content would, according to Bateman, make plants more susceptible to *Rhizoctonia* infection. A 50-100% increase in damping-off of sugarbeets occurred in plants grown in infested and herbicide-treated soil compared to infested but nontreated controls. The increases in glucose exudates and electrical conductivity and decreased calcium levels are additional evidence that herbicides may predispose plants to increased disease.

Effect of Benomyl on Pathogens Causing Foot Rot in Rye

G. J. BOLLEN and E. P. VAN DER HOEVEN—*Laboratory of Phytopathology, Wageningen, The Netherlands.*

In field trials, benomyl was used for control of pathogens of foot and haulm-base in rye on a sandy soil (pH 4.6; organic matter 3.2%). In 1970 the crop had been sprayed at stages 4, 7, and 10 (Feekes-scale) with 1.6, 1.3, and 1.9 kg/ha of the fungicide (50% W.P.). The last spray was applied to the base of the crop. Two months after this spray was applied the crop was harvested. At that time the treatment resulted in about a 50% decrease in discoloration of the haulm-base, mainly caused by *Fusaria.* However, incidence of sharp eyespot caused by *Rhizoctonia solani* had increased from 2.9% in the unsprayed crop to 35.0% in the sprayed one. In 1971 and 1972, benomyl again significantly favored incidence of *R. solani.* In vitro, isolates of *R. solani* from rye were less sensitive to benomyl (ED_{50} 2.1-3.1 μg/ml a.i.) than those of *Fusaria* (ED_{50} 0.5-1.2 μg/ml a.i.) and of *Cercosporella*

($ED_{50} < 0.5$ μg/ml a.i.) obtained from haulm-bases of the same crop. Thus, in a crop sprayed with the fungicide, *R. solani* can have an advantage over the other pathogens mentioned. The increase in yield as a result of the spray will at least partly be due to the inhibitory effect on *Fusaria* and *Cercosporella.* How much of the additional yield will be lost by the increase of sharp eyespot remains questionable. No significant difference in sensitivity towards benomyl was found between isolates of *R. solani* originating from sprayed plots and unsprayed ones. The antagonism toward *R. solani* exerted by the microflora of the rhizosphere soil of the crop was also determined. This antagonism was significantly less for the sprayed crop than for the untreated one.

Biological Effects of Panogen PX in Soil on Common Root Rot of Wheat Seedlings

STANLEY H. F. CHINN—*Research Branch, Agriculture Canada, Saskatoon, Saskatchewan.*

Many similarities exist between our study on the mechanism responsible for the control of common root rot of wheat caused by *Cochliobolus sativus* in soil treated with Panogen PX and that of numerous studies on root rot of citrus caused by *Armillaria mellea* in soil treated with carbon disulfide. Panogen PX at 5 ppm had little direct effect on conidia of *C. sativus*. Nevertheless, the treatment reduced disease ratings of the seedlings from 29.3% in the control to 3.2% in the treated soil. Control in Panogen PX-treated soil may be attributed to a number of mechanisms: (a) the increase in *Penicillium* spp. and certain other microorganisms may have been paralleled by a concomitant increase in antibiotic production, (b) a cross-protection mechanism operating within the seedlings may have been involved as *Penicillium* spp. invaded and became established in the subcrown internodes of the seedlings, (c) Panogen PX at 5 ppm may have acted as a fungistatic agent or it may have augmented naturally occurring fungistatic substances, thus minimizing conidial germination and infection, and (d) conidia may have been slightly injured, resulting in their gradual eradication from soil and contributing to their inability to invade plant tissue.

Control of Armillaria mellea by an Integrated Chemical and Biological Method

HOWARD D. OHR and DONALD E. MUNNECKE—*U.S. Department of Agriculture, Agricultural Research Service, Southern Weed Science Research Laboratory, Stoneville, Mississippi, and Department of Plant Pathology, University of California, Riverside.*

The involvement of *Trichoderma* in the control of *Armillaria mellea* following fumigation has long been noted. Control of the pathogen was initially attributed to the increase of *Trichoderma* inoculum due to the elimination of competitors. Later investigations demonstrated that an increased population of *Trichoderma* was not adequate and that *Armillaria* must also be stressed. Following fumigation, a lag period is present before growth resumes in *Armillaria* but not in *Trichoderma*. Investigations into events occurring during this lag period have demonstrated that *Trichoderma* is more tolerant to methyl bromide fumigation than is *Armillaria* and that a negative correlation exists between isolations of *Trichoderma* and *Armillaria* from fumigated *Armillaria*-infested citrus roots stored in nonsterile soil. Methyl bromide fumigation causes a reduction in the amount of antibiotic substances produced by *Armillaria* in vitro, and more ionic substances leak from fumigated than from nonfumigated *Armillaria*. It is reasoned, therefore, that sublethal fumigation with methyl bromide stresses the *Armillaria* by reducing antibiotic production and by causing cellular leakage and possible cellular damage. These events give the advantage to *Trichoderma*, which is then able to kill *Armillaria*.

Biological Control Through Chemical Treatment

LLOYD T. RICHARDSON—*Research Institute, Research Branch, Agriculture Canada, London, Ontario.*

Biological control of seed- and soil-borne pathogens can be achieved by manipulating the microbial ecology of the soil by means of fungicides with appropriate selectivity. This was demonstrated with thiram treatments to protect pea seedlings against *Pythium ultimum*. A fungicide that is too highly specific may control one disease but permit others to increase. An extreme case of specificity was found in the toxicity of benomyl and thiabendazole against *Fusarium* spp., *F. solani* being tolerant and other species being sensitive. Application of selective fungicides in combination may be necessary when more than one pathogen is involved. Certain combinations may be advantageous even with a single pathogen and no competitors present. Synergistic effects were observed when pea seeds treated with various combinations of thiram and chloroneb or of captan and chloroneb were planted in soil infested with *P. ultimum*.

Integrated Control of Sclerotium rolfsii Using Sublethal Treatments With Fumigants to Promote Biological Control of Sclerotia

A. M. SMITH—*Biological Chemical Research Institute, Rydalmere, New South Wales, Australia.*

Fumigation of sclerotia with sublethal doses of carbon bisulfide causes leakage of small quantities of sugars and amino acids. Similar treatments with propylene oxide result in virtually no leakage. If fumigation is followed by a brief nonlethal drying treatment, both carbon bisulfide- and propylene oxide-treated sclerotia leak massive quantities of nutrients. When placed in soil, sclerotia subjected to combined treatments of fumigation and drying are rapidly colonized and rotted by soil microorganisms, whereas sclerotia subjected to either treatment alone remain intact, viable, and infective. A principle emerges from this work that may have important implications in the control of plant pathogens: biological control of fungal pathogens can be promoted by altering the permeability properties of their propagules until nutrient leakage is stimulated above a certain threshold. Death of the propagules results from the subsequent activity of other microorganisms, not from the loss of nutrients per se. These results also provide a possible new method for predicting the effectiveness of a particular chemical as a fumigant.

Studies Into the Biology and Control of Ganoderma Root Disease of Tea in Malaysia

G. VARGHESE and P. S. CHEW—*Faculty of Agriculture, University of Malaya, Kuala Lumpur, Malaysia.*

A root disease of tea caused by a species of *Ganoderma* has lately become a serious problem. Comparative studies on the morphology and physiology showed that the species concerned was closely related to *G. pseudoferreum*, a confirmed parasite of Hevea rubber. The tea fungus, however, could be distinguished on the basis of spore morphology. The species recorded on tea was found to be distinctly different from *Ganoderma* species recorded on oil palm. The disease spreads from scattered patches in the field. Tests were carried out to control the disease by inactivating inoculum sources in the patches. Among various treatments, fumigation of the patches by methyl bromide gave promising results. Fumigated patches showed increased activity of soil microflora and colonization of tea stumps by *Trichoderma, Aspergillus, Penicillium*, and *Spicaria. Ganoderma* could not be isolated from stumps colonized by the above fungi. Addition of mulch with NPK significantly increased microflora activity following fumigation. The practical application of this treatment under estate conditions is under investigation.

Index

Acer. See Maple
Acetaldehyde
−alfalfa volatiles, 92
−soil fungi inhibited, 92, 96
−Trichoderma volatiles, 96
Achromobacter spp., seed inoculant, 130
Actinomyces
−antagonistic to Rhizoctonia solani, 130
−colonizing conidia, 202; oospores, 200
−ethylene tolerant (foreword)
−seed inoculant, 130
−soils suppressing Phytophthora cinnamomi, 154
−take-all decline, role, 142
−volatiles, 96
Aeration
−plants, 52
−soil: effect on potato scab, 126; indices, 50-51
Aerobes, inhibition by ethylene, 1
Agaricus bisporous, bacteria stimulate sporophores, 96
Agnotobiotic, definition, 11
Alfalfa (Medicago)
−bacterial wilt, 17
−medicarpin, 189
−root rot resistance: breeding, 177; effect of clipping, 17; induced, 17
−volatiles: chemical nature, 92; toxicity, 92
−volatiles, effect on: antagonists, 94-95; microsphere, 95; soil microflora, 91-94; Sclerotium rolfsii, germination of sclerotia, 93
Alkyl sulfides, from Allium, 90-91
Allium. See Onion
Allyl alcohol, fungistatic role, 194, 197
Almond (Prunus), Verticillium wilt, 167
Ambrosiella, toxin produced, 32
Amendments, inorganic
−availability, effect of soil, 102
−effects on: control of disease, 104; disease, 100-106; host, 102-103; pathogen, 101-102; susceptibility, 103; virulence, 102
Amino acids
−absorption by axenic roots, 14
−diffusates cause "frenching" in tobacco, 16
−germination of sclerotia, 200
−leakage from sclerotia, 205
Ammonia
−decomposing debris, source, 88
−effects on: antagonists in root zone, 57, 103; soil-borne pathogens, 56-57, 102; soil pH, 55; spore germination, 58
−Phytophthora cinnamomi suppressed, 86
Ammonium sulfamate, controls: Fomes annosus, 162; wood rotters, 162
Amygdalin, toxic hydrolysates, 191
Anaerobiosis, in soil microsites, (foreword), 51
Ananas. See Pineapple
Antagonism, 127
−alfalfa volatiles, 93-95
−nitrogen effect, 59
−persistence in steamed soil, 156
−pine stumps, 159-160

−potential, 153
−resistance role, root rot of wheat, 117-120
−systemic fungicides reduce, 204
−take-all decline, 139-141; specificity, 140; elimination from soil, 140
Antagonists
−Fomes annosus in pine stumps, 159-161
−grain yield effect, 130
−resistance role, wheat root rot, 117-120
−rye leaves, 199
−seed application, 130
−survival in steamed soil, 156
−Verticillium, 169
−wheat variety effect, 117-120
Antibiotics
−antifungal compounds, 185-193
−fumigation effect on, 205
−mycorrhizal fungus sources, 112-113
−residue colonization role, 80-81
−soil moisture and pH effects, 81
Aphanomyces cochloides
−barley leaf fragments colonized, 79
−crucifer debris suppresses, 86
Aphanomyces euteiches
−crucifer volatiles toxic to, 86, 91
−field infection, index, 179
−mineral nutrients and reproduction, 102
−races, 179
Apricot (Prunus), Verticillium wilt, 167
Arabidopsis thaliana, sugar utilization, axenic, 15
Arachis. See Peanut
Armillaria mellea
−broad leaf trees, 69
−carbon dioxide effect, 63
−conifers, 69
−control, integrated, 205
−dieback, 31-32
−ethanol stimulates rhizomorphs, 97
−food base and pathogenicity, 69-71
−methyl bromide effects, 204, 205
−pathogenic strains, 69, 71
−rhizomorph production, 69-71
Arthrobacter, pathogens lysed, 16
Asparagus (Asparagus), toxic glucoside produced, 191
Aspergillus
−methyl bromide effect, 206
−tea stumps colonized, 204, 206
Aspergillus flavus
−conidia germination, 200
−peanut colonized, 16
Apsergillus niger, onion, 185
Aureobasidium pullulans, Helminthosporium sativum inhibited, 199
Autoclaved soil, fungistasis effects, 199-200
Avocado (Persea)
−phytoalexin produced, 186, 190
−Phytophthora cinnamomi pathogen, 152
Avena. See Oats